The Practical Approach Series

SERIES EDITOR

B. D. HAMES
Department of Biochemistry and Molecular Biology
University of Leeds, Leeds LS2 9JT, UK

★ **indicates new and forthcoming titles**

Affinity Chromatography

★ Affinity Separations

Anaerobic Microbiology

Animal Cell Culture
(2nd edition)

Animal Virus Pathogenesis

Antibodies I and II

★ Antibody Engineering

Basic Cell Culture

Behavioural Neuroscience

Biochemical Toxicology

Bioenergetics

Biological Data Analysis

Biological Membranes

Biomechanics—Materials

Biomechanics—Structures and
Systems

Biosensors

★ Calcium-PI signalling

Carbohydrate Analysis
(2nd edition)

Cell–Cell Interactions

The Cell Cycle

Cell Growth and Apoptosis

Cellular Calcium

Cellular Interactions in
Development

Cellular Neurobiology

Clinical Immunology

★ Complement

Crystallization of Nucleic
Acids and Proteins

Cytokines (2nd edition)

The Cytoskeleton

Diagnostic Molecular Pathology
I and II

Directed Mutagenesis

★ DNA and Protein Sequence
Analysis

DNA Cloning 1: Core
Techniques (2nd edition)

DNA Cloning 2: Expression
Systems (2nd edition)

★ DNA Cloning 3: Complex
Genomes (2nd edition)

★ DNA Cloning 4: Mammalian
Systems (2nd edition)

Electron Microscopy in
Biology

Electron Microscopy in
  Molecular Biology

Electrophysiology

Enzyme Assays

★ Epithelial Cell Culture

Essential Developmental
  Biology

Essential Molecular Biology I
  and II

Experimental Neuroanatomy

★ Extracellular Matrix

Flow Cytometry (2nd edition)

★ Free Radicals

Gas Chromatography

Gel Electrophoresis of Nucleic
  Acids (2nd edition)

Gel Electrophoresis of Proteins
  (2nd edition)

Gene Probes 1 and 2

Gene Targeting

Gene Transcription

Glycobiology

Growth Factors

Haemopoiesis

Histocompatibility Testing

HIV Volumes 1 and 2

Human Cytogenetics I and II
  (2nd edition)

Human Genetic Disease
  Analysis

★ Immunochemistry 1

★ Immunochemistry 2

Immunocytochemistry

In Situ Hybridization

Iodinated Density Gradient
  Media

★ Ion Channels

Lipid Analysis

Lipid Modification of Proteins

Lipoprotein Analysis

Liposomes

Mammalian Cell Biotechnology

Medical Bacteriology

Medical Mycology

Medical Parasitology

Medical Virology

★ MHC Volume 1

★ MHC Volume 2

Microcomputers in Biology

Molecular Genetic Analysis of
  Populations

Molecular Genetics of Yeast

Molecular Imaging in
  Neuroscience

Molecular Neurobiology

Molecular Plant Pathology
  I and II

Molecular Virology

Monitoring Neuronal Activity

Mutagenicity Testing

★ Neural Cell Culture

Neural Transplantation

★ Neurochemistry (2nd edition)

Neuronal Cell Lines

NMR of Biological
  Macromolecules

Non-isotopic Methods in
  Molecular Biology

Nucleic Acid Hybridization

Nucleic Acid and Protein
  Sequence Analysis

Oligonucleotides and
  Analogues

Oligonucleotide Synthesis

PCR 1
PCR 2
Peptide Antigens
Photosynthesis: Energy Transduction
Plant Cell Biology
Plant Cell Culture (2nd edition)
Plant Molecular Biology
Plasmids (2nd edition)
★ Platelets
Pollination Ecology
Postimplantation Mammalian Embryos
Preparative Centrifugation
Prostaglandins and Related Substances
Protein Blotting
Protein Engineering
★ Protein Function (2nd edition)
Protein Phosphorylation
Protein Purification Applications

Protein Purification Methods
Protein Sequencing
★ Protein Structure (2nd edition)
★ Protein Structure Prediction
Protein Targeting
Proteolytic Enzymes
Pulsed Field Gel Electrophoresis
Radioisotopes in Biology
Receptor Biochemistry
Receptor–Ligand Interactions
RNA Processing I and II
★ Subcellular Fractionation
Signal Transduction
Solid Phase Peptide Synthesis
Transcription Factors
Transcription and Translation
Tumour Immunobiology
Virology
Yeast

Immunochemistry 1

A Practical Approach

Edited by

ALAN P. JOHNSTONE

Division of Immunology, St George's Hospital Medical School
London

and

MALCOLM W. TURNER

Immunobiology Unit, Institute of Child Health
London

OXFORD UNIVERSITY PRESS
Oxford New York Tokyo

Oxford University Press, Great Clarendon Street, Oxford OX2 6DP

Oxford   New York
Athens   Auckland   Bangkok   Bogota   Bombay   Buenos Aires
Calcutta   Cape Town   Dar es Salaam   Delhi   Florence   Hong Kong
Istanbul   Karachi   Kuala Lumpur   Madras   Madrid   Melbourne
Mexico City   Nairobi   Paris   Singapore   Taipei   Tokyo   Toronto
and associated companies in
Berlin   Ibadan

Oxford is a trade mark of Oxford University Press

Published in the United States
by Oxford University Press Inc., New York

A catalogue record for this book is available from the British Library

Library of Congress Cataloging in Publication Data
(Data available)

ISBN 0 19 963606 0 (Hbk)
ISBN 0 19 963605 2 (Pbk)

Two volume set
ISBN 0 19 963608 7 (Hbk)
ISBN 0 19 963607 9 (Pbk)

Typeset by Footnote Graphics, Warminster, Wilts
Printed in Great Britain by Information Press, Ltd, Eynsham, Oxon.

100 1182913

Preface

Immunochemistry is of immense importance to virtually all areas of modern biology and medicine. In its broadest sense, the term has come to mean the structure and function of all molecules of immunological importance. The wide scope of the subject caused our provisional contents list to quickly outgrow the standard Practical Approach size and we were delighted that Oxford University Press agreed to publish the work as two volumes. This space has allowed coverage of a wide range of topics, commensurate with the importance of immunochemistry. However, we realize that many aspects remain uncovered, some of which are covered by other volumes in the Practical Approach series.

The availability of high quality antibodies has been an essential prerequisite for progress in most areas of immunochemistry and various strategies currently used in their production are covered in the first five chapters of *Immunochemistry 1*: G. T. Stevenson and K. S. Kan describe how antibodies may be chemically engineered and R. J. Owens discusses the use of genetic engineering in this field, whilst M. D. Melamed, J. Sutherland, M. D. Thomas, and J. Newton address the continuing difficulties associated with the preparation of human monoclonal antibodies. In contrast, J. Chr. Jensenius and C. Koch illustrate the simplicity of production and utility of applications of chicken antibodies. *Immunochemistry 1* also covers some of the uses of antibodies in immunoassays: R. Thorpe, M. Wadhwa, and T. Mire-Sluis give advice on the standardization of antibodies and antigens, whereas D. M. Kemeny and C. H. Self, D. L. Bates, and D. B. Cook cover the latest ELISA techniques in two further chapters. The application of enzyme immunoassays to cells (the cell–ELISA and ELISPOT procedures) are detailed by N. W. Pearce and J. D. Sedgwick, and I. A. Hemmilä presents the latest refinements of photoluminescence immunoassays. *Immunochemistry 1* also describes the incorporation of antibodies and their fragments into immunotoxins (E. J. Derbyshire, C. Gottstein, and P. E. Thorpe) and the use of synthetic peptides in mapping the epitopes of both antibodies and T cells (S. J. Rodda, G. Tribbick, and N. J. Maeji).

Immunochemistry 2 begins with two chapters on affinity and avidity measurements (by H. Saunal, R. Karlsson, and M. H. V. van Regenmortel and by D. Goldblatt). Next, E. Claassen and F. van Iwaarden describe the use of liposomes to modulate immune responses. The use of antibodies in histo- and cytochemistry is presented in three chapters: P. Brandtzaeg, T. S. Halstensen, H. S. Huitfeldt, and K. N. Valnes discuss comprehensively immunofluorescence and immunoenzyme histochemistry; G. D. Johnson explores immunological applications of confocal microscopy; and G. Damgaard, C. H. Nielsen, and R. G. Q. Leslie review various aspects of flow cytometry. *Immunochemistry 2* ends with three chapters focusing on immunochemical aspects of some immunologically important molecules, namely: soluble adhesion molecules (by M. G. Bouma, M. P. Laan, M. A. Dentener, and W. A. Buurman), complement components (by R. Würzner, T. E. Mollnes, and B. P. Morgan); complement receptors (by I. Bartók and M. J. Walport).

We would like to thank all of the authors for their hard work in producing such high quality contributions when they all have so many other competing responsibilities. We are sure that their efforts will contribute significantly to the increasing use of immunochemistry in so many diverse biological fields.

London   A. P. J.
March 1997   M. W. T.

Contents

Immunochemistry 1

Contributors   xxi

Abbreviations   xxv

1. Chemically engineered antibodies   1

G. T. Stevenson and K. S. Kan

1. Introduction   1

2. Some essential features of IgG molecules   2
    Interchain SS bonds   2
    Proteolytic dissections   3

3. Some further reactions of SH groups and SS bonds   3
    Properties of pyridyl disulfides   3
    SS-interchange on reduced Ig modules   6
    Alkylation of SH groups for blocking or cross-linking   7

4. Buffers and equipment   8
    Buffers   8
    Equipment   10

5. The human Fcγ module   12
    Precautions when handling human material   12
    Basic IgG   12
    Preparation of Fcγ1   14
    Preparation of Fc–SS–pyridine and Fc–maleimide   16

6. Murine Fab'γ modules   18
    IgG antibodies from ascites or culture fluid   18
    Preparation of $F(ab'\gamma)_2$ 18
    Fab'–SS–pyridine and Fab'(–SH)$_1$ 20

7. Some derivatives of Fcγ and Fab'γ modules   21
    Chimeric FabFc$_2$ 21
    Thioether-linked $F(ab')_2$ with potential hinge SH groups   22
    Bispecific chimeric derivatives from thioether-linked $F(ab')_2$ 23

References   24

2. Genetically engineered antibodies   27

Raymond J. Owens

1. Introduction   27

2. PCR cloning of immunoglobulin variable domains   28
     Synthesis of first strand cDNA   28
     Primer design   30
     Problems with the method   33

3. The construction of engineered human variable domains   35
     Designing the sequence   35
     Assembling the sequence   38

4. Construction of Fvs and single-chain Fvs   41
     Single-chain Fvs   41
     Disulfide-linked Fvs   42

5. Expression of recombinant antibodies   44
     Mammalian cells   44
     E. coli   48

Acknowledgements   52

References   52

3. Production of human monoclonal antibodies from B lymphocytes   55

M. D. Melamed and J. Sutherland

1. Introduction   55

2. Equipment and media   56
     Basic requirements   56
     Media   57
     Supplements   57

3. Assays for detection of human monoclonal antibodies   58
     ELISA   59
     Haemagglutination assays   61

4. Donor selection   63

5. Isolation and proliferation of human B lymphocytes   64
     Introduction   64
     Isolation of PBMC from human blood   65
     Production and use of Epstein–Barr virus   65
     T cell depletion   66

6. Cell fusion   69
     Introduction   69

Fusion of LCLs 70
Electrofusion 71

7. Strategies for production of human Mabs 73
EBV only 73
Fusion of selected LCLs 73
Fusion of bulk LCLs 74
Use of anti-CD40 75

8. Cloning 75
Limiting dilution 76
Other techniques 78
Points to note 78

9. Scale up of production 79

Acknowledgements 81

References 81

4. Generation and selection of human monoclonal antibodies against melanoma-associated antigens: a model for production of anti-tumour antibodies 83

M. D. Thomas, M. D. Melamed, and J. Newton

1. Introduction 83

2. Production of hybridomas from lymph node lymphocytes 83

3. Identification of tumour-specific Mabs 84

References 88

5. Antibodies packaged in eggs 89

Jens Christian Jensenius and Claus Koch

1. Chicken antibodies 89

2. Antibodies in eggs 90

3. Chicken husbandry 91

4. Immunization 91

5. The antibody response 93

6. Purification of IgG from yolk 95

7. Use of chicken antibodies 101

References 106

6. Standardization of immunochemical procedures and reagents   109

Robin Thorpe, Meenu Wadhwa, and Tony Mire-Sluis

1. Introduction   109

2. Quality of immunochemical reagents   109

3. Standardization of immunochemical procedures   110
Use of standards   110
Nature and availability of standard preparations   111
Unitage of standard preparations   112

4. Problems with the standardization of immunochemical procedures   115
Matrix effects   115
Use of multiple standards   115
Problems due to the epitope specificity of antibodies   119

Acknowledgements   119

References   119

7. Synthetic peptides in epitope mapping   121

Stuart J. Rodda, Gordon Tribbick, and N. Joe Maeji

1. Introduction   121

2. Multiple peptide synthesis on pins   121
Cleaved or non-cleavable?   122
Choice of ending   122

3. Peptide synthesis   122
Storage and handling of peptides   126

4. Antibody (B cell) epitope mapping with synthetic peptides   127
Direct binding of antibodies on pins   127
Direct binding on biotinylated peptides   130
Confirmation of relevance of binding   132
Analoguing for detailed epitope analysis   134

5. T cell epitope mapping with synthetic peptides   134
Design of peptide sets   135
Epitope mapping with T cell clones   138
Mapping of T helper epitopes with peripheral blood mononuclear cells   141

Acknowledgements   145

References   145

8. Enzyme-linked immunoassays   147

D. M. Kemeny

1. Introduction   147

2. Assay format   147
Non-competitive ELISA   147
Competitive ELISA   149

3. Enzymes and substrates   152
Alkaline phosphatase (EC 3.1.3.1)   152
Horse-radish peroxidase (EC 1.11.1.7)   153
β-D-Galactosidase (EC 3.2.1.23)   154

4. Purification of conjugate   154

5. Assay optimization   155

6. Coating of the solid phase   155

7. Sample   157

8. Labelled detector   158

9. Equipment   159

10. Buffers and substrates   160
Buffers   160
Enzyme substrates   161

11. Enzyme labelling of proteins   162

12. Assays   165

13. Quantification   170
Quality controls   172
Coefficient of variation   173

References   173

9. Enzyme amplification systems in ELISA   177

Colin H. Self, David L. Bates, and David B. Cook

1. Introduction   177

2. Enzyme-amplified assay with colorimetric determination   179

3. Application of fluorescence to enzyme amplification   184
Water sources and buffers   186
Preparation of enzymes   187
Alkaline phosphatase substrate   187
Balance of cycling enzyme concentrations   187

4. Concluding remarks   189

References   190

10. Photoluminescence immunoassays 193

I. A. Hemmilä

1. Introduction 193

2. Lanthanides as probes 195

3. DELFIA as a label system for bioanalytical assays 195
Microtitration plates as solid phase matrix 197
Labelling of immunoreagents with lanthanide ions 199
Dissociative fluorescence enhancement of lanthanides and their
chelates 202
Time-resolved fluorometry of lanthanides 202
DELFIA multilabel measurement 204
DELFIA FIA and IFMA: assay optimization 205
Receptor binding assay with DELFIA technology 209
DELFIA cytotoxic release assays 209

4. Time-resolved fluorometric assays with stable fluorescent
chelates 211

5. Conclusion 212

References 213

11. Enzyme immunoassays for detection of cell surface molecules, single cell products, and proliferation 215

N. W. Pearce and J. D. Sedgwick

1. Introduction 215

2. Cell–ELISA 216
Background to the technique 216
Protocol for cell–ELISA 216
Advantages and disadvantages 221
Applications of the technique 221

3. ELISPOT 222
Background to the technique 222
Protocol for ELISPOT 223
Protocol for reverse ELISPOT 226
Advantages and disadvantages 229
Applications of the technique 230

4. Cell proliferation 230
Background to the technique 230
Protocol for cell proliferation using BrdU 231
Advantages and disadvantages 231
Applications of the technique 236

References 236

Contents

12. Immunotoxins   239

Elaine J. Derbyshire, Claudia Gottstein, and Philip E. Thorpe

1. Introduction   239

2. Preparation of chemically coupled immunotoxins   241
General considerations   241
Conjugation methods   242

3. Purification of chemically coupled immunotoxins   253
Gel permeation chromatography   253
Affinity chromatography   254

4. Calculation of the composition of chemically prepared ITs   255

5. Preparation of ITs by genetic engineering   255
Extraction of RNA from B cell hybridoma cell line   256
Preparation of DNA coding for variable regions   258
Cloning of DNA and assembly of DNA encoding scFv and
scFv–immunotoxin   260
Expression and purification of fusion proteins   266

6. Biochemical characterization of ITs   269

Acknowledgements   270

References   271

A1 *List of suppliers*   275

Index   283

xv

Immunochemistry 2

Contributors   xxi

Abbreviations   xxv

1. Antibody affinity measurements   1

Hélène Saunal, Robert Karlsson, and
Marc H. V. van Regenmortel

1. Introduction   1

2. General theory   2

3. Equilibrium constant *K* measurement   2
    Equilibrium dialysis   3
    Ammonium sulfate precipitation   4
    Measurement of free Ab at equilibrium by ELISA titration   4
    Measurement of free Ab at equilibrium using BIAcore   9

4. Measurement of the kinetic rate constants k_a and k_d 11
    Principles and technology of real time biosensor measurements   11
    Principle of the method   12
    Mass transfer during BIAcore experiments   14
    More complex interactions   17
    Range of kinetic rate constants measurable with the BIAcore system   23
    Experimental aspects of the determination of rate and equilibrium
        constants   24
    Example of kinetic measurements in BIAcore analyses and
        interpretation of experimental data   25

References   29

2. Simple solid phase assays of avidity   31

David Goldblatt

1. Introduction   31
    The measurement of avidity   31

2. Competition inhibition assays   33

3. Elution assays   35
    Thiocyanate elution assays   37
    Urea elution assays   42
    Diethylamine elution assays   44

4. Comparison of elution techniques   46

References   49

3. Liposomes   53

Eric Claassen and Freek van Iwaarden

   1. Introduction   53

   2. Lipids   54

   3. Multilamellar vesicles   55

   4. Unilamellar liposomes   58

   5. Size   59

   6. Content   59

   7. Giant liposomes   60

   8. Cytotoxic T cells   61

   9. Stealth (PEG; SL) liposomes   62

   10. Enhanced entrapment   63

   11. Carbocyanin labelling of liposomes   64

   12. Conclusions   68

   References   68

4. Immunofluorescence and immunoenzyme histochemistry   71

P. Brandtzaeg, T. S. Halstensen, H. S. Huitfeldt, and K. N. Valnes

   1. Introduction   71

   2. Immunostaining for fluorescence or light microscopy   71
      Reagents   71
      Immunohistochemical staining methods   80

   3. Multicolour immunostaining and its evaluation   86
      Primary antibodies from different species   87
      Primary antibodies from the same species   87
      Computerized image analysis of multicolour fluorescence   101
      Confocal microscopy   110

   4. Methodological considerations   110
      Tissue preparation methods   110
      Choice of immunostaining method   115

   5. Immunohistochemical reliability criteria   117
      Specificity criteria   117
      Detection accuracy   120
      Detection precision   121
      Detection sensitivity   121
      Detection efficiency   122

6. General conclusions 123

Acknowledgements 123

References 123

5. Confocal laser scanning microscopy 131

G. D. Johnson

1. Introduction 131
Principle 132

2. Practical considerations relating to the equipment 133
Microscope specification 133
Supplementary equipment 133
Operating adjustments 134

3. Application of CLSM in immunological studies 136
Improved definition 136
Image intensity enhancement 137
Optical sectioning and 3D reconstruction 137
Multichannel analysis 139
Quantitative image analysis 142

4. Summary 145
Future prospects for CLSM in immunology 146

References 148

6. Flow cytofluorimetry 149

G. Damgaard, C. H. Nielsen, and R. G. Q. Leslie

1. Introduction 149

2. Applications of flow cytofluorimetry 150

3. Investigation of membrane glycoprotein expression and
molecular interactions at the cell surface 152
Quenching and pH dependence 152
Calibration 153
Measurement of a membrane glycoprotein on blood leucocytes 154
Investigation of molecular interactions at the cell surface 159

4. Measurement of intracellular proteins and mRNA 159
Detection of intracellular proteins 150
Measurement of intracellular enzymatic activity 161
Measurement of mRNA 162
General remarks regarding the investigation of intracellular
components 162

5. Investigation of cellular activation 164
   Measurement of intracellular calcium ions 164
   Measurement of surface activation markers 164

6. Investigation of cellular replication 165
   PCNA and BrdU approaches 166
   The macromolecular dilution approach 166

7. Measurement of cellular uptake and ingestion of soluble
and particulate materials 168
   Binding of opsonized immune complexes to whole blood cells 169
   Measurement of binding and ingestion by phagocytes 171

8. Measurement of phagocytic oxidative processes 172

9. Measurement of cell-mediated cytotoxicity and apoptosis 173
   Cell-mediated cytotoxicity 173
   Measurement of apoptosis 177

References 178

7. Analysis of soluble adhesion molecules   181

M. G. Bouma, M. P. Laan, M. A. Dentener, and W. A. Buurman

1. Introduction 181

2. E-selectin 182

3. ICAM-1 183

4. VCAM-1 184

5. Potential roles of soluble adhesion molecules in health and
disease 185

6. ELISA for sICAM-1 185

7. Potential pitfalls 193

Acknowledgements 193

References 193

8. Immunochemical assays for complement components   197

R. Würzner, T. E. Mollnes, and B. P. Morgan

1. Introduction 197
   The complement system 197
   Clinical importance of the complement system 197
   Assays for the assessment of complement activation 200
   Application of specific complement assays to human body fluids 201

2. Collection and preservation of samples   201

3. EIAs for the assessment of complement activation   203

4. Native-restricted and neoepitope-specific mAbs   207
     The classical pathway   207
     C3 and the alternative pathway   209
     The terminal pathway   211

5. mAbs directed against complement control proteins   215

6. Allospecific mAbs   218

References   220

9. Assays for complement receptors   225

István Bartók and Mark J. Walport

1. Introduction   225

2. Identification of complement receptors using fluorescence-labelled antibodies   227
     Isolation of human neutrophils   228
     Assay of cell surface receptors   229

3. Radioligand binding assay using C3b dimers to enumerate CR1   230
     Binding of radiolabelled ligand to CR1   231
     Calculation of data   233
     Scatchard plot analysis   234

4. Complement receptor production and surface expression   237
     Immunoprecipitation of surface CR1   237
     Biosynthetic labelling of CR1   239

5. Rosette formation assays   242
     Generation of fixed fragments on sheep erythrocytes   242
     Enumeration of complement fragments on sheep erythrocytes   245
     Rosette formation   248

6. Complement receptor-dependent phagocytosis   253
     Preparing C3bi-coated particles   253
     Phagocytosis of C3bi-coated fluorescent microspheres   254

7. Conclusion   255

References   255

A1 *List of suppliers*   257

Index   265

Contributors

ISTVÁN BARTÓK
Department of Medicine, Royal Postgraduate Medical School, DuCane Road, London W12 0HS, UK.

DAVID L. BATES
Dako Diagnostics Ltd., Denmark House, Angel Grove, Ely, Cambridgeshire CB7 4ET, UK.

M. G. BOUMA
Department of General Surgery, Maastricht University, PO Box 616, Maastricht, The Netherlands.

P. BRANDTZAEG
LIIPAT, Institute of Pathology, Rikshospitalet, N-0027 Oslo, Norway.

W. A. BUURMAN
Department of General Surgery, Maastricht University, PO Box 616, Maastricht, The Netherlands.

ERIC CLAASSEN
Institute for Animal Science and Health, ID-DLO, POB 65 8200 AB Lelystad, The Netherlands.

DAVID B. COOK
Deparment of Clinical Biochemistry, The Medical School, Framlington Place, University of Newcastle, Newcastle upon Tyne NE2 4HH, UK.

G. DAMGAARD
Department of Medical Microbiology, Institute of Medical Biology, University of Odense, J. B. Winslowsvej 19, 5000 Odense C, Denmark.

M. A. DENTENER
Department of General Surgery, Maastricht University, PO Box 616, Maastricht, The Netherlands.

ELAINE J. DERBYSHIRE
Department of Pharmacology, University of Texas South Western Medical Center, Dallas 75235, USA.

DAVID GOLDBLATT
Immunobiology Unit, Institute of Child Health, 30 Guilford Street, London WC1N 1EH, UK.

CLAUDIA GOTTSTEIN
Department of Pharmacology, University of Texas South Western Medical Center, Dallas 75235, USA.

Contributors

T. S. HALSTENSEN
Department of Environmental Medicine, The National Institute of Public Health, University of Oslo, Oslo, Norway.

I. A. HEMMILÄ
Wallac Oy, PO Box 10, SF-20101 Turku 10, Finland.

H. S. HUITFELDT
Institute of Pathology, Rikshospitalet, N-0027 Oslo, Norway.

JENS CHRISTIAN JENSENIUS
Department of Immunology, Institute of Medical Microbiology, The Bartholin Building, DK-8000 Aarhus C, Denmark.

G. D. JOHNSON
Department of Immunology, University of Birmingham Medical School, Vincent Drive, Birmingham B15 2TJ, UK.

K. S. KAN
Department of Biological Sciences, King Alfred's College, Winchester SO22 4NR, UK.

ROBERT KARLSSON
BIA core AB, Pty Ltd, Uppsala, Sweden.

D. M. KEMENY
Department of Immunology, King's College Hospital School of Medicine, London SE5 9PJ, UK.

CLAUS KOCH
Statens Serum Institut, Artillerivej 5, 2300 Copenhagen S, Denmark.

M. P. LAAN
Department of General Surgery, Maastricht University, PO Box 616, Maastricht, The Netherlands.

R. G. Q. LESLIE
Department of Medical Microbiology, Institute of Medical Biology, University of Odense, J. B. Winslowsvej 19, 5000 Odense C, Denmark.

N. JOE MAEJI
Chiron Technologies Pty Ltd., 11 Duerdin Street, Clayton, Victoria 3169, Australia.

M. D. MELAMED
Antibody and Cell Culture Research Unit, Faculty of Science, University of East London, London E15 4LZ, UK.

TONY MIRE-SLUIS
Immunobiology Division, National Institute for Biological Standards and Control, Blanche Lane, Potters Bar EN6 3QG, UK.

Contributors

T. E. MOLLNES
Department of Immunology and Transfusion Medicine, University of Tromsø, Nordland Centre Hospital, N 8017 BODØ, Norway.

B. P. MORGAN
Department of Medical Biochemistry, University of Wales College of Medicine, Health Park, Cardiff CF4 4XN, UK.

J. NEWTON
Antibody and Cell Culture Research Unit, Faculty of Science, University of East London, London E15 4LZ, UK.

C. H. NIELSEN
Department of Medical Microbiology, Institute of Medical Biology, University of Odense, J. B. Winslowsvej 19, 5000 Odense C, Denmark.

RAYMOND J. OWENS
Celltech Ltd., 216 Bath Road, Slough SL1 4EN, UK.

N. W. PEARCE
Centenary Institute of Cancer Medicine and Cell Biology, Building 93, Royal Prince Alfred Hospital, Missenden Road, Camperdown, Sydney, NSW, Australia.

STUART J. RODDA
Chiron Technologies Pty Ltd., 11 Duerdin Street, Clayton, Victoria 3169, Australia.

HÉLÈNE SAUNAL
UPR 9021, CNRS, IBMC, 15 rue Descartes, Strasbourg, France.

J. D. SEDGWICK
Centenary Institute of Cancer Medicine and Cell Biology, Building 93, Royal Prince Alfred Hospital, Missenden Road, Camperdown, Sydney, NSW, Australia.

COLIN H. SELF
Deparment of Clinical Biochemistry, The Medical School, Framlington Place, University of Newcastle, Newcastle upon Tyne NE2 4HH, UK.

G. T. STEVENSON
Tenovus Research Laboratory, Southampton University Hospitals, Southampton SO16 6YD, UK.

J. SUTHERLAND
Antibody and Cell Culture Research Unit, Faculty of Science, University of East London, London E15 4LZ, UK.

M. D. THOMAS
Antibody and Cell Culture Research Unit, Faculty of Science, University of East London, London E15 4LZ, UK.

Contributors

PHILIP E. THORPE
Department of Pharmacology, University of Texas South Western Medical Center, Dallas 75235, USA.

ROBIN THORPE
Immunobiology Division, National Institute for Biological Standards and Control, Blanche Lane, Potters Bar EN6 3QG, UK.

GORDON TRIBBICK
Chiron Technologies Pty Ltd., 11 Duerdin Street, Clayton, Victoria 3169, Australia.

K. N. VALNES
Institute of Pathology, Rikshospitalet, N-0027 Oslo, Norway.

FREEK VAN IWAARDEN
Department of Cellbiology, Medical Faculty, Vrije Universiteit, Amsterdam, The Netherlands.

MARC H. V. VAN REGENMORTEL
UPR 9021, CNRS, IBMC, 15 rue Descartes, Strasbourg, France.

MEENU WADHWA
Immunobiology Division, National Institute for Biological Standards and Control, Blanche Lane, Potters Bar EN6 3QG, UK.

MARK J. WALPORT
Department of Medicine, Royal Postgraduate Medical School, DuCane Road, London W12 0HS, UK.

R. WÜRZNER
Institut für Hygiene, University of Innsbruck, Fritz Pregl Strasse 3, A-6020 Innsbruck, Austria.

Abbreviations

2-IT	2-iminothiolane hydrochloride
A/C	alternating current
AAF	2-actylaminofluorene
ABC	avidin–biotin complex
ABTS	2,2′-azino-di[3-ethylbenzthiazoline sulfonate]
ADCC	antibody-dependent cell-mediated cytotoxicity
AEC	3-amino-3-ethylcarbazole
AET	amino ethyl thiouronium bromide
AF	aminofluorene
AMCA	aminomethylcoumarin acetic acid
AP	alkaline phosphatase
APAAP	alkaline phosphatase anti-alkaline phosphatase
APC	antigen-presenting cell
ARDS	acute respiratory distress syndrome
ART-tips	aerosol resistant tips
Az	sodium azide
BALF	broncho-alveolar lavage
BCIP	bromochloroindolyl phosphate
BDHC	benzidine dihydrochloride
BNHS	N-hydroxysuccinimidobiotin
BrdU	bromodeoxyuridine
BrdUrd	5-bromodeoxyuridine
BSA	bovine serum albumin
CCD	charge coupled device
CDR	complementarity determining region
CFSE	5-(and-6)-carboxy-2′,7′ dichlorofluorescein diacetate succinimidylester
CHO	Chinese hamster ovary cell
cm	chloramphenicol resistance gene
CN	4-chloro-1-naphthol
CPD	citrate, phosphate, dextrose
CR	complement receptor
CSA	cyclosporin A
CTL	cytotoxic T lymphocyte
Cy	indocarbocyanine/ide
D/C	direct current
DAB	diaminobenzidine
DABCO	1,4-diazobicyclo(2,2,2)-octane
DAF	decay accelerator factor
DAPI	4′6-diamidino-2-phenylindole

DCFH-DA	dichlorofluorescein-diacetate
DEPC	diethyl pyrocarbonate
DHFR	dihydrofolate reductase
DHR	dihydrorhodamine
DIC	diisopropylcarbodiimide
DKP	diketopiperazine
DMEM	Dulbecco's modified Eagles medium
DMF	dimethylformamide
DMSO	dimethyl sulfoxide
dsFv	disulfide linked Fv
DT	diphtheria toxin
DTPA	diethylenetriaminepenta acetic anhydride
DTT	dithiothreitol
DTTA	diethylenetriaminetetra acetic anhydride
EBV	Epstein–Barr virus
ECLIA	enzyme-linked chemiluminescence immunoassay
EDT	ethanedithiol
EDTA	ethylenediaminetetraacetic acid
EF2	elongation factor-2
EIA	enzyme immunoassay
ELFIA	enzyme-linked fluoroimmunoassay
ELISPOT	enzyme-linked immunospot
EPOS	enhanced polymer one-step staining
F:P:	fluorochrome protein
FC	flow cytofluorimetry
FCS	fetal calf serum
FDNB	fluorodinitrobenzene
FI	fluorescence intensity
FIA	fluoroimmunoassay
FISH	fluorescent *in situ* hybridization
FITC	fluorescein isothiocyanate
FMLP	f-met-leu-phe
FOAM	fluorescent overlay antigen mapping
F:P	fluorochrome-to-protein ratio
FR	framework region
FSC	forward light scatter
Fv	variable domain fragment
GAP	glycine acid peptide
gpt	guanosine phosphoribosyl transferase gene
GS	glutamine synthetase
GTC	guanidinium thiocyanate
H/B	HBSS with 0.01% bovine serum albumin
HAT	hypoxanthine aminopterin thymidine
HBSS	Hanks balanced salt solution

hCMV	human cytamegalovirus major immediate-early gene
HGPRT	hypoxanthine guanine phosphoribosyl transferase
HIFCS	heat inactivated FCS
HOBt	1-hydroxybenzotriazole
HRP	horse-radish peroxidase
HT	hypoxanthine thymidine
IC	immune complex
ICAM-1	intercellular adhesion molecule-1
IFN-γ	interferon-γ
Ig	immunoglobulin
IL-1β	interleukin-1β
IOD	integrated optical density
IPTG	isopropyl-β-D-thiogalactopyranoside
IRMA	immunoradiometric assay
ISC	immunoglobulin-secreting cell
IT	immunotoxin
KLH	keyhole limpet haemocyanin
LAK	lymphokine-activated killer cell
LCL	lymphoblastoid cell line
Leu-leuO-Me	leucyl-leucine-methyl ester
LIA	luminoimmunoassay
LISS	low ionic strength saline
LPS	lipopolysaccharide
LSAB	labelled streptavidin–biotin
mAb	monoclonal antibody
MNC	mononuclear cell
MPS	multiple peptide synthesis
MRI	magnetic resonance imaging
MSX	methionine sulfoximine
MTX	methotrexate
NBT	nitroblue tetrazolium
neo	neomycin resistance gene
NPG	n-propyl gallate
NPP	p-nitrophenyl phosphate
OD	optical density
ONPG	ortho nitrophenyl-β-D-galatopyranoside
PAP	peroxidase anti-peroxidase
PBMC	peripheral blood mononuclear cell
PBS/BSA	PBS containing 0.5% (w/v) bovine serum albumin
PBS	phosphate-buffered saline
PBSe	PBS containing 1 mM EDTA
PBST	PBS with 0.1% (w/v) added Tween 20
PBSTA	PBST with 0.1% (w/v) added sodium azide
PCNA	proliferating cellular nuclear antigen

PCR	polymerase chain reaction
PE	pseudomonas exotoxin
PEG	polyethylene glycol
PFC	plaque-forming cell
PHA	phytohaemagglutinin
PI	propidium iodide
PLP	periodate–lysine–paraformaldehyde
PMSF	phenylmethylsulfonyl fluoride
PMT	photomultiplier tube
PPD	paraphenylenediamine
PRINS	primed *in situ* labelling
PVA	polyvinyl alcohol
PWM	pokeweed mitogen
Px	peroxidase
rDNA	recombinant DNA
rbs	ribosome binding site
RB200SC	lissamine rhodamine B sulfonyl chloride
RGB	red, green, and blue
RIP	ribosome-inactivating protein
R-PE	R-phycoerythrin
RPMI	Roswell Park Memorial Institute
sE-selectin	soluble E-selectin
SFE	specific fluorescence equivalent
sFv	single chain Fv
sICAM-1	soluble intercellular adhesion molecule-1
SIF	specificity interval factor
SIT	silicone-intensified tube
SMCC	N-succinimidyl-4-(N-maleimidomethylcyclohexane)-1-carboxylate
SMPT	succinimidyloxycarbonyl-α-methyl-α-(2-pyridyldithio)-toluene
SNARF	carboxy-seminaphthorhodafluor
SPDP	N-succinimidyl-3(2-pyridyldithio) propionate
SRBC	sheep red blood cell
SSC	side light scatter
sVCAM-1	soluble vascular cell adhesion molecule-1
TBS	tris-buffered saline
Tc	cytotoxic T cell
TCA	trichloroacetic acid
TCM	tissue culture media
TCR	T cell receptor
Th	helper T cell
TNF-α	tumour necrosis factor-α
TOPO	tri-n-octylphosphineoxide

TRITC	tetramethylrhodamine isothiocyanate
U	relative units
VCAM-1	vascular cell adhesion molecule-1
V_H	heavy chain variable domain
V_L	light chain variable domain

1

Chemically engineered antibodies

G. T. STEVENSON and K. S. KAN

1. Introduction

The chemical engineering described in this chapter will utilize just two immunoglobulin (Ig) modules: mouse monoclonal Fab'γ from individual antibodies, and human Fcγ from normal IgG1. These can be combined in various ways to provide antibody derivatives exhibiting combinations of bi-specificity, chimerism, and multiple Fc regions. Our derivatives have been designed to kill neoplastic cells, *in vitro* or *in vivo*, usually by the recruitment of natural effectors such as complement and cells bearing Fcγ receptors. After describing the preparation of the Ig modules we shall give three examples of derivatives which may be prepared from them. Once the principles of construction are understood the range of accessible derivatives will encompass far more than these examples, and may be extended to other purposes such as immunohistology. The technology offers a rapid and economic prediction of the performance of some constructs, such as chimeric antibodies, which could then be engineered genetically to avoid the need for repeated chemical preparation. But it is capable also of yielding derivatives with geometries virtually impossible to obtain by genetic means.

We manipulate the Fab'γ and Fcγ modules by taking advantage of the sulfhydryl (SH) groups released by the reduction of interchain disulfide (SS) bonds in IgG molecules. For the most part these SH groups are clustered at the IgG hinge. The approach had its beginnings in 1962 when Nisonoff and Mandy (1) described the preparation of rabbit bispecific F(ab')$_2$, in which the single SS bond which normally links identical Fab'γ in an IgG molecule was used instead to link Fab'γ of two different antibody specificities. We shall describe a similar derivative in which two Fab'γ modules are thioether-linked rather than SS-linked in order to provide stability *in vivo*, and to which additional modules can be attached; but the use of endogenous cysteine residues to provide the necessary S atoms at defined locations remains an essential feature. With continuing improvements in genetic engineering of antibody molecules, particularly in the expression systems, we can look forward to a ready availability of modules in which the number and position of cysteine residues can be specified at will for particular tasks in chemical engineering.

2. Some essential features of IgG molecules

2.1 Interchain SS bonds

The heavy and light chains of IgG (and of Ig molecules in general) are joined by sets of non-covalent bonds, supplemented by SS bonds (*Figure 1*). The interchain SS bonds are somewhat variable in number and position, so that if one is to utilize them for protein engineering it is essential to identify the Ig molecule by animal species, and by Ig class, subclass, and even allotype. As a source of antibody Fab′γ we deal here with mouse IgG1 and IgG2a from hybridomas of BALB/c origin, which all have one SS bond between each heavy (γ) chain and light (κ or λ) chain, and three SS bonds between the two γ chains. (Other strains of mouse can present an IgG2a allotype with only two inter-γ SS.) As a source of human Fcγ for chimeric molecules we use the subclass IgG1, which makes up about 70% of the total IgG population. It and its Fcγ fragment have two inter-γ SS bonds, regardless of allotype.

Consider an interchain SS in an IgG molecule represented by IgG(–SS–). Exposure to a thiol (R–SH) under suitable conditions leads to an initial reduction or SS-interchange:

$$IgG(-SS-) + R-SH \rightleftharpoons IgG(-SH, -SS-R)$$

The formula for the partially reduced protein is written in a way intended to indicate that the constituent chains do not dissociate as a result of the re-

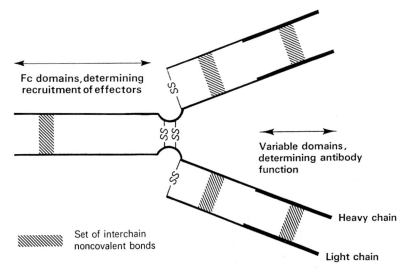

Figure 1. Non-covalent and SS interchain bonds in human IgG1.

duction: the sets of non-covalent bonds maintain the unions of the heavy and light chains. If the reduction is driven by a sufficient surplus of thiol the reduction of the interchain SS goes to completion:

$$IgG(-SS-) + 2R\text{–}SH \rightarrow IgG(-SH, -SH) + R\text{–}SS\text{–}R$$

Certain dithiols such as dithiothreitol (DTT) will drive this reaction to the right without being in large molar excess, due to their tendency to cyclize by forming an intramolecular SS:

$$IgG(-SS-) + T(-SH)_2 \rightarrow IgG(-SH, -SH) + T(-SS-)$$

An IgG molecule with all its interchain SS bonds thus reduced retains its four-chain structure. Its antibody activity, at least for small haptens, also appears unimpaired (2). However the hinge opens out, so that the Y-shaped molecular outline seen on electron microscopy of antigen–antibody complexes is altered so as to suggest a C-terminal transposition of the hinge to about the junction of the $C_\gamma2$ and $C_\gamma3$ domains (3). The molecule shows an accompanying loss of the ability to activate the classical complement pathway (4). It should be noted that the protein's intrachain SS bonds, which are an essential feature of each Ig domain, are not susceptible to reduction by thiol unless there is some protein-unfolding agent such as urea present: they are shielded by lying amid the conjoined faces of β-pleated sheets.

2.2 Proteolytic dissections

Enzymatic cleavages of IgG by papain and pepsin, first reported more than three decades ago, were of great importance in elucidating the structure of Ig molecules and remain one of the day-to-day tools of the immunochemist.

Papain cleaves each γ chain just N-terminal to the inter-γ SS bonds to yield two Fabγ fragments and one Fcγ fragment (*Figure 2*). The figure also depicts the next manipulation we undertake on human Fcγ1, reduction of the inter-γ SS bonds to yield Fc(-SH)$_4$: the two half-γ chains of the Fc are now held together by a set of non-covalent bonds between their C-terminal ($C_\gamma3$) domains.

Figure 3 depicts the cleavage of mouse IgG1 by pepsin at pH ~ 4.0. The γ chains are cut on the C-terminal side of the inter-γ SS bonds, and the Fcγ is extensively degraded. Upon reduction with a thiol the two Fab'γ separate— there are no significant inter-γ non-covalent bonds at the hinge—and each Fab'γ then exhibits five SH groups.

3. Some further reactions of SH groups and SS bonds

3.1 Properties of pyridyl disulfides

Although SS-interchange reactions are conventionally written with the thiol depicted as R–SH, in reality they entail a nucleophilic attack by the thiol ion

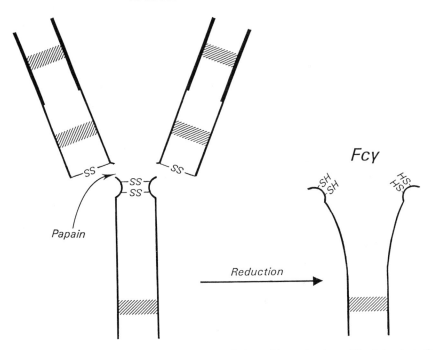

Figure 2. Digestion of human IgG1 by papain, followed by reduction of the interchain SS in its Fc fragment.

R–S⁻, so that a reduction of an interchain SS bond in IgG is better represented by:

$$IgG(-SS-) + R-S^- \rightarrow IgG(-S^-, -SS-R)$$

Thus these reactions are impeded as the thiol ion, with a pK usually ~ 9, becomes increasingly protonated with a lowering of the pH: we find, for example, that exposure of human IgG to 1 mM DTT at pH 5.0 for 30 min at 37 °C yields little or no cleavage of protein SS bonds. However a notable and useful exception to this rule is provided by SS bonds attached to the 2 or 4 carbon of a pyridyl ring (5–7), especially in dipyridyls such as 4,4′-dipyridyl disulfide (Py–SS–Py) (*Figure 4*). Protonation (including low fractional protonation) of the pyridyl N renders the SS unusually electrophilic at low pH: thus these bonds are readily reduced by a thiol R–SH at a pH near 5, the reaction apparently funnelling through the minute proportion of R–S⁻ present. When the pyridyl ring is SS-bonded to protein, as in Fab′γ–SS–Py, we find the bond again abnormally susceptible to reduction at pH 5.0, despite the fact that in the general case of mixed disulfides R–SS–Py the reactivity of the SS can be reduced considerably by the group R introducing a proximal negative charge or a degree of steric hindrance (7).

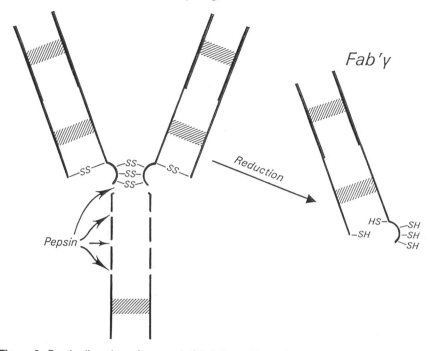

Figure 3. Peptic digestion of mouse IgG1, followed by reduction of the interchain SS in the F(ab')$_2$ fragment. Cleavage of mouse IgG2a is essentially the same, but in that molecule the γ-light SS bonds extend from a C-terminal (or next to C-terminal) Cys of the light chain to a Cys in C$_\gamma$1 near its junction with the V$_H$ domain.

Further useful properties of SS–Py groups are seen upon considering a reaction such as the introduction of SS–Py into a protein molecule by SS-interchange:

$$\text{Py–SS–Py + Prot–SH} \rightarrow \text{Prot–SS–Py + Py–SH}$$

The released 4-pyridylthiol (Py–SH) tautomerizes to yield 4-thiopyridone (Py=S) (*Figure 4*), a reaction which goes close to completion and in turn pulls the SS-interchange essentially to completion. As a further bonus the Py=S gives a strong chromophoric signal (ϵ_{mM} at 324 nm = 19.8), indicating the extent of reaction: on this basis we have found the both the 4,4'- and 2,2'-

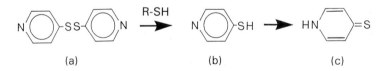

(a)   (b)   (c)

Figure 4. (a) Reduction of 4,4'-dipyridyldisulfide by a thiol R–SH. (b) The released pyridylthiol is in tautomeric equilibrium with (c) 4-thiopyridone. The equilibrium lies overwhelmingly to the right and in turn pulls the SS-interchange to the right.

dipyridyl disulfides to be valuable titrating agents for SH groups. Because of its somewhat greater reactivity at pH 5.0 (7) we prefer the former for our protein engineering.

3.2 SS-interchange on reduced Ig modules

Equilibria of SS-interchange reactions involving SH groups and SS bonds in proteins can be altered profoundly when the cysteine residues involved are held in close proximity by the conformation of the protein. Consider for example an attempt to convert the four SH groups in the hinge of reduced Fcγ1 to SS–Py groups by SS-interchange:

$$Fc(-SH)_4 + 4Py-SS-Py \rightarrow Fc(-SS-Py)_4 + 4Py-SH\ (\rightarrow 4Py{=}S)$$

No matter how high the concentration of Py–SS–Py used to drive this reaction rapidly to the right it will usually be found that the number of SS–Py groups inserted into the Fc is less than one. The explanation is seen in *Figure 5*, in

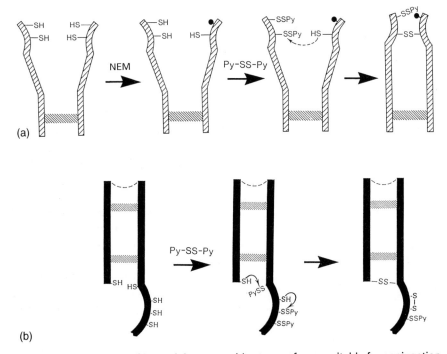

Figure 5. Manipulations of Ig modules to provide storage forms suitable for conjugation. (a) Reduced Fcγ1 has one of its SH groups alkylated by NEM; SS-interchange with Py–SS–Py then yields a closed-hinge Fc displaying a single SS–Py group. It is not known if there is any bias towards one of the Cys residues in the alkylation. (b) Fab'(–SH)$_5$ undergoes SS-interchange with Py–SS–Py to re-form the γ-light SS bond, to form an intrachain SS at the hinge, and to leave one SS–Py group attached to the hinge. It is not known which two of the three SH from the inter-γ SS bonds tend to form the intrachain SS under these circumstances.

the context of Py–SS–Py meeting Fc(–SH)$_3$: an initial formation of Fc–SS–Py is apt to be followed, wherever a contralateral SH exists, by an intramolecular SS-interchange which reconstitutes the inter-γ SS bond. This intramolecular interchange is apparently rapid compared with the reaction of the contra-lateral SH with Py–SS–Py. We meet this reaction again on returning to the subject of closing the Fc hinge.

3.3 Alkylation of SH groups for blocking or cross-linking

Of all groups occurring naturally in proteins SH are the most readily alky-lated. For simple blocking of SH activity iodoacetate or iodoacetamide is often used. However these reagents also show some reactivity towards amino groups and methionine, and greater selectivity for SH is available from maleimides, where the SH adds to the double bond:

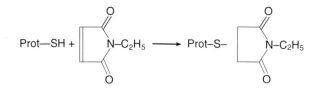

Again the reaction involves Prot–S$^-$ rather than Prot–SH, so the rate falls with decreasing pH. However, as will be seen in our protocols, we find the rate at pH 5.0 to be adequate, and it is often convenient to work at this pH as it hinders both untoward SS-interchange and oxidation of SH. For simple blocking of SH groups by alkylation, *N*-ethylmaleimide (NEM), depicted in the above equation, is highly suitable.

Bismaleimide linkers, a simple example of which is *o*-phenylenedi-maleimide (PDM), provide a highly satisfactory means of linking SH-display-ing protein modules by tandem thioether bonds. First one of the protein modules reacts with the linker, which is present in sufficient molar excess to avoid homologous cross-linking, to yield protein–S–(*o*-succinimidylphenyl)-maleimide (e.g. Fc–maleimide) (*Figure 5*). The Prot–maleimide is then allowed to react with the second protein partner to yield Prot–S–R–S–Prot′, where R is *o*-phenylene-disuccinimidyl (shown between the two Fab′γ modules in *Figure 6*).

During the reaction of PDM with a first partner which displays more than one SH group there will be a strong tendency to form intramolecular cross-links. This need not be a problem if there is an odd number of SH present and only a single maleimide is required on the protein: thus Fab′γ(–SH)$_5$ plus *o*-PDM yields Fab′γ(–maleimide)$_1$ (8). However if starting with an even number of SH, the reaction is apt to yield no active maleimide group attached to the protein.

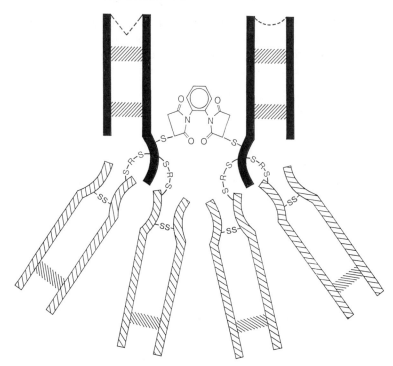

Figure 6. Fab_2Fc_4 prepared by adding human $Fc\gamma1$ to mouse bispecific $F(ab'\gamma)_2$ as described in Section 7.3. The diagram is idealized: reaction of Fc–maleimide to the eight SH groups released upon reducing bispecific $F(ab'\gamma)_2$ is constrained only by steric factors, so it is likely, for example, that some Fc are linked to light chains. The links all involve tandem thioether bonds; the intervening R group is shown in full for the inter-Fab' link.

4. Buffers and equipment

The preparations described in this chapter are on a scale appropriate for antibody therapy of human lymphoma. For most experimental purposes this is far too large, and considerable scaling down of both equipment and reagents will be necessary. We know of no particular difficulties involved in scaling either down or up.

4.1 Buffers

All formulae refer to 1 litre of buffer. Any buffer used in the presence of SH groups should be degassed and/or N_2 bubbled before use. The disodium salt of ethylenediaminetetraacetic acid (Na_2EDTA) is always present in order to complex heavy metals and thereby impede oxidation of SH groups to disulfides. Subsequently in the chapter each buffer will be designated by the bracketted abbreviation in the headings below. Note that the concentration dependence of apparent pK (pK') mean that acetate buffers show little change in pH on being diluted, but the pH of formate and phosphate buffers rises. The pK' of Tris falls as the temperature rises: fomulae given here are for buffers at room temperature.

(a) 2 M Tris–HCl pH 8.0 (2 M T8):
- 242 g Tris[hydroxymethyl]aminomethane (Tris)
- 37.2 g Na$_2$EDTA.2H$_2$O
- 200 ml 5 M HCl

0.2 M T8 is a simple dilution with water.

(b) 0.5 M NaCl, 0.5 M Tris pH 8.7 (0.5 M TBS8.7):
- 29.2 g NaCl
- 60.6 g Tris
- 25 ml 5 M HCl
- 3.72 g Na$_2$EDTA.2H$_2$O

(c) 0.5 M Na acetate pH 4.3 (0.5 M A4.3):
- 12.7 g Na acetate
- 346 ml 1 M acetic acid
- 1.86 g Na$_2$EDTA.2H$_2$O

(d) 0.5 M Na acetate pH 5.0 (0.5 M A5):
- 27.3 g Na acetate
- 167 ml 1 M acetic acid
- 1.86 g Na$_2$EDTA.2H$_2$O

0.1 M A5 and 0.05 M A5 are simple dilutions with water.

(e) 0.2 M NaCl, 0.05 M Na acetate pH 5.8 (acetate-buffered saline, 0.2 M ABS5.8):
- 11.7 g NaCl
- 4.10 g Na acetate
- 1.49 g Na$_2$EDTA.2H$_2$O (doubling as a source of acid in this buffer)

(f) 1.5 M Na formate pH 3.8 (1.5 M F3.8)
- 102 g Na formate
- 92.5 ml 5 M HCl
- 0.93 g Na$_2$EDTA.2H$_2$O

0.05 M F4.0 is a 30-fold dilution with water.

(g) 0.1 M glycine–HCl pH 3.0 (0.1 M Gly3)
- 7.51 g glycine
- 0.37 g Na$_2$EDTA.2H$_2$O
- 4.05 ml 5 M HCl

(h) 0.072 M imidazole–HCl pH 6.7 (0.072 M Im6.7)
- 1.23 g imidazole
- 1.20 ml 5 M HCl
- 1.49 g Na$_2$EDTA.2H$_2$O

(i) 0.03 M imidazole–HCl pH 6.9 (0.03 M Im6.9)
  - 2.04 g imidazole
  - 0.62 g $Na_2EDTA.2H_2O$
  - 3.0 ml 5 M HCl

(j) 0.018 M imidazole–HCl pH 7.2 (0.018 M Im7.2)
  - 1.23 g imidazole
  - 0.37 g $Na_2EDTA.2H_2O$
  - 1.2 ml 5 M HCl

(k) 1.8 M phosphate pH 6.2 (1.8 M P6.2)
  - 72.4 g Na_2HPO_4
  - 176 g KH_2PO_4
  - 7.44 g $Na_2EDTA.2H_2O$

It is important to note that in aqueous buffers Py–SS–Py and any protein displaying –SS–Py groups have a tendency to bind to chromatographic gels, and to an even greater extent to plastic column fittings. This problem is overcome by adding dimethylformamide (DMF) to 1.0 M to the buffer. In addition, whenever Py–SS–Py has been passed through a column, it is our practice to rinse the column with a dilute solution of thiol (e.g. 2 mM 2-mercaptoethanol in a buffer of pH 8), thereby rinsing off any residual bound dithiopyridine after converting it to the more water soluble thiopyridone. Failure to do this can lead to problems associated with unexpected SS-interchange on the columns.

4.2 Equipment

4.2.1 Chromatography columns and circuits

A suitable range of columns fitted with jackets to permit water cooling is available from Pharmacia. PTFE tubing, tubing connections, and manually and electrically operated flow valves are obtained from PhaseSep. We have the columns organized into circuits each capable of running and monitoring two columns simultaneously according to an automated program. This is a convenient arrangement for requirements such as recycling gel chromatography, where the species of interest is purified progressively by passage from column A to column B, and thence back to A, and so on, for as many passes as are required. After being thus purified the product may be concentrated, on the same chromatography rig, by passage from one of the size-exclusion columns onto an ion exchange column for a cycle of adsorption and elution. If the running buffer on the size-exclusion columns is not suitable for binding the protein molecules to the ion exchange column it may be modified, in molarity and/or pH, by blending the size-exclusion outflow with a modifying buffer at a T-piece before leading it onto the ion exchange column. It is important that all connections are tight to prevent leakage or (a more common problem) the entry of bubbles. The occurrence of bubbles can also be minimized by routinely degassing the buffers, and by the simple expedient of having the output of the fluid stream at the top of the fluid circuitry, putting the stream under a pressure which discourages entry of air by Venturi effects at tubing connections. When carrying out chromatography of molecules containing free SH or maleimide groups, cooling water at 5°C is pumped through the column jackets from a suitable thermocirculator (e.g. Conair Churchill 0.2 series). Under other circumstances Ig molecules, including the chemically engineered derivatives we shall describe, are sufficiently stable to permit prolonged chromatographic separations at room temperature.

4.2.2 Chromatography gels

Three different types of gel are used, all obtained from Pharmacia.

(a) Gel (size-exclusion) chromatography: Sephadex G-25 Fine, Sephadex G-50 Fine, and Superdex S200 Fast Flow.

(b) Ion exchange chromatography: SP–Sepharose Fast Flow and DEAE–Sepharose Fast Flow.

(c) Affinity chromatography: Protein A–Sepharose, prepared by conjugating protein A (Sigma) to CNBr-activated Sepharose 4B (Pharmacia) using the manufacturer's recommended methods. About 10 mg of protein A is coupled to every millilitre of swollen gel.

4.2.3 Pumps

Most of our work is carried out with piston pumps having a capacity for accurate delivery of non-pulsatile flow over a range of 10–500 ml/h. The Pharmacia LKB model 2248 is suitable. For coarser duties such as washing and equilibration of columns one can avoid undue wear on the piston pumps by using a peristaltic pump such as the Watson-Marlow model 5058.

4.2.4 Monitors and recorders

Two UV monitors, suitable for measuring the 10 mm absorbancy of the fluid stream at 280 nm while avoiding turbulent mixing (e.g. Single Path Monitor UV-1, Pharmacia), are used on each circuit. A variable wavelength monitor has only occasionally proved useful. The Kipp and Zonen two-channel pen recorder (supplied by Semat Technical Ltd.) has proved suitable for charting the output of the monitors.

4.2.5 Electronic control

To control each circuit we employ electronic sequence controllers made by Scott Smith Electronics. These employ a 'patch' of sockets in rows and columns (usually 20 × 20), with each row representing an electrical output (for flow valves, pumps, recorders, event markers, and fraction collectors; and commands for hold and return-to-start), and each column representing a time stage. An output in a given time stage is activated by the presence of a diode pin in the socket, and the array of pins thus represents an analogue display of the program. (Control of the rig by microcomputer would require less expensive hardware but suitable programs are complex and not necessarily available off-the-shelf. The sequence controllers have proved highly reliable and are well suited to testing rig components and trouble-shooting.) For many runs we also use hard-wired peak detectors (custom-built by the Department of Electronics and Computer Science, University of Southampton) for collecting samples. Each detector receives inputs from the UV monitors and sends its 5 volt outputs to the sequence controller. It actuates the sequence controller to start collection at a selected absorbancy on the leading edge of a protein peak and to terminate it at another selected absorbancy on the trailing edge. To minimize the chances of activation by transient disturbances (such as bubbles) the peak detector does not start its surveillance until a selected time into the program, and the levels of absorbancy which activate it must be maintained for times (often about 1% of the anticipated peak collection time) which are also set on the control panel. The detector will shut down the program if the protein peak fails to appear within a stipulated time, or if it then over-runs a stipulated maximum delivery time.

4.2.6 High-performance liquid chromatography (HPLC)

An HPLC circuit is a useful adjunct to the main chromatography rigs for monitoring reactions such as peptic digestions and thioether bonding of proteins. A variety of highly satisfactory Waters systems are available from Millipore UK Ltd. We employ a system with a model 510 HPLC pump and a 486 Tunable Absorbance Detector to carry out size-exclusion chromatography on Zorbax GF250 (DuPont) or Superdex 30 (Pharmacia).

5. The human Fcγ module

5.1 Precautions when handling human material

Human blood or any of its products should be screened, by the laboratory or by the supplier, for hepatitis B, hepatitis C, and HIV viruses. Other potentially dangerous agents, such as HTLV viruses, T. pallidum, and malarial parasites, do not represent a significant risk in the UK but may need to be considered in other countries. It is sensible to regard all human material as potentially infectious, and to decontaminate any site of spillage with a suitable disinfectant. Frequently the local rules governing this problem are more than adequate.

5.2 Basic IgG

We obtain bulk preparations of human normal IgG as Fraction II precipitate from Bio Products Laboratory. Much of Fraction II is destined for clinical use, but a surplus is available relatively cheaply because the quantity of plasma treated is presently driven by needs for albumin and Factor VIII. Similar preparations of total IgG may be prepared without difficulty from human normal serum as detailed in standard immunochemical texts: one popular method involves a combination of precipitation with ammonium sulfate followed by anion exchange chromatography. We further purify the IgG obtained from either of these sources by ion exchange column chromatography which removes small amounts of non-Ig material together with the most acidic 30% or so of IgG, to yield basic IgG (see *Protocol 1*). Although this basic fraction is not rigorously defined—normal IgG presents a broad continuum of charge and the associated electrophoretic and chromatographic properties—it possesses two notable advantages: it is more completely digested by papain than is total IgG, and from the digest the Fabγ and residual IgG are readily separable from Fcγ because of their basicity.

Fraction II precipitate of UK origin is prepared by cold ethanol fractionation from the pooled plasma of unpaid volunteers who have been screened for hepatitis B surface antigen and for antibodies to HIV-1, HIV-2, and hepatitis C virus. We arrange for the Fraction II as received to be tested additionally for the hepatitis C genome by nested polymerase chain reaction for the untranslated 5′ region: this step is taken because occasional cases of transmission of hepatitis C by therapeutic intravenous Ig preparations have been reported, but it appears unlikely that the viral infectivity can long survive *in vitro* at room temperature.

Protocol 1. Preparation of basic IgG

Equipment and reagents
- See Section 4

Method

1. Dissolve about 18 g of the precipitated Fraction II protein in 0.018 M Im7.2 and make up to a final volume of 500 ml.

2. Adjust the final $E_{280, 1\,cm}$ of the solution to 52 $\pm$ 2 ($\sim$ 36 mg/ml of protein) to allow for the variable moisture content of the precipitate.

3. Clarify the solution by centrifuging and then pass it through a DEAE–Sepharose column 10 cm $\times$ 76 cm, equilibrated with the same buffer, at $\leq$ 1200 ml/h. The basic IgG is unretarded or only minimally retarded.

4. Transfer the IgG peak from the DEAE column at 400 ml/h, after blending at a T-piece with 0.072 M Im6.7 at 200 ml/h (to yield 0.036 M Im6.9), onto a column of SP–Sepharose 5 cm $\times$ 22 cm, equilibrated initially with 0.03 M Im6.9.[a]

5. Continue for 7.0 h, thereby passing to the SP–Sepharose all the protein regarded as 'minimally retarded' on the DEAE column.

Protocol 1. *Continued*

6. Rinse the SP–Sepharose with two column volumes (860 ml) of 0.03 M
    Im6.9 to remove non-bound material (although the baseline need not
    be reattained).

7. Harvest the basic IgG by elution with 0.5 M TBS8.7 at 300 ml/h.

8. Collect the eluted peak above $E_{280,\ 1\,cm} = \sim 1$ to yield a final concentra-
    tion usually > 22 mg/ml.

9. Store the preparation at 5°C.

[a] At this stage any acidic IgG which has escaped the DEAE column passes through the SP column
without binding, i.e. there is back-up selectivity. The SP column also serves to concentrate the
basic IgG by the cycle of binding and elution, finally yielding about half the protein present in
the Fraction II precipitate.

5.3 Preparation of Fcγ1

Some 70% of human IgG, and possibly more of the basic IgG fraction
described above, consists of the subclass IgG1. The conditions described in
Protocol 2 for digesting IgG with papain and harvesting the Fc yield essentially only the Fc from IgG1, i.e. Fcγ1. The only other class detectable by
radial immunodiffusion has been Fcγ2, at < 1% of the concentration of Fcγ1.
(The brief digestion with little thiol present should leave most of the IgG2
intact, Fcγ3 appears to be lost during ion exchange chromatography, and
there is negligible IgG4 present in the basic IgG preparation.)

Protocol 2. Preparation of Fcγ1

Equipment and reagents
• See Section 4

A. *Activation of papain*

1. Obtain papain as a 100 mg crystalline suspension in ~ 4 ml 0.05 M
    sodium acetate pH 4.5 (preparation P 3125 from Sigma).

2. To this add 4 ml 4 mM DTT in 0.2 M T8, and incubate for 15 min at
    37°C.

3. Bring the papain to a pH suitable for chromatographic separation by
    adding 2 ml 1.5 M FE3.8 ($\rightarrow$ pH ~ 4.2).

4. Change the enzyme medium by passing the solution through a
    0.45 μm filter onto a column of Sephadex G-50 2.6 cm × 94 cm, equi-
    librated with 5.5 mM 2-mercaptoethanol in 0.05 M A4.3 (0.38 ml pure
    2-mercaptoethanol to l litre), and elute at 100 ml/h.

5. Collect the predominant protein peak above $E_{280,\ 1\,cm} = 0.8$.

6. Read E_{280} in the spectrophotometer in 5 mm cells (versus the equilibrating mercaptoethanol solution), and dilute with the mercaptoethanol solution to 0.55 mg papain/ml ($E_{278, 5\,mm} = 0.69$).

7. Deep-freeze in glass containers any material which is not to be used within 24 h.

B. *Digestion*

1. To the solution of basic IgG in 0.5 M TBS8.7 add 0.2 vol. of 1.8 M P6.2 ($\rightarrow$ pH ~ 6.7).

2. Add water if necessary to bring the protein concentration down to 20 mg/ml.

3. Pre-warm the IgG and papain solutions to 37 °C for about 10 min, and then add to the protein 0.1 times its volume of the papain solution.

4. Incubate at 37 °C for 20 min, then chill and inactivate the papain by adding 0.05 vol. of cold 40 mM NEM in DMF, and leave on ice for 20 min.

5. Change the buffer by passing the digest ($\leq$ 840 ml) through Sephadex G-25 (10 cm $\times$ 64 cm) equilibrated with 0.03 M Im6.9, and lead the emergent protein at 1000 ml/h directly onto SP–Sepharose (5 cm $\times$ 22 cm) equilibrated with the same buffer. Papain, Fabγ, and residual IgG are retained on this column.

6. When the concentration of emerging protein has reached about 0.1 mg/ml reduce the flow rate to 200 ml/h, blend the effluent at a T-piece with 50 mM acetic acid flowing at the same rate, and lead the confluent stream onto a smaller SP–Sepharose column (2.6 cm $\times$ 15 cm) equilibrated with 0.05 M AE4.3.

7. Continue this process for 8.0 h.[a]

8. Obtain a purified, concentrated solution of Fcγ1 by eluting the small SP–Sepharose column with 0.5 M TBS8.7 at 50 ml/h.

[a] Any attempt to raise the yield of Fcγ1 by running longer is apt to result in contamination by other proteins beginning to emerge from the first SP–Sepharose column. Finally a purified, concentrated solution of Fcγ1 is obtained upon eluting the small SP–Sepharose column with 0.5 M TBS8.7.

To minimize damage to Fcγ1 the digestion is carried out for the minimum time yielding > 80% cleavage of the IgG1 hinge, and the concentration of thiol used to maintain papain in its SH form is kept too low to cause any detectable cleavage of the Fcγ1 hinge SS at the prevailing pH. The digest is fractionated by passing first through an SP–Sepharose column on which the Fcγ1 is unretarded but Fabγ and residual IgG are bound. The emergent Fc is then led onto a second smaller SP–Sepharose column, where it is concentrated by a cycle of binding and elution. Typically we digest individual lots of about 12 g of basic IgG and obtain about 2.7 g of Fcγ1. Each batch of papain obtained from the supplier (100 mg) usually serves three such digestions, so we activate the entire batch and store frozen that which is not to be used immediately. Note that activation takes place with the highly efficient reducing agent DTT, but that the papain is then retained in the reduced form with a concentration of 2-mercaptoethanol which is finally too low to break interchain SS bonds of IgG during the digestion.

5.4 Preparation of Fc–SS–pyridine and Fc–maleimide

We now describe (*Protocol 3*) the conversion of the Fcγ module to Fc–maleimide, which can be conjugated via thioether bonds to any molecule expressing a free SH group (9). The yield is only about 33%, but this is of little consequence with the current availability and cheapness of human IgG. As the maleimide group is unstable it is generally convenient to store the intermediate form Fc–SS–Py, which is readily converted to Fc–maleimide when required.

The steps involved in this preparation are summarized in *Figure 5*. The overall strategy is to sacrifice one of the hinge SS bonds, freeing a cysteine residue which is converted sequentially from the –SH form to –SS–Py, back to –SH, and finally to –S–R–maleimide. These manipulations require that the other cysteine which participated in the sacrificed SS bond first be blocked permanently by alkylation, to prevent re-formation of the bond by SS-interchange. The second, non-utilized hinge SS bond may be left intact (giving a closed-hinge configuration), or reduced and alkylated (open-hinge configuration). We describe the preparation of the 'normal' closed derivative. A protein conjugate such as chimeric FabFc (mouse Fab′γ/human Fcγ1) may be converted from closed to open configuration, with an accompanying loss of complement-activating properties, simply by reducing the SS hinge: the reduction will not affect the tandem thioether link between the two protein modules.

Alkylation of Fc(–SH)$_4$ by NEM is carried out stochastically on the following basis. We require molecules with one SH group alkylated. Molecules with zero or two groups alkylated will have the even number of remaining SH joined either by SS-interchange or by intramolecular linking with the bis-maleimide reagent and, without a free maleimide group, will act simply as inert carrier, as will molecules with four SH alkylated. Molecules with three groups alkylated will form active Fc–maleimide with a permanently open hinge, which is undesirable. An average level alkylation of 0.6 SH per molecule has therefore been selected as suitable: this is expected to yield 37% of the molecules with one alkylated group and only 1% with three, because a random distribution means that the proportion P of molecules alkylated at n sites is given by:

$$P = \frac{4!}{(4 - x)!} \left(\frac{x}{4}\right)^n \left(\frac{4 - x}{4}\right)^{4-n}$$

Protocol 3. Preparation of Fc–SS–pyridine and Fc–maleimide

Equipment and reagents
• See Section 4

A. *Reduction and alkylation of human Fcγ1[a]*

1. To 3.2 g of Fcγ1 in 200 ml 0.5 M TBS8.7 add 20 ml of 22 mM DTT in water, and incubate at 25°C for 15 min.
2. Pass onto a column of Sephadex G-25 (5 × 92) equilibrated with 0.1 M A5 and run at 1000 ml/h at 5°C.
3. Harvest the emergent Fc–$(SH)_4$ above $E_{280,1\ cm} = 1.0$.
4. NEM is now added to a molarity 0.8 times that of the Fc.[b]
5. Measure carefully the $E_{280,1\ cm}$ of a 16-fold dilution of the Fc–$(SH)_4$ solution, and let this value = y.
6. Take (3.3 × y) ml of 40 mM NEM in DMF and bring to 150 ml with 0.1 M A5.
7. With the Fc solution stirring rapidly at 5°C, add slowly to it 0.2 times its volume of the dilute NEM solution.
8. Raise the temperature to 37°C, incubate for 2 h, and then return to 5°C in preparation for the next stage.

B. *Conversion to Fc–SS–pyridine[c]*

1. Add to the cold Fc solution 0.1 times its volume of 12 mM Py–SS–Py in 40% DMF/60% 0.1 M A5 (v/v).
2. Incubate at 5°C for 30 min.
3. Pass the solution onto two columns of Superdex 200 (each 10 cm × 89 cm) in series, equilibrated with 1.0 M DMF in 0.1 M A5 buffer, and run at 1000 ml/h.
4. Concentrate the emergent Fc–SS–Py by passing it at 500 ml/h, blended with 1 M DMF in water at 500 ml/h, onto a column of SP–Sepharose (5 cm ×10 cm) equilibrated with 1 M DMF in 0.05 M A5.
5. Elute from this column a concentrated solution of Fc–SS–Py, using 1 M DMF in 0.5 M NaCl, 0.5 M A5.[d]

C. *Conversion to Fc–maleimide*

1. To the required quantity of Fc–SS–Py with its accompanying Fcγ1, at a total protein concentration adjusted to 15 mg/ml in buffer at pH 5.0, add 0.05 times its volume of 21 mM DTT in water.
2. Incubate at 25°C for 15 min.[e]

17

Protocol 3. *Continued*

3. Separate the resulting Fc(–SH)$_1$ by passage through Sephadex G-25 equilibrated with 0.05 M A5.

4. With the protein at 5°C add 0.2 times its volume of 6 mM phenylened-imaleimide in 50% DMF/50% 0.05 M A5 (v/v).[f]

5. Incubate at 5°C for 30 min, and then pass again through G-25, again equilibrated with 0.05 M A5, to collect the Fc–maleimide.

6. Use immediately for the conjugation reaction (Section 7).

[a] Our instructions refer to the therapeutic quantities of Fc used in our laboratory; there are no problems in scaling down.

[b] With attrition of the maleimide groups and failure of the reaction to proceed fully to completion this is found to yield the desired level of approx. 0.6 alkylated SH group per Fc.

[c] The addition of Py–SS–Py now first closes the Fc hinge by an SS-interchange which reconstitutes SS bonds wherever both homologous SH groups are still available; and secondly converts unpaired SH to SS-Py. Where alkylation of two SH by NEM has left a pair of non-homologous SH, an 'unnatural' SS might form to yield either an intrachain SS or an interchain SS between non-homologous cysteine residues. The extent to which this occurs is probably immaterial: if the non-homologous cysteine pairs do not become SS-bonded at this stage, they will almost certainly be linked intramolecularly by the bismaleimide linker in the next (8).

[d] From our initial 3.2 g of Fcγ1 we recover about 1 g of Fc–SS–Py, accompanied by about 1.8 g of Fcγ1 without an active group, at a total protein concentration > 15 mg/ml. In this form the protein may be stored in glass containers for at least two months at 5°C.

[e] Under these conditions only the electrophilic pyridine-linked SS bond is reduced.

[f] Add the aqueous component of this solution shortly before use: maleimides should not be left for any prolonged period in contact with water.

6. Murine Fab′γ modules

6.1 IgG antibodies from ascites or culture fluid

Murine IgG1 and IgG2a antibodies are purified from ascitic fluid or cell culture fluid by combinations of ion exchange chromatography and affinity chromatography on protein A–Sepharose columns, amounting to minor variants of standard techniques. Antibody is eluted from the protein A–Sepharose with 0.1 M Gly3. If it is shortly to be digested with pepsin the pH is adjusted (to 4.4 ± 0.3) by addition to the eluate of 0.1 vol. of 0.6 M sodium formate. If lengthy storage is planned the antibody is instead transferred to a suitable neutral buffer (e.g. 0.2 M TBS8) by leading the eluate from the protein A–Sepharose directly to a Sephadex G-25 column equilibrated with this buffer; antibody so stored is transferred later to 0.05 M A5 for peptic digestion. The antibody is concentrated if necessary by ultrafiltration so as to be at ≥ 15 mg/ml for the next stage.

6.2 Preparation of F(ab′γ)$_2$

F(ab′γ)$_2$ is produced by limited proteolysis of mouse IgG1 or IgG2a antibody (14 mg/ml) with pepsin (1 mg/ml), usually at pH 3.9 for about 6 h (see

Protocol 4). In order to maximize the yield of this valuable material it is essential that the progress of the digestion be followed at hourly intervals by HPLC; Zorbax GF250 and Superdex 30 have both proved satisfactory gels for this purpose.

Protocol 4. Preparation of F(ab'γ)$_2$ fragments

Equipment and reagents
- See Section 4

Method

1. Dilute the antibody solution, at pH ~ 4.3, to 15 mg/ml with 0.05 M A4.3, and adjust the pH to 3.9 by addition of 50 mM formic acid.[a]

2. Add to the antibody 0.05 vol. of a solution of pepsin (porcine, preparation P 6887 from Sigma), 2.1 mg/ml in 0.05 M A4.3.

3. Incubate the mixture at 37°C.

4. At hourly intervals apply aliquots of 20 μl to the HPLC column equilibrated with 0.2 M phosphate pH 7.0.[b]

5. Halt the digestion when it is about 90% complete by adding 0.05 vol. 2 M T8.[c]

[a] It is essential that this step be carried out meticulously as the rate of digestion is critically related to pH. Differences in susceptibility to pepsin, perhaps related to extents of glycosylation, may dictate that the pH be varied for an occasional antibody.
[b] A progressive fall in the IgG peak will be seen while the F(ab')$_2$ peak rises.
[c] An attempt to pursue the digestion to completion is likely to be accompanied by a disproportionate degradation of the F(ab')$_2$, and sometimes the beginning of a fall in the F(ab')$_2$ peak dictates an earlier termination of the digestion.

F(ab')$_2$ is separated from IgG, digestion products, and enzyme by recycling chromatography on Superdex S200. A digestion of 1 g of IgG requires passage through the equivalent to 7200 cm^3 of gel. Two columns A and B, each 5 cm × 92 cm, are used in the sequence A → B → A → B. When the leading compound protein peak (containing IgG and F(ab')$_2$) has first passed from A onto B, the program stops the flow in the latter column while all material of lower molecular weight in column A is flushed to waste. The separation then proceeds with column B → A → B → collection. It is convenient to interpolate an ion exchange step before the collection in order to concentrate the F(ab')$_2$. This can be carried out by having the Superdex equilibrated with 0.1 M A5 (the lowest molarity which avoids significant binding to the gel at this pH), blending the final outflow containing the F(ab'γ)$_2$ zone with water at an equal flow rate, and leading the confluent stream onto SP–Sepharose equilibrated with 0.05 M A5. Elution with 0.5 M TBS8.7 yields a concentrated protein solution (up to about 20 mg/ml, depending upon the column and tubing geometries). The yield of $F(ab')_2$ from IgG should be 400–500 mg.

6.3 Fab'–SS–pyridine and Fab'(–SH)$_1$

A simple reduction of $F(ab'\gamma)_2$ with DTT at pH 8 yields Fab'(–SH)$_5$, which for some purposes is a suitable module for conjugation—for example in preparing FabFc$_2$ as described below. On other occasions, and in particular for preparing bispecific $F(ab'\gamma)_2$ which retains potential hinge region SH groups for addition of further modules, one requires Fab'(–SH)$_1$, reached through SS-interchange as depicted in *Figure 5*. In this series of reactions (see *Protocol 5*) the intermediate product Fab'–SS–Py is suitable for storage. One begins the manipulations by diluting $F(ab'\gamma)_2$ to 5 mg/ml, as at higher concentrations there is too great a tendency for the dimer to re-form during the SS-interchanges.

Protocol 5. Preparation of Fab'–SS–pyridine and Fab'(–SH)$_1$

Equipment and reagents
- See Section 4

A. *Fab'–SS–pyridine*

1. To the $F(ab'\gamma)_2$ at 5 mg/ml in buffer of pH 7.5–8.5 add 0.05 vol. of 21 mM DTT in water.

2. Incubate at 25°C for 15 min, then separate the Fab'(–SH)$_5$ by passing through Sephadex G-25 equilibrated with 0.05 M A5 at 5°C.

3. Stir the harvested Fab'(–SH)$_5$ under N_2 at 5°C and add to it, over a period of about 1 min, 0.05 vol. of 10.5 mM Py–SS–Py in 50% DMF/ 50% 0.5 M A5.

4. Store the resulting Fab'–SS–Py in glass in this form, in the presence of the surplus Py–SS–Py.

B. *Fab'(–SH)$_1$[a]*

1. To the stirring Fab–SS–Py, under N_2 at 5°C, add 0.1 vol. of 21 mM DTT in water.

2. Incubate at 5°C for 30 min.

3. Separate the Fab'(–SH)$_1$ by passage at 5°C through Sephadex G-25 equilibrated with 0.05 M A5.

[a] The single dithiopyridine group on Fab–SS–Py is now converted to SH by addition of DTT, without any need to remove first the surplus Py–SS–Py. As in the manipulations of Fc, those SS bonds not attached to the pyridyl group are not susceptible to reduction at the prevailing pH of 5.0.

---

7. Some derivatives of Fcγ and Fab'γ modules

7.1 Chimeric FabFc₂

In this derivative (see *Protocol 6*) a single mouse Fab'(–SH)₅ is linked to two human Fcγ1, each link involving tandem thioether bonds. The derivative is univalent so as to minimize antigenic modulation (10), and has dual Fc regions to improve the recruitment of effectors (11). It possesses also the usual advantages of chimeric antibodies bearing Fc regions of host type: efficient effector recruitment, a prolonged metabolic survival, and a reduction in immunogenicity. By using mouse Fab'γ one continues to display immunogenic constant region epitopes on C_L and $C_\gamma 1$, a situation which will be remedied with the greater availablity of recombinant human Fab', but the problem of idiotypic epitopes will persist.

There are no SS-interchanges in this preparation which risk dimerization of Fab'(–SH)₅, so one may begin with an arbitrarily high concentration of F(ab'γ)₂. Note however that with the protein at 20 mg/ml a full reduction of the interchain SS will lower the DTT concentration by 1 mM. The ratio of Fc–maleimide to Fab'(–SH)₅ of 2:1 maximizes the yield of FabFc₂; to obtain FabFc as the main product use a ratio of 1:1. The yield of chimeric antibody will be higher if one is permitted to harvest from the sizing run on Superdex 200 those chimeric species adjoining the definitive one: if for example in the present preparation one may present a mix of FabFc₃, FabFc₂, and FabFc, while specifying that at least 60% of the protein is in the form of FabFc₂, and that the average chain composition is $\text{FabFc}_{2 \pm 0.2}$. Tests *in vitro* have not revealed any significant loss of titre for complement lysis or ADCC as a consequence of this wider harvesting.

Protocol 6.   Preparation of chimeric FabFc₂

Equipment and reagents
• See Section 4

Method

1. To a solution of antibody F(ab'γ)₂ at ≤ 20 mg/ml in buffer of pH 7.5–8.5 add one-sixth volume of 21 mM DTT in water, and incubate for 15 min at 25 °C.

2. Separate the Fab'(–SH)₅ by passing through Sephadex G-25 equilibrated with 0.05 M A5 at 5 °C.

3. Keep cold under N₂ until the addition of Fc–maleimide.[a]

4. Add the cold Fc–maleimide to cold, stirring Fab(–SH)₅ under N₂.

5. Incubate at 25 °C for 18 h.

Protocol 6. *Continued*

6. Stop the reaction by adding 0.1 vol. of 110 mM cystamine[b] in 0.3 M Tris base (→ pH 7.6–8.0) and incubating at 25°C for 30 min.

[a] The Fc–maleimide prepared as described in Section 5.4 displays about 0.33 maleimide group per molecule. On this basis a ratio of total Fc to Fab of 6:1 is chosen to yield an average ratio of 2:1 in the conjugated products.
[b] Disulfide-interchange with the cystamine removes residual SH groups and at the same time reconstitutes many γ-light chain SS bonds.

To separate the *Protocol 6* products on Superdex 200 the protein solution must first be concentrated. Again this can conveniently be carried out by cation exchange chromatography on SP–Sepharose (equilibrated with 0.05 M A5) after adjusting the pH and ionic strength of the solution by adding an equal volume of 30 mM acetic acid. For a preparation involving 400 mg F(ab'γ)$_2$ and 2.4 g Fcγ1, a column of SP–Sepharose 5 cm × 10 cm is suitable. It is loaded at 1000 ml/h and eluted with 0.5 M TBS8.7 at 250 ml/h, leading the eluate directly onto two recycling Superdex 200 columns, each 5 cm × 92 cm, equilibrated with 0.2 M ABS5.8. (We have chosen this buffer rather than the usual phosphate-buffered saline for therapeutic preparations because it has been associated with less protein binding to the polymyxin column used as a back-up for removal of pyrogens.) The Superdex run now proceeds at 500 ml/h with a total of five column passes before harvesting the FabFc$_2$ zone. For therapeutic preparations sterile, pyrogen-free water is used for the buffers, the columns are pre-sterilized with 0.2 M NaOH, and protected by bacteriological filters, and the protein is passed finally through a column of polymyxin–Sepharose (12) and 0.45 μm and 0.22 μm filters in series into a tared sterile transfusion bag (Baxter). Aliquots are taken from the bag under sterile conditions for estimating concentration (E_{280}), dispersion (HPLC), sterility (cultures for bacteria and fungi), and pyrogens (limulus amebocyte test). The volume in the bag is calculated either from the weight increment due to filling (preferred), or from the chromatography trace obtained in combination with an accurately metering pump.

7.2 Thioether-linked F(ab')$_2$ with potential hinge SH groups

The F(ab')$_2$ will usually be bispecific, capable of linking two different epitopes on the surface of the target cell, or of linking two different cells, or of recruiting an agent such as a toxin to a target cell surface (13). It may be used as a derivative in its own right without adding any further modules. Its preparation differs from our previous account of the preparation of thioether-linked F(ab'γ)$_2$ (8) in that those hinge region cysteine residues which are not involved in the intermodule thioether link are all tied up in SS bonds, one an unnatural intrachain bond (*Figure 5*). Apart from using the F(ab')$_2$ by itself we now have the important option of reducing the SS bonds (without affecting the thioether bonds) to provide SH groups to which other modules may be linked: for example a ribosome-inactivating protein such as saporin can be SS-linked to yield a bispecific immunotoxin; or, as shown in our next section, human Fc can be thioether-linked to yield a bispecific chimeric antibody.

We shall assume that this construct is to be bispecific, and that each Fab'γ partner has been stored in the form Fab–SS–Py (Section 6.3). Equal amounts of each partner are used. One is converted to Fab'(–SH)$_1$ as described in Section 6.3 and kept cold, under N$_2$, at pH 5 pending conjugation. The second partner is similarly converted to Fab'(–SH)$_1$. Immediately upon delivery from the Sephadex G-25 column add to it 0.2 times its volume of 6 mM PDM in 50% DMF/50% 0.05 M A5 (v/v) and incubate at 5 °C for 30 min. Separate the resulting Fab'–maleimide on Sephadex G-25, leading the emergent protein into stirring Fab'(–SH)$_1$. Incubate the Fab'–maleimide + Fab'(–SH)$_1$ under N$_2$ at 25 °C for 18 h.

Subsequent purification of the bispecific F(ab'γ)$_2$ depends upon plans for its use, or for any subsequent modular additions to it. The immediate addition of Fcγ1 modules is described in the next section. If the F(ab'γ)$_2$ is to be used in its own right one must decide whether any homodimers Fab'γ–SS–Fab'γ which have formed by oxidation or untoward SS-interchange need to be removed: this entails reduction of the protein, preferably before purification of the dimer on a sizing gel. If the reduction is carried out a further decision is required regarding the released SH: should these groups simply be blocked by alkylation, or should the intramodular SS bonds be reconstituted by SS-interchange with cystamine in dilute protein solution (as in the preceding section)? Favouring the latter more complicated course is the retention of the option to add further modules, and the possibility that the function of some antibodies might be enhanced by reconstitution of many of the γ-light chain SS bonds. The simplest course is to block residual SH groups by alkylation at the end of the incubation period: to the reaction mixture add 0.05 vol. of 40 mM NEM in 50% DMF/50% 0.05 M A5 and incubate at 25 °C for 15 min. The protein can then be concentrated by a cycle of absorption and elution on SP–Sepharose, and the dimer purified by passage through Superdex 200 as described in previous sections. The extent of homodimer formation may be checked at any time by comparing HPLC patterns given by the purified construct and by a reduced and alkylated aliquot; remedial action can then be taken in the light of the above discussion.

7.3 Bispecific chimeric derivatives from thioether-linked F(ab')$_2$

When F(ab'γ)$_2$ prepared as in the preceding section is reduced one has available eight SH groups for the attachment of further modules. We have prepared a variety of such derivatives with the modular formula Fab$_2$Fc$_4$ (*Figure 6*) which have proved strikingly effective in invoking complement cytotoxicity against human neoplastic B lymphocytes. *Protocol 7* describes the preparation of one such derivative which is bispecific for the lymphocytic surface antigens CD20 and CD38.

Protocol 7. Preparation of bispecific chimeric derivatives from thioether-linked F(ab′)$_2$

Equipment and reagents
- See Section 4

Method

1. Starting with equal quantities of Fab′–SS–Py (anti-CD20) and Fab′–SS–Py (anti-CD38), proceed to prepare the bispecific dimer as described in Section 7.2.

2. At the end of the 18 h incubation of Fab′(–SH)$_1$ and Fab′–maleimide, concentrate the protein by leading it onto SP–Sepharose equilibrated with 0.05 M A5.[a]

3. Elute with 3 mM DTT in 0.5 M TBS8.7, leading the eluate onto Superdex 200 which has been equilibrated with 0.1 M A5 but has then had 0.2 times its volume of the DTT/eluting buffer pre-introduced.[b]

4. Add immediately Fc–maleimide in 0.05 M A5 (Section 5.4) and incubate under N$_2$ at 25°C for 18 h.[c]

5. Add 0.1 vol. of 110 mM cystamine in 0.4 M Tris base ($\rightarrow$ pH 7.8–8.0) and incubate at 25°C for 30 min.[d]

6. Dilute the reaction mixture and acidify to permit concentration on SP–Sephadex as described previously.

7. Separate Fab$_2$Fc$_4$ on Superdex 200 as described for FabFc$_2$.

[a] The elution proceeds with simultaneous reduction.
[b] In this way the F(ab′γ)$_2$ is separated from other products and residual monomer while in the presence of DTT, but the last part of the column passage sees it separated from the DTT and delivered as F(ab′)$_2$(–SH)$_8$ in the acetate buffer at pH 5.0.
[c] To obtain Fab$_2$Fc$_4$ as the predominant product the total amount of Fc (carrier plus reactive protein) should be six times the total amount of F(ab′γ)$_2$.
[d] This ends the reaction by carrying out SS-interchange at pH ~ 8.0.

References

1. Nisonoff, A. and Mandy, W. J. (1962). *Nature*, **194**, 355.
2. Fuijo, H. and Karush, F. (1966). *Biochemistry*, **5**, 1856.
3. Seegan, G. W., Smith, C. A., and Schumaker, V. N. (1979). *Proc. Natl. Acad. Sci. USA*, **76**, 907.

4. Isenman, D. E., Dorrington, K. J., and Painter, R. H. (1975). *J. Immunol.*, **114**, 1726.
5. Grassetti, D. R. and Murray, J. F. (1967). *Arch. Biochem. Biophys.*, **119**, 41.
6. Brocklehurst, K. (1979). *Int. J. Biochem.*, **10**, 259.
7. Grimshaw, C. E., Whistler, R. L., and Cleland, W. W. (1979). *J. Am. Chem. Soc.*, **101**, 1521.
8. Glennie, M. J., McBride, H. M., Worth, A. T., and Stevenson, G. T. (1987). *J. Immunol.*, **139**, 2367.
9. Stevenson, G. T., Glennie, M. J., and Kan, K. S. (1993). In *Protein engineering of antibody molecules for prophylactic and therapeutic applications in man* (ed. M. Clark), pp. 127–41. Academic Titles, Nottingham, UK.
10. Lane, A. C., Foroozan, S., Glennie, M. J., Kowalski-Saunders, P., and Stevenson, G. T. (1991). *J. Immunol.*, **146**, 2461.
11. Stevenson, G. T., Pindar, A., and Slade, C. J. (1989). *Anti-Cancer Drug Design*, **3**, 219.
12. Issekutz, A. C. (1983). *J. Immunol. Methods*, **61**, 275.
13. Glennie, M. J., Brennand, D. M., Bryden, F., McBride, H. M., Stirpe, F., Worth, A. T., *et al.* (1988). *J. Immunol.*, **141**, 3662.

2

Genetically engineered antibodies

1. Introduction

Monoclonal antibody technology has produced immunoglobulins with specificities to a wide range of clinically relevant antigens, including tumour-associated markers, viral coat proteins, and lymphocyte cell surface glycoproteins. However, the use of rodent antibodies in human diagnosis or therapy poses a number of problems. Foremost amongst these is the immunogenic nature of the immunoglobulins (1). This in turn limits the use of repeated doses of antibody essential for chronic treatment. The availability of human monoclonal antibodies would greatly reduce this problem but such antibodies are generally restricted in their range of specificities. Further, human hybridomas are often unstable and/or poor producers preventing the production of large amounts of the human immunoglobulin.

Intrinsic properties of immunoglobulins, whether rodent or human derived, may also limit their effective use. For example, for certain applications particular effector functions of the antibody may be required. However, the appropriate isotype may not be readily obtained with the correct specificity. The relatively large molecular size of immunoglobulins may also militate against their effectiveness *in vivo*. Thus antibody fragments may be preferred for some clinical applications, for example tumour imaging, since they have much shorter half-lives *in vivo* compared with whole antibody (1).

Some of the limitations of monoclonal antibodies as therapeutic agents have been addressed by genetic engineering. A new generation of engineered human antibodies have been developed in which much of the rodent-derived sequence of an antibody has been replaced with sequences from human immunoglobulins. Such antibodies represent an alternative to human monoclonal antibodies and are proving to be less immunogenic in man (1).

Furthermore genetic truncation has been used to produce antibody fragments, including the smallest antigen-binding unit or Fv. These in turn have formed the building blocks of novel antigen-binding molecules. To produce these antibodies for clinical use has required the development of high level expression systems. Mammalian cells generally provide the highest yields of whole recombinant antibodies, whereas production in *E. coli* using secretion vectors has emerged as the system of choice for antibody fragments. The cloning and expression technology can also be used to rescue otherwise unstable or poorly produced human monoclonal antibodies for re-expression in high yielding mammalian or microbial systems.

In this chapter the experimental methods for genetically engineering antibodies will be described. For the most part the techniques will be exemplified by reference to the modification of mouse antibodies for use in human therapy.

2. PCR cloning of immunoglobulin variable domains

The use of hybridization methods for isolating antibody variable region genes (see for example refs 2 and 3) has been superceded by strategies based on the polymerase chain reaction (PCR). The method depends upon amplification of the required gene through repeated rounds of extension between two primers that hybridize to the 5′ and 3′ ends of the sequence. The primers need not match the template sequence exactly, except at the very 3′ end and often include restriction enzyme sites to permit subsequent cloning of the amplified gene.

Two general PCR strategies have been reported for the cloning of both human and mouse variable domains. The approaches differ in the locations of the 5′ PCR primers. In the first method, a set of upstream primers has been designed that hybridize to the DNA sequence encoding the signal peptides of the heavy and light chains of mouse (4) and human antibodies (5). A potential weakness of this approach is the possibility that for any new antibody there may not be a primer that recognizes or is sufficiently similar to its leader to allow amplification. The second method avoids this since the 5′ primers bind to relatively conserved sequences in the first framework region of heavy and light chain variable domains. Again sets of primers for cloning both mouse (6) and human (7) variable regions have been designed.

Both the above type of 5′ primers have been combined with 3′ primers that recognize either the mouse or human constant regions or the J segments of the variable domains (*Figure 1*). Choosing the position of the 3′ primer depends upon the form of recombinant antibody that is required. For example, to produce a mouse Fab fragment the 3′ primers should bind to the 3′ ends of the $C_{kappa/lambda}$ and C_H1 gene segments respectively. The use of appropriate J region primers enables chimeric whole antibody genes and single-chain Fv fragment genes to be constructed directly.

2.1 Synthesis of first strand cDNA

The starting point for cDNA cloning is the isolation of mRNA from the antibody-producing cells. The guanidium thiocyanate/phenol chloroform extraction method of Chomcynski and Sacchi (8) is probably the most widely used. A commercially available reagent RNAzol (Cinnabiotecx) provides the

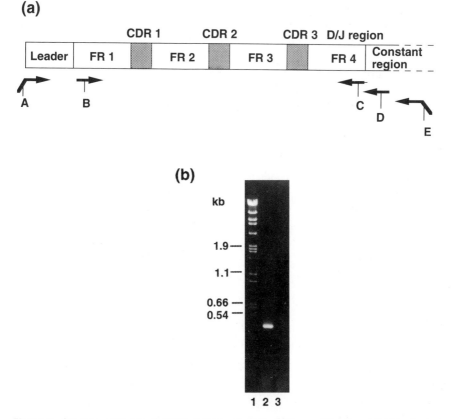

Figure 1. Cloning antibody variable region genes by PCR amplification. (a) The diagram shows the alternative positions of oligonucleotide primers and restriction sites which enable the PCR amplified gene to be cloned. Restriction enzyme sites are either integral to the sequence recognized by the primer (B, C, and D) or are incorporated at the 5' ends of the priming sites (A and E). FR, framework region; CDR, complementarity determining region. (b) The photograph shows the results of PCR amplification of a variable region gene (V_H) analysed on a 1.2% agarose gel stained with ethidium bromide and viewed under UV light. The V_H gene (track 2) was amplified according to the procedure in *Protocol 1* using leader sequence 5' primers (*Table 1*). Track 3 is a control without template and DNA size markers are shown in track 1.

simplest version of the method and typically yields 1 mg total RNA from 10^8 hybridoma cells. Further purification of the mRNA by oligo(dT) affinity chromatography is not necessary for subsequent PCR amplification of the variable domains. Two options are available for the production of first strand cDNA. Reverse transcription may be non-specifically primed with random hexamers in which case only a single reaction is required to produce the PCR template. Alternatively the first strand cDNA may be generated via specific constant region primers with separate reactions for heavy and light chains.

Liu *et al.* (9) have taken the approach to the ultimate limit and report the PCR cloning of variable domains from individual B cell hybridoma cells. The protocol requires two rounds of PCR amplification to produce sufficient template for sequencing.

2.2 Primer design

The design of PCR primers discussed in detail in this section is restricted to the cloning of mouse variable domains. However, the principles will apply to the isolation of variable region genes of other species.

2.2.1 5′ primers

Sets of leader sequence primers for the amplification of both mouse V_H and V_L sequences are shown in *Tables 1* and 2. The primers have been designed to hybridize to all possible known variable domain leader sequences (10). In view of the relatively large number of primers required it is advisable to synthesize them as mixtures with low redundancy. These can then be pooled such that each individual sequence is present in equal amounts for the PCR reaction. A restriction enzyme site and Kozak sequence (11) is included at the 5′ end of each primer. Restriction sites that rarely occur in variable domain sequences are shown in *Table 3* (12). The final choice of site is generally determined by the availability of restriction sites in the vector used to clone the V region sequence. Restriction sites close or at the 5′ ends of DNA fragments are often less efficiently digested than internal sites. Therefore a short sequence should be added to the 5′ end of the restriction site. The inclusion of a Kozak motif ensures that translation initiation is optimized. A typical mouse V_L primer (*Table 2*, No. 4) is shown below.

5′ GCGCGC<u>TTCGAA</u>GCCGCCACCATGAGGTGCGCTCACTCAGGTCCT 3′
        *Bst*bI   Kozak   Leader

The sets of primers that have been designed to recognize the amino terminal framework 1 region of mouse V_H and V_L sequences are shown in *Tables 1* and *2*. Much fewer and shorter primers are required to specify the framework region consensus sequence compared to the leader sequences. The choice of 5′ cloning site is dictated by the consensus sequence and hence is highly constrained compared to the leader primers. Since the cloned variable regions lack signal sequences, these must be added in order to permit expression. A potential problem with the framework primers is that the consensus sequence becomes incorporated into the variable region sequence and may result in changes to amino terminal residues compared to the authentic sequence. By definition any changes due to somatic mutation are not accommodated by these primers. This may affect the antigen-binding activity of the resulting antibody since the N-termini of both variable domains are contiguous with the complemenatarity determining region (CDR) surface (13).

Table 1. Mouse V$_H$ PCR primers[a]

5′ leader sequence primers

 1. ATGAAATGCAGCTGGGTCATSTTCTT
 2. ATGGGATGGAGCTRTATCATSYTCTT
 3. ATGAAGWTGTGGTTAAACTGGGTTTT
 4. ATGRACTTTGGGYTCAGCTTGRT
 5. ATGGACTCCAGGCTCAATTTAGTTTT
 6. ATGGCTGTCYTRGSGCTRCTCTTCTG
 7. ATGGRATGGAGCJGGRTCTTTMTCTT
 8. ATGAGAGTGCTGATTCTTTTGTG
 9. ATGGMTTGGGTGTGGAMCTTGCTATT
10. ATGGGCAGACTTACATTCTCATTCCT
11. ATGGATTTTGGGCTGATTTTTTTTATTG
12. ATGATGGTGTTAAGTCTTCTGTACCT

5′ framework primers

        *Pst*I
AGGTSMAR<u>CTGCAG</u>SAGTCWGG
        *Xho*I
AGGTSMARCTJ<u>CTCGAG</u>TCWGG

3′ J region primers I

      *Apa*I
 1. GCAGAT<u>GGGCC</u>CTTCGTTGAGGCTGMRGAGACDGTGASTGA
 2. GCAGATGGGCCCTTCGTTGAGGCTGMRGAGACDGTGASCAG
 3. GCAGATGGGCCCTTCGTTGAGGCTGMRGAGACDGTGASMGT

3′ J region primer II

        *Bst*EII
TGAGGAGACGGTGACCGT<u>GGTCCC</u>TTGGCCCCAG

[a] All primers are shown 5′–3′.
Key: R = A/G, Y = T/C, W = A/T, J = T/G, M = A/C, S = C/G, D = A/G/T.

2.2.2 3′ primers

Primers that will bind to all known J region sequences have been designed for both mouse V$_H$ and V$_L$ domains (*Tables 1* and *2*). These primers may be modified at their 5′ ends to permit the direct construction of mouse/human chimeric light and heavy chains. Two options for constructing the chimeric genes are possible. In the first case the 3′ primer is used to introduce an in-frame restriction site that is present at the 5′ end of constant region, for example *Apa*I for V$_H$ and *Spl*I for V$_L$ (*Tables 1* and *2*). This permits the direct fusion of the variable domains to the constant regions. In the second option the 3′ primer introduces a restriction site at the 3′ end of the variable domain, for example *Bst*EII for V$_H$ and *Bgl*II for V$_L$. This enables the variable

Table 2. Mouse V$_L$ PCR primers[a]

5′ leader sequence primers

1. ATGAAGTTGCCTGTTAGGCTGTTGGTGCT
2. ATGAATTTGCCTGTTCATCTCTTGGTGCT
3. ATGGAGWCAGACACACTCCTGYTATGGGT
4. ATGAGTGTGCTCACTCAGGTCCT
5. ATGAGGRCCCCTGCTCAGWTTYTTGG
6. ATGGATTTWCAGGTGCAGATTWTCAGCTT
7. ATGGATTTWCARGTGCAGATTWTCAGCTT
8. ATGGATTTTCAATTGGTCCTCATCTCCTT
9. ATGAGGTJCYYTGYTSAGYTYCTGRG
10. ATGAGGTGCCTARCTSAGTTCCTGRG
11. ATGAAGTACTCTGCTCAGTTTCTAGG
12. ATGAGGCATTCTCTTCAATTCTTGGG
13. ATGGGCWTCAAGATGGAGTCACA
14. ATGTGGGGAYCTJTTTYCMMTTTTTCAAT
15. ATGGTRTCCWCASCTCAGTTCCTT
16. ATGTATATATGTTTGTTGTCTATTTC
17. ATGGAAGCCCCAGCTCAGCTTCTCTT
18. ATGRAGTYWCAGACCCAGGTCTTYRT
19. ATGGAGAGACACATTCTCAGGTCTTTGT
20. ATGGATTCACAGGCCCAGGTTCTTAT
21. ATGATGAGTCCTGCCCAGTTCCTGTT

5′ framework one primer

*Pvu*II
GACATT<u>CAGCTG</u>ACCCAGTCTCCA

3′ J region primer I

*Sp*II
GGATACAGTTGGTGCAGCATC<u>CGTACG</u>TTT
CC<u>CGTACG</u>TTTGATYRCCAGCTTGGTSCC

3′ J region primer II

*Bg*III
GTTA<u>GATCT</u>CCAGCTTGGTCCC

[a] All primers are shown 5′–3′.
Key: R = A/G, Y = T/C, W = A/T, J = T/G, M = A/C, S = C/G, D = A/G/T.

region to be joined via a 3′ splice site to the intron between the variable and constant regions. Vectors have been produced for the expression of both forms of the antibody.

The J segment primers may also be extended to include a linker to facilitate construction of single-chain Fv fragments (see Section 4).

Alternatively if the authentic J region sequence is required 3′ PCR primers that bind to either the C$_{kappa}$ or C$_H$1 sequences may be used. Subsequently

Table 3. Frequency of restriction enzyme sites in the variable domains of mouse V_H and V_L sequences

Restriction site	Frequency		Restriction site	Frequency	
	Light chain[a]	Heavy chain[b]		Light chain[a]	Heavy chain[b]
AatI	0	1	HindIII	1	3
AftI	0	1	HpaI	0	0
AvaIII	5	0	KpnI	17	2
ApaI	0	0	MluI	0	2
ApaLI	1	1	NarI	0	8
AseI	9	16	NcoI	ND	10
BamHI	12	3	NdeI	2	1
BclI	1	1	NheI	0	0
BglI	0	3	NotI	ND	0
BstBI	0	0	NruI	0	0
BspH1	0	2	PflMI	13	0
BstEII	6	13	PstI	29	46
BstXI	19	3	PvuII	0	35
Bsu36I	2	8	SacI	0	5
DrdI	ND	33	SacII	0	1
ClaI	0	0	SalI	0	0
EagI	0	0	SmaI	0	0
EcoRI	0	6	SpII	2	0
EcoRV	0	7	XbaI	ND	0
EspI	0	32	XhoI	ND	0

ND: not determined.
[a] 50 sequences from Kabat's data base were analysed.
[b] 100 sequences from Kabat's data base were analysed.

the cloned V regions can be manipulated to produce the required engineered antibody or fragment.

2.3 Problems with the method

Two problems are frequently encountered when using PCR to amplify variable region genes. The first is a general one associated with the use of *Taq* polymerase, which lacks a proof-reading function. This results in an error frequency of approximately 1/1000 bases which means that for the most part error-free versions of variable region sequences are obtained. Errors in amplified V_H and V_L sequences can be readily identified by sequencing several clones from two independent amplifications.

The second problem is specific to antibody gene cloning. All myeloma fusion partners derived from the original MOPC21 tumour (X63-Ag8, NS-1, P3X63Ag8.653) express an aberrant V_L mRNA (14). This sequence is characterized by a 4 bp deletion at the V/J junction and the substitution of the invariant cysteine at residue 23 with tyrosine (Genbank No. 05184). Thus the mRNA will not produce a functional light chain. However, this sequence is readily cloned using either the framework or leader sequence primer sets. With respect to the leader sequence primers the pseudolight chain is recognized by primer set 3. It is suggested that a separate PCR amplification is carried out using this primer to distinguish the B cell-derived light chain from the aberrant one.

Whilst the pseudo light chain is usually co-amplified with the authentic V_L sequence we have consistently obtained only one V_H sequence using the V_H primers. However, it has been reported that the P3-X63 Ag8.653 also produces an aberrant V_H transcript (Genbank No. X58634) which is distinguished by a frameshift mutation in framework four leading to a nonfunctional heavy chain.

Protocol 1.   PCR amplification of variable regions

Equipment and reagents

- Thermal cycler (e.g. Perkin Elmer)
- 5 × reverse transcriptase buffer: 500 mM Tris pH 8.3, 375 mM KCl, 15 mM MgCl$_2$
- 10 × PCR buffer: 100 mM Tris pH 8.3, 15 mM MgCl$_2$, 0.1% (w/v) gelatin, 500 mM KCl

- RNasin (Promega)
- Random primers (Pharmacia)
- Reverse transcriptase (Moloney murine leukemia virus RNase H⁻ Gibco-BRL)
- *Taq* polymerase (Perkin Elmer).

A. *Synthesis of first strand cDNA[a]*

1. Mix 10 μg of total RNA dissolved in H$_2$O with 100 pmoles random primers or light/heavy chain specific primer in total volume of 10 μl.

2. Heat to 70°C for 10 min and then place immediately on ice.

3. Add the following reagents to the primed RNA:
    - 5 × reverse transcriptase buffer   4 μl
    - 0.1 M DTT   2 μl
    - 10 mM dNTPs   1 μl
    - RNasin   1 μl
    - Reverse transcriptase (200 U/μl)   1 μl

4. Mix and incubate at 42°C for 1 h.

B. *PCR amplification*

1. Mix the following reagents:
    - First strand cDNA   5 μl
    - 10 × PCR buffer   10 μl
    - 2.5 mM dNTPs   10 μl
    - 10 pmoles of each 5′ and 3′ primers
    - *Taq* polymerase   0.2 μl
    - H$_2$O to a final volume of 100 μl

2. Overlay the reaction mixture with 40 μl of paraffin oil and amplify using the conditions 94°C 1 min, 55°C 1 min, 72°C 1 min, for 30 cycles.

3. Remove the paraffin oil and extract the amplified mix twice with an equal volume of phenol/chloroform (1:1, v/v) and once with chloroform. At this point the yield of amplified product may be assessed by agarose gel electrophoresis. If necessary the sample may be concentrated by ethanol precipitation prior to restriction enzyme digestion and gel purification.

[a] First strand cDNA synthesis kits are available in which all components are provided, e.g. Pharmacia.

3. The construction of engineered human variable domains

The construction of engineered human variable domains offers a viable alternative to isolating human antibody variable regions. In many cases it provides the only way of obtaining high affinity antibodies to clinically relevant human self-antigens, for example, cytokines and tumour-associated markers. The concept involves the production of hybrid variable regions which combine the antigen-binding site of the rodent-derived antibody with human framework sequences. A balance has to be struck between maximizing the human sequence content of the engineered variable regions whilst at the same time maintaining the antigen-binding activity. Two general approaches have been described for achieving this which are discussed in the following section.

3.1 Designing the sequence
3.1.1 CDR replacement
The CDR regions of an antibody are defined as that part of the variable domains that show hypervariability in sequence. Each variable region contains three hypervariable sequences which together represent the antigen-binding site of the antibody (10). In general, the CDRs include the structural loops that connect the β-stranded frameworks at the top of the variable region (*Table 4*). The construction of engineered human antibodies to a variety of complex antigens has been reported (e.g. refs 15–17). In all cases it has been shown that simply introducing the CDR sequences into the human antibody background is not sufficient to obtain full binding activity. Therefore the process of CDR replacement or grafting requires the manipulation of both the CDR and framework sequences. A general approach to designing such an antibody is outlined below.

Starting with the sequence of the mouse light and heavy chain variable

Table 4. The six variable domain sequences which contribute to the antibody combining site, defined by either sequence hypervariability (complementarity determining regions, CDR) (10) or structural loops (18)

Variable domain	Sequence[a] (CDR)	Structure[a]
V_H		
1	31–35	26–32
2	50–65	53–55
3	95–102	96–101
V_L		
1	24–34	26–32
2	50–56	50–52
3	89–97	91–96

[a] Residues are numbered according to ref.10.

regions, the first step in the design process is to identify the boundaries of the CDRs. Most commonly these are defined according to Kabat *et al.* (10) and, with the exception of CDR2 on the heavy chain (H2), these include all the sequences required to determine the local conformation of the apical loops (*Table 4*). In the case of H2, residues adjacent to the Kabat CDR are important in determining the loop conformation (see below).

Next, all framework residues which could influence the disposition of CDRs and hence the structure of the combining site are considered. These include those which are in contact with the CDRs and those which are at the interface of the two variable regions. A number of not necessarily mutually exclusive strategies have been described for identifying the important framework residues for designing a CDR-grafted antibody. In some cases a molecular model of the Fv has been constructed to guide the sequence design (16,20). However, given the high level of structural conservation amongst antibody Fv regions this is not a prerequisite for successful CDR replacement (15,17). In fact examination of known antibody structures provides important information. For example, Tramantano (19) argued on structural grounds alone that residue 71 in framework 4 of V_H is the major determinant of the position of H2. A combination of site-specific mutagenesis studies and examination of molecular structures has led to the identification of a subset of framework residues that directly or indirectly interact with the CDR loops and determine their packing (*Table 5*). It can be seen from *Tables 4* and *5* that many of these important framework residues flank the CDRs. Among these flanking positions are most of the framework residues that are involved in both contacting the opposite domain and the CDRs. Moreover, all framework residues which have been observed to participate in antigen binding are

Table 5. Variable domain framework residues that make contact with CDRs and/or the opposite domain

Variable domain	Framework residue[a]
V_H	1, 2, 4, 24, 27–30, 36–40, 43–49, 66–69, 71, 73, 78, 80, 82, 86, 91–94, 103, 105
V_L	1–5, 7, 22, 35, 36, 38, 43–46, 48, 49, 58, 60, 62, 66, 67, 69, 70, 71, 85, 87, 98, 100

[a] Residues numbered according to ref. 10.

found in the residues adjacent to the CDRs (13). Another group of framework residues that contact the CDRs are the amino termini of both V_L and V_H. The inclusion of some or all of these residues from the donor mouse antibody, providing they differ in the acceptor human framework, has been shown to be necessary to retain full binding activity.

In practice, the number of framework changes required to design the CDR-grafted variable domain depends upon the degree of homology between the framework of the donor mouse antibody and acceptor human antibody. Two ways of selecting the acceptor human sequence have been described. In the first, a generic human framework is chosen, usually based upon the availability of an X-ray structure, e.g. KOL for V_H and REI for V_L (15,17). This framework is then modified to accept the CDR loops of a specific mouse antibody. In the second approach, the human frameworks are chosen from a database of known antibody sequences that show greatest homology to the framework of the mouse antibody. The V_L and V_H frameworks are not necessarily from the same antibody. It is important to screen the chosen frameworks for residues not usually found in the consensus for the human antibody subgroup from which they have been derived. In practice this may be achieved by first matching the mouse sequence to the human subgroup consensus and then selecting the best matched member of that group as the human acceptor framework. Some fine-tuning may still be necessary (20,21).

A further design consideration is the packing of the two domains. For the most part contact residues at the V_H/V_L interface are conserved between mouse and human antibodies. However if the human frameworks are not from the same antibody, those residues, not already considered as CDR contacts, should be checked to ensure compatability across the V_H/V_L interface (13).

3.1.2 Surface residue replacement

An alternative to CDR replacement for the construction of engineered human variable domains has been described. Padlan (22) suggested that it may be possible to reduce the antigenicity of non-human derived variable

Table 6. Variable domain framework residues exposed or partially exposed on the antibody surface (from ref. 13)

Variable domain	Framework residue[a]
V_H	1, 3, 5, 7, 8, 10, 11, 13, 14, 15, 16, 17, 19, 21, 23, 25, 26, 28, 30, 40–44, 46, 68, 70, 72, 73–77, 79, 81, 82a,b, 83–85, 87–89, 105, 108, 110, 112, 113
V_L	1–3, 5, 7–10, 12–18, 20, 22, 37, 39–43, 45, 57, 59, 60, 63, 65–70, 72, 74, 76, 77, 79–81, 83, 100, 101, 103, 105, 106, 107

[a] Residues numbered according to ref. 10.

regions by simply replacing those surface exposed residues in its frameworks with the equivalent residues from human antibodies. Since many of the CDR contacting and all interdomain contacting residues are relatively buried, any effect of the substitutions on the antigen-binding characteristics would be minimized. Examination of the exposure patterns for the variable domains of a number of human and mouse Fv structures reveals a high degree of similarity (22). This level of conservation indicates that the solvent exposure of the framework residues of an antibody of unknown structure may be predicted with a high degree of confidence. The framework residues whose side chains are at least partly exposed are listed in *Table 6*. The procedure that was proposed by Padlan (22) was to match the target non-human V_H and V_L sequences to the most homologous human frameworks and then replace those residues from the list of exposed residues with the human equivalent where this differs from the mouse sequence.

This approach does not take into account exposed residues that may contact the CDRs, for example the amino terminus of V_L. Therefore reference should also be made to the list in *Table 5* in producing the final design.

3.1.3 Worked example

To illustrate the two approaches to producing engineered human variable region sequences the design of a specific sequence is shown in *Figure 2*. The antibody B72.3 recognizes the tumour antigen TAG-72 (3). The sequences of both V_L and V_H designed using either CDR replacement or surface replacement are shown. As can be seen only a small number of mouse framework residues are retained in the engineered human sequences. For the most part the same residues are identified by both design methods.

3.2 Assembling the sequence

The manipulation of antibody variable regions to produce engineered versions has been carrried out in a number of ways. Early methods based on site-directed mutagenesis have largely been replaced with total gene synthesis

(a) V_L

```
                                                            43 45  48
B72.3    DIQMTQSPAS LSVSVGELVT ITC cdr1 WYQQKQG KSPQLLVY cdr2 GV PSRFSGSGS
Eu       DIQMTQSPST LSASVGDRVT ITC cdr1 WYQQKPG KAPKLLMY cdr2 GV PSRFIGSGS
hB72.3a  DIQMTQSPSS LSASVGDRVT ITC cdr1 WYQQKPG KSPQLLVY cdr2 GV PSRFIGSGS
hB72.3b  DIQMTQSPSS LSASVGDRVT ITC cdr1 WYQQKPG KAPKLLVY cdr2 GV PSRFIGSGS


         70,71           84,85         101
B72.3    G TQYSLKINSL QSEDFGSYYC cdr3 FGGGTRLEI KR
Eu       G TEFTLTISSL QPDDFATYYC cdr3 FGQGTKVEI KR..FR4 is human subgroup 1
hB72.3a  G TQYTLKISSL QSEDFATYYC cdr3 FGGGTKVEI KR  6 mouse B72.3 frameworks
hB72.3b  G TEYTLKISSL QSEDFGSYYC cdr3 FGQGTKVEI KR  4 mouse B72.3 frameworks
```

(b) V_H

```
              12       20        27          37,38 42       48        66,67 69
B72.3    QVQLQQSDAE LVKPGASVKI SCKASGYTFT cdr1 WAKQKPEQGL EWIG cdr2  KATLTA
Eu       QVQLVQSGAE VKKPGSSVKV SCKASGGTFS cdr1 WVRQAPGQGL EWMG cdr2  RVTITA
hB72.3a  QVQLVQSGAE VKKPGSSVKV SCKASGYTFT cdr1 WAKQKPEQGL EWIG cdr2  KATLTA
hB72.3b  QVQLVQSGAE VVKPGSSVKI SCKASGYTFS cdr1 WAKQKPGQGL EWIG cdr2  KATLTA

         73        82c        93,94
B72.3    DKSSSTAYMQ LNSLTSEDSA VYFCKR cdr3 WGQG TLVTVSS
Eu       DESTNTAYME LSSLRSEDTA FYFCAG cdr3 WGQG TLVTVSS  FR4 is human subgroup 1
hB72.3a  DKSTNTAYME LSSLRSEDTA FYFCKR cdr3 WGQG TLVTVSS..11 B72.3 mouse frameworks
hB72.3b  DESTNTAYME LNSLRSEDTA FYFCKR cdr3 WGQG TLVTVSS..12 B72.3 mouse frameworks
```

Figure 2. Design of the human engineered variable domains of the antibody B72.3 (3). Aligned are the framework (i.e. non-CDR) amino acid sequences of the variable regions (V_H and V_L) of the parent mouse antibody B72.3, the human antibody, EU (10), and the human engineered (hB72.3) antibody designed by either CDR replacement (hB72.3a) or surface residue replacement (hB72.3b). The mouse framework residues retained in the EU framework sequences are shown in bold type and numbered above the sequences according to Kabat *et al.* (10). CDR, complementarity determining region.

using PCR. Lewis and Crowe (23) have adapted the PCR strand overlap method of mutagenesis to replace the CDRs in a human variable domain sequence with those from a rodent antibody. Although several rounds of PCR amplification are required to build the CDR replaced variable domain, correctly assembled sequences are obtained in high yield. As an alternative Daugherty *et al.* (24) have used PCR to synthesize engineered variable regions in one step using six overlapping 80 bp oligonucleotides (*Figure 3*). This is by far the most rapid method for producing the sequences but suffers from a relatively high error rate. Consequently several variable region genes may have to be sequenced to obtain an error-free product.

The sequences of the assembly oligonucleotides are designed in the following steps.

(a) The amino acid sequence designed by either CDR replacement or resurfacing is backtranslated into a DNA sequence. Codon usage for mouse antibodies is used.

(b) The sequence is checked for the presence of the restriction sites which will be used to clone the sequence or any subsequent manipulation of vectors. These (if any) are removed by altering the sequence.

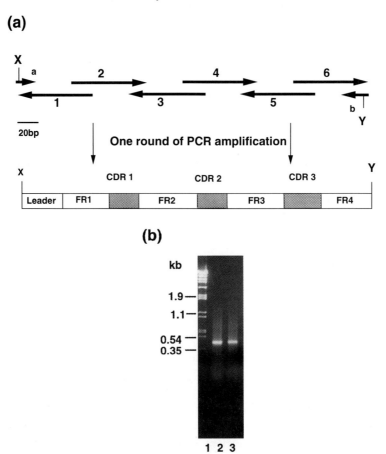

Figure 3. Synthesis of CDR-grafted variable region genes by PCR amplification. (a) The diagram shows the strategy for assembling variable region genes from overlapping oligonucleotides by the polymerase chain reaction. Six 80 bp oligonucleotides (1–6: 1 pmol each) are mixed with two 20 bp flanking primers (a and b: 10 pmol each) using the conditions in *Protocol 2*. (b) Samples were analysed on an agarose gel following one round of PCR amplification (tracks 2 and 3). DNA size markers are shown in track 1 of the photograph.

(c) Any internal repeats are eliminated by changing the sequence.
(d) The final sequence is divided into 80mer fragments according to the scheme shown in *Figure 3*.

Protocol 2. Construction of engineered variable region genes

Equipment and reagents
• For PCR reagents see *Protocol 1*

40

Method

1. Mix the following reagents:
   - 10 × PCR buffer   10 μl
   - 2.5 mM dNTPs   10 μl
   - 1.0 pmole each assembly oligonucleotide
   - 10 pmole each amplifying primer
   - *Taq* polymerase   0.2 μl
   - dH$_2$O to final volume of 100 μl

2. Incubate using the conditions of *Protocol 1* and process product as described.

3. If necessary, re-amplify the primary product. To do this set-up the following reaction:
   - 10 × PCR buffer   10 μl
   - 2.5 mM dNTPs   10 μl
   - First round PCR mix   1 μl
   - 10 pmole each amplifiying primer
   - *Taq* polymerase   0.2 μl
   - dH$_2$O to final vol of 100 μl

4. Repeat step 2.

4. Construction of Fvs and single-chain Fvs

Recombinant Fv fragments have been produced by either isolating the variable region gene segments or by inserting stop codons into heavy and light chain genes (25,26). The resulting Fv fragments have been expressed in both mammalian and microbial cell systems (see Section 5) and shown to retain the antigen-binding activity of the parent molecule. However, the variable domains of a Fv are only held together by relatively weak non-covalent interactions. It has been observed that at low concentrations certain Fvs dissociate. Two strategies have been described for increasing the utility of Fvs. In the first, a peptide linker is used to join the variable domains into a single polypeptide to produce single-chain Fvs (sFv). In the second, cysteine residues are inserted into each domain in such a way that a disulfide bridge is spontaneously formed, covalently joining the V_H to V_L (dsFv). The construction of these two types of Fv variant are described below. It must be noted that both represent experimental approaches whose success for any given antibody in terms of retention of full binding activity cannot be predicted.

4.1 Single-chain Fvs

A number of different linkers have been used to join the variable domains, including synthetic oligomers of (Gly$_4$Ser) (27), natural sequences derived from proteins of known three-dimensional structure (28), and a combination of the two (29). The composition of the linkers is dictated by the need for the sequence to be relatively flexible, i.e. to lack pronounced secondary structure and to be soluble. The path length of the linker has been estimated to be 0.35 nm from an examination of the X-ray structures of Fab fragments. Consistent with this, linker lengths of 14 or 15 residues appear to be optimal (30,31). Although longer sequences up to 25 residues are tolerated with some loss of binding activity, linkers shorter than 14 residues result in reduced stability and increased aggregation (30,31).

sFv constructs have been made using a variety of monoclonal antibodies to both small molecular weight haptens and complex polypeptide antigens. Although detailed analyses of affinity have not always been carried out, the results indicate that in most cases the sFv retains the monovalent binding activity of the parent antibody. By far the most common linker used is the $(Gly_4Ser)_3$ and this would be the first choice of sequence for the construction of a new sFv. Next, the order of the variable domains has to be addressed. Examples of both V_H–linker–V_L (29) and V_L–linker–V_H (31) being the most active configuration, have been reported. Differences between antibodies may reflect the relative contribution of the amino terminus of V_H and V_L to the positioning of the CDRs (13). It is recommended that both orientations of the domains are evaluated. The construction of such an sFv by PCR is outlined in *Figure 4*. The reagents and conditions for assembling the sFv are the same as given in *Protocol 1*.

4.2 Disulfide-linked Fvs

A major limitation of single-chain Fvs appears to be their tendency to aggregate which leads to precipitation on storage. Therefore, alternative ways of stabilizing Fv fragments have been investigated. For a limited number of examples the covalent assembly of V_H and V_L through an introduced disulfide bond has been reported (32,33). In each case the following criteria have been used to identify suitable positions to introduce the S-S bridge:

(a) Separation of C-alpha atoms should be similar to those in natural disulfide bonds (0.56–0.68 nm).

(b) Both residues should be completely buried at the V_H–V_L interface.

(c) Residues should permit disulfide bond formation with little or no strain.

(d) Both residues should be in the framework to minimize affects on CDR position, and provide for a generally applicable strategy.

Applying these criteria to an examination of either X-ray or model built Fv structures has led to the identification of three possible pairs of framework residues (numbered according to ref. 10):

- V_L44–V_H103 and V_L198–V_H45 (33)
- V_L100–V_H44 (32)

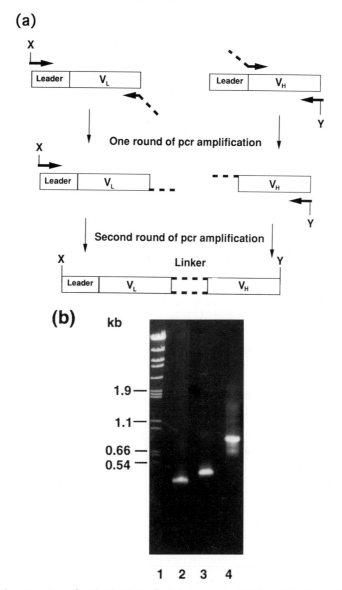

Figure 4. Construction of a single chain Fv (sFv) gene by PCR amplification. (a) The diagram shows the strategy for linking two variable domain genes together via a short oligonucleotide sequence to produce an sFv gene. (b) The results of one round of PCR amplification using the conditions in *Protocol 1* to produce the V_H (track 2) and V_L (track 3) fragments are shown in the photograph of an agarose gel. A further round of amplification using the products from the first round and primers a and b leads to the production of the sFv gene (track 4). DNA size markers are shown in track 1.

Only one example of a dsFv has been constructed using the first two possible positions. In this case a disulfide-linked Fv was produced by secretion into the *E. coli* periplasm with approximately 30% activity of the parent Fv (33). For the third position three active dsFvs have been constructed but in these cases the dsFvs were assembled by renaturation of the two domains separately expressed intracellularly in *E. coli* (32). The generality of this disulfide bond strategy awaits further exemplification.

5. Expression of recombinant antibodies

5.1 Mammalian cells

The earliest examples of the expression of immunoglobulins in mammalian cells involved the re-introduction of antibody genes into lymphoid cells with gene expression being directed by immunoglobulin promoters/enhancers. However, antibody production from recombinant myelomas (10–100 mg/litre) rarely approached that obtained from the original hybridoma from which the light and heavy chain genes were cloned. Subsequently, improvements in expression levels have been obtained by using heterologous promoters (for example, major immediate–early gene promoter–enhancer of human cytomegalovirus, hCMV (34) and beta-actin (35)) in both myeloma and non-myeloma cell hosts. The use of such vector systems to produce antibodies in both transient and stable mammalian cell lines is described below.

5.1.1 Low copy vectors

It has been found that the hCMV promoter is one of the most efficient promoters for expression in a number of cells including chinese hamster ovary (CHO), COS, and NSO myeloma. Vectors incorporating this promoter have been developed for expression of antibody genes (3). Similar vectors are commercially available, e.g. pcDNA3 (In Vitrogen). Apart from the hCMV promoter/enhancer these plasmids contain an SV40 replication origin which enables the vectors to be used for transient expression in COS cells (*Protocol 3*). COS cells cotransfected with heavy chain and light chain containing vectors yield up to 1–2 µg antibody/ml culture medium in four to five days.

Protocol 3. Transient expression of antibodies in COS cells

Equipment and reagents

- COS-1 (ECACC No. 880317; ATCC: CRL1650) or COS-7 (ECACC No. 87021302; ATCC CRL1651) cells
- Dulbecco's modified Eagle's medium (DMEM; Gibco)
- TBS (Tris-buffered saline): 25 mM Tris–HCl pH 7.4, 137 mM NaCl, 5 mM KCl, 0.7 mM $CaCl_2$, 0.5 mM $MgCl_2$, 0.6 mM Na_2HPO_4
- Non-essential amino acids (NEAA; Gibco): 5 ml per 500 ml DMEM
- Penicillin streptomycin (P/S; Gibco): 5 ml per 500 ml DMEM
- HBS (Hepes-buffered saline): 21 mM Hepes pH 7.1, 137 mM NaCl, 5 mM KCl, 0.7 mM Na_2HPO_4, 6 mM glucose
- 0.25 M Tris–HCl pH 7.5

- DMSO
- Fetal calf serum (FCS)—heat inactivated at 56°C for 30 min
- Trypsin/EDTA (Gibco)
- DEAE–dextran (Pharmacia): 1 mg/ml in TBS

- Heavy and light chain plasmid vectors with SV40 origin of replication
- Suitable tissue culture equipment including humidified CO_2 incubator set at 37°C and 5% CO_2

Method

1. Start with a T75 flask of subconfluent COS-1 cells growing in DMEM plus 10% FCS, P/S, and NEAA.

2. Remove the medium from cells by aspiration and wash cells with 20 ml serum-free DMEM.

3. Remove DMEM and wash the cells briefly with 3 ml trypsin/EDTA.

4. Trypsinize cells by adding a second 3 ml aliquot of trypsin/EDTA to the cells and leave at 37°C for 3–5 min.

5. Dislodge the cells, add 30 ml DMEM plus 10% FCS, P/S, NEAA to inhibit trypsin, and disperse into a single cell suspension by pipetting up and down.

6. Count cells (a T75 flask typically gives approx. 2×10^7 cells). Plate cells at 2×10^5/ml per well in DMEM plus 10% FCS, P/S, NEAA in a 24-well Falcon plate. Leave cells overnight in the tissue culture incubator.

7. The following day wash the cells twice with 1 ml serum-free DMEM.

8. Prepare transfection buffer by mixing serum-free DMEM with 0.25 M Tris–HCl (4:1, v/v).

9. Mix 10 μg plasmid DNA with 1 ml transfection buffer and then add 0.25 ml DEAE–dextran (1 mg/ml).

10. Remove serum-free medium from the cells and add 1 ml DEAE–dextran/ DNA complex to each well of the 24-well plate and leave cells for 6–8 h in the tissue culture incubator.

11. Remove complex and add 1 ml 10% DMSO in HBS and leave for 2–5 min on cells.

12. Remove DMSO, wash twice with 1 ml serum-free DMEM, and replace with 1 ml DMEM plus 10% FCS, P/S, NEAA.

13. Culture the cells for 48–72 h and then harvest the medium for analysis of product secretion.

In order to establish stable cell lines selectable markers are incorporated into the vectors. Heavy and light chain genes are expressed from separate plasmids each containing a different selectable marker to enable coselection of antibody-producing cells. A possible combination is neomycin resistance (*neo*) and guanosine phosphoribosyl transferase (*gpt*) genes (34). An effective way of generating antibody-producing cell lines is to construct a light chain-producing cell line first and then to re-transfect this with the heavy chain vector. By this means, CHO cell lines producing up to 50 mg/litre antibody have been obtained (36). The approach is particularly useful if a number of different antibodies are to be made in which the same light chain is matched to a number of heavy chain variants. Details of media and methods for using *neo/gpt* selection are given in ref. 34.

5.1.2 Vector amplification

Significant improvement in expression levels can be achieved by vector amplification. An amplifiable marker usually encodes an enzyme that is both essential for cell survival and one for which an inhibitor with high affinity binding is available. By applying increasing amounts of the inhibitor to the cells expressing the desired gene product variant clones that are resistant to higher levels of inhibitor are selected. Resistance is usually the result of an increase in the number of copies of the marker gene integrated into the genome of the cells (37). The region that is amplified is usually larger than that occupied by the marker gene. Hence associated sequences, for example the antibody genes, become co-amplified, leading to higher expression levels.

The most commonly used system of vector amplification is based on the enzyme dihydrofolate reductase (DHFR) which is inhibited by methotrexate (MTX). Expression vectors for this system carry the DHFR gene and are

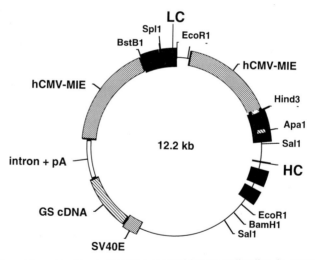

Figure 5. Plasmid vector for expressing recombinant antibodies in mammalian cells. This double gene construct is based on the Celltech vector pEE12 (41). The positions of some key restriction sites are shown. GS, glutamine synthetase selectable marker gene; hCMV MIE, human cytomegalovirus major immediate–early gene promoter/enhancer; LC, light chain gene; HC, heavy chain gene; pA, polyadenylation signal; SV40E, SV40 early promoter.

transfected into a *dhfr⁻* CHO cell line (38). A generic DHFR expression vector (pSV2dhFR) in which expression of the amplifiable marker is under the control of SV40 promoters has been described (39). High levels of antibody expression have been reported using this system (40).

A number of other amplifiable markers are available including the enzyme, glutamine synthetase (GS) which catalyses the formation of glutamine from glutamate and ammonia and is inhibited by L-methionine sulfoximine (MSX). GS can be used as a dominant selectable marker in a variety of different cell types, including myelomas, and appears to be maximally amplified in a single round by increasing the concentration of MSX applied to the transfected cells (40,41).

In order to adapt a vector amplification system for the production of antibodies both heavy and light chain genes must be integrated in close proximity to the amplifiable marker. This is most simply achieved by placing all the genes on the same plasmid (*Figure 5*). Using this approach NSO myeloma cell lines (42, and unpublished data) have been constructed expressing high levels of a recombinant antibody in suspension culture ($\geq$ 400 mg/litre). The procedure for generating such cell lines is given in *Protocol 4*.

Protocol 4. Construction of amplified antibody-producing NSO cell lines using GS

Equipment and reagents

- NSO myeloma cells (ECACC No. 85110503)
- Iscove's modification of DMEM without glutamine (IMDM; Sigma)
- Dialysed FCS (Gibco) and FCS
- PBS (Dulbecco's phosphate-buffered saline; Gibco)
- 200 mM glutamine (e.g. Gibco)
- 100 × glutamate plus asparagine (G + A): 40 mM glutamic acid and asparagine monohydrate (Sigma) in distilled water, filter sterilized, and stored at 4°C
- Gene pulser (Bio-Rad)

- 100 mM L-methionine sulfoximine (Sigma) in PBS
- 50 × nucleosides: 1.3 mM adenosine, 1.3 mM guanosine, 1.3 mM cytidine, 1.3 mM uridine, 0.5 mM thymidine (Sigma) in 100 ml distilled water, filter sterilized, and stored at –20°C
- pEE12 GS vector incorporating heavy and light chain genes (available from Celltech Therapeutics Ltd.) linearized by restriction digest
- Electroporation cuvettes (0.4 cm; Bio-Rad)

Method

1. Grow up NSO cells (10^7 for each transfection) in IMDM containing 10% FCS and 2 mM glutamine. Cells should be $\geq$ 90% viable.

2. Harvest cells by centrifugation at 500 *g* for 5 min at room temperature.

3. Wash cells once in cold PBS by centrifuging and place on ice.

4. Resuspend in PBS at 10^7 cells/ml and add 1 ml to an electroporation cuvette, on ice containing 40 μg of the linearized plasmid to be transfected.

5. Leave cell–DNA mixture for 5 min.

Protocol 4. *Continued*

6. Wipe the outside of the cuvette dry and electroporate plasmid into the cells using two consecutive pulses at 1500 V, 3 µFd according to manufacturers instructions.

7. Return cuvette to ice for 5 min.

8. Plate 50 µl cells/well in IMDM containing 10% FCS and 2 mM glutamine over 96-well tissue culture plates (e.g. 1×10^5 cells/ml over 20 plates). Place in tissue culture incubator overnight.

9. Add 150 µl IMDM plus 10% dialysed FCS, $1 \times G + A$, and $1 \times$ nucleosides to each transfected well and return to incubator.

10. Leave for 12–18 days until colonies of glutamine-independent cells appear (frequency approx. $2–5/10^5$ cells plated).

11. Pick colonies, expand, and assay cell supernatants for production of antibody.

12. Amplify positive cell lines by replating over 96-well plates (at approx. 10^5 cells/well) and add 10–200 µM MSX.

13. Repeat steps 10 and 11.

5.2 *E. coli*

5.2.1 Vector design

Fabs, Fvs, and sFvs have all been produced by intracellular expression in *E. coli*. In general, the polypeptides accumulate as inclusion bodies and require cycles of denaturation and refolding *in vitro* to recover functional fragments (42). Although levels of expression are generally very high (20–30% total cell protein) yields of active protein are variable and often require optimization for each fragment. By contrast the production of antibody fragments as secreted products greatly simplifies the recovery of the material in an active form (43,44).

The approach involves using bacterial signal sequences to direct the export of the antibody fragment to the periplasm. This offers three main advantages over intracellular expression:

(a) The amino terminal residue of the secreted product is authentic compared to the cytoplasmic product which generally has a methionine residue.

(b) Protease activity is considered to be much less in the periplasmic space than in the cytoplasm thereby increasing product stability.

(c) The product is only folded in an active form in the periplasm, thus any product dependent toxicity is avoided.

A number of different secretion cloning vectors for *E. coli* have been developed. The most commonly used bacterial leader sequence is from the *E.*

coli outer membrane protein A (OmpA) (43,45). Other leaders that have been used for antibody fragments include the signal peptide of pectate lyase B (pelB) from *Erwinia carotovara* (46) and enterotoxin II (sr-II) from *E. coli* (44). For secretion of antibody fragments comprising two polypeptide chains, e.g. Fab′, the same leader sequence can be used for each subunit.

Systems for the regulated production of heterologous proteins in *E. coli* are well established. For the inducible expression of antibody fragments two have been reported. LacZ operator/promoter (26,43) and the alkaline phosphatase promoter (phoA) (44). Expression from the lacZ operator/promoter is induced by the addition of isothiopropylgalactoside (IPTG) or lactose to the culture media. Conveniently the lacZ repressor gene, *lacI*, is incorporated into the expression vector (*Figure 6*). This permits regulated expresion even in *lacI⁺* hosts. In the case of the phoA promoter, expression is switched on by phosphate starvation. These promoter systems have been used to construct dicistronic operons comprising the V_H V_L genes or Fd heavy chain and light chain genes, to direct the expression of Fvs of Fabs respectively.

An example of a bacterial secretion vector for producing antibody fragments is shown in *Figure 6* (47). This plasmid (pACtac) is derived from the pUC based vector, pTTQ9 (48) available from Amersham International and pACYC184 (49). It contains the strong hybrid trp–lac (*tac*) promoter, 5S ribosomal gene (*rrnB*) terminators, and *lacI* gene. Efficient translation is ensured by insertion of tandem ribosome binding sites (rbs) between the promoter/operator and downstream OmpA leader sequence. The presence of an extra rbs appears to be a feature of the mRNA leader regions of some major

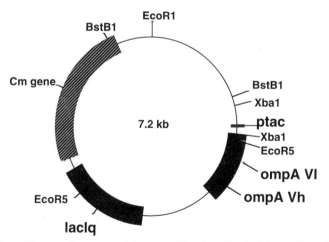

Figure 6. Plasmid vector for expressing recombinant antibody fragments in *E. coli*. This dicistronic construct for production of Fv fragments is based on the Celltech vector pACtac (47). The positions of some key restriction sites are shown. OmpA, ompA leader sequence; cm, chloramphenicol; V_H, heavy chain variable domain gene; V_L, light chain variable domain gene; *lacIq*, lac repressor gene; ptac, hybrid *trp/lac* promoter.

E. coli outer membrane proteins, including OmpA. The first rbs is followed by a short coding sequence that terminates close to the second rbs (45). In the case of pACtac vector this sequence is derived from the 5′ end of the lacZa protein, as shown below:

rbs1   lacZ start
AGGAAACAGCG**ATG**AGCTTGGCTGCATCCGTCGAGTTCTAGATAACG
rbs2   lac2   OmpA leader start
AGGCG**TAA**AAAA**ATG**AAA

Two features of the vector aim to minimize plasmid loss. The chloramphenicol resistance gene (cm) is used since the antibiotic cannot be destroyed by broken cells as in the case of β-lactams. The origin of replication is derived from the pACYC plasmid which results in a moderate vector copy number per cell.

5.2.2 Factors affecting the level of expression

While the use of bacterial secretion vectors has been successful, this strategy alone does not ensure high level production of functional antibody fragments in *E. coli*. Yield improvements are observed between cells grown in shake flask culture and the same strain grown to higher density in a bioreactor (26). The choice of *E. coli* host strain does not appear to play a major part, since a number of different K12 strains have been used equally well, e.g. W3110 (ATCC No. 2735), RV308 (ATCC No. 31608). Anecdotal observations indicate that yield is principally determined by the compatibility of different amino acid sequences with the bacterial secretory and/or protein folding pathways. Thus Carter *et al.* (44) reported that a particular set of human consensus framework sequences supported a tenfold higher expression level (1–2 g/litre) of engineered human Fab′ fragments compared to the parent mouse framework sequences. Mackenzie *et al.* (50) showed that improvements in the yield of functional mouse Fab′ fragments were obtained by switching from mouse C_{kappa} to C_{lambda} light chain constant regions. There appears to be no generic approach to overcoming the potential problem of misfolding. However, growing cultures at a reduced temperature (generally 20–30 °C) is a simple way of improving product solubility. Careful optimization of the induction and harvest time will also contribute to obtaining the maximum yield of a functional antibody fragment from *E. coli*. Growth and product accumulation kinetics are likely to be different for each fragment that is expressed and should be monitored with reference to a non-induced culture run in parallel. In general antibody fragments secreted into the periplasm of *E. coli* accumulate there, although a proportion will leak into the culture medium. Thus both periplasm and media should be sampled to ensure that the maximum available product is recovered. *Protocol 5* describes the culture and induction procedures to establish optimal harvest time for antibody fragment production in *E. coli*.

Protocol 5. Production of antibody fragments in *E. coli*

Reagents

- *E. coli* strain, e.g. W3110 (ATCC strain 2735) transformed with pACtac (or equivalent lacZ promoter/operator based vector) containing antibody fragment sequences (e.g. V_H and V_L)
- IPTG (100 mM stock: Sigma)

- LB medium: 10 g tryptone, 5 g yeast extract, 5 g NaCl per litre
- Chloramphenicol (34 mg/ml in EtOH stock: Sigma)
- Periplasmic extraction buffer: 100 mM Tris–HCl pH 7.4, 10 mM EDTA

Method

1. Thaw a 1 ml glycerol stock of the *E. coli* strain (harvested and frozen at $OD_{600\ nm}$ = 0.5–0.75) into 100 ml LB containing chloramphenicol (25 μg/ml).

2. Incubate on an orbital shaker at 30°C and 250 r.p.m. taking optical density measurements every hour until the culture reaches $OD_{600\ nm}$ = 1.0 (approx. 6 h).

3. Inoculate two 2 litre baffled Erlenmeyer flasks each containing 270 ml LB medium with 30 ml of the above culture.

4. At a cell density of $OD_{600\ nm}$ = 0.5 take samples (X ml) from each flask and induce expression in one of the cultures by addition of IPTG to a final concentration of 200 μM. Optical density measurements should be made and samples taken at 30 min and 60 min post-induction, thereafter hourly up to 6 h.

5. The volume of each sample is normalized for cell number using the following formula:

   $$\text{Volume} \times = \frac{9\ ml}{\text{culture } OD_{600\ nm}}$$

6. Harvest the cells by centrifugation at 12 000 *g* for 15 min at 4°C. Decant supernatant and retain a sample for assay (media fraction).

7. Resuspend the cell pellet in 300 μl periplasmic extraction buffer. Remove and store a 50 μl aliquot (total cell fraction) for analysis by SDS–PAGE.

8. Incubate the remaining 250 μl cell suspension overnight at 30°C with shaking.

9. After overnight incubation clarify the suspension by centrifugation at 12 000 *g* for 15 min at 4°C. Remove and store supernatant (periplasmic fraction) for analysis.

10. Expression of antibody fragment in each of the cell, media, and periplasmic fractions is evaluated by ELISA and immunoblotting methods.

Once the optimum time and method (periplasmic extraction or culture supernatant) of harvest has been established, the culture can be repeated to produce a single batch of material. This is prepared as described above except that the whole culture is harvested at the point which has been shown to be optimal for yield or fragment quality. As a general rule Fvs and sFvs will normally be harvested from the culture supernatant approximately 10 h post-induction and Fabs will be harvested from the periplasm between 1–4 h post-induction.

Acknowledgements

I am grateful to Chris Bebbington and Neil Weir for their advice on antibody expression. I would also like to thank Maureen Day for her expert assistance in preparing this chapter.

References

1. Adair, J. R. (1992). *Immunol. Rev.*, **130**, 1.
2. Morrison, S. L., Johnson, M. J., Herzenberg, L. A., and Oi, V. T. (1984). *Proc. Natl. Acad. Sci. USA*, **81**, 6851.
3. Whittle, N., Adair, J. R., Lloyd, C., Jenkins, E., Devine, J., Schlom, J., *et al.* (1987). *Protein Eng.*, **1**, 499.
4. Jones, S. and Bendig, M. M. (1991). *Bio/Technology*, **9**, 88.
5. Larrick, J. W., Danielsson, L., Brenner, C. A., Wallace, E. F., Abrahamson, M., Fry, K., *et al.* (1989). *Bio/Technology*, **7**, 934.
6. Orlandi, R., Güssow, D. H., Jones, P. T., and Winter, G. (1989). *Proc. Natl. Acad. Sci. USA*, **86**, 3833.
7. Marks, J. D., Tristem, M., Karpas, A., and Winter, G. (1991). *Eur. J. Immunol.*, **21**, 985.
8. Chomczynski, P. and Sacchi, N. (1987). *Anal. Biochem.*, **162**, 156.
9. Liu, A. H., Creadon, G., and Wysocki, L. J. (1992). *Proc. Natl. Acad. Sci. USA*, **89**, 7610.
10. Kabat, E. A., Wu, T. T., Reid-Miller, M., Perry, H. M., and Gottesman, K. S. (1987). *Sequences of proteins of immunological interest*, 4th edn. US Department of Health and Human Services. Washington, DC.
11. Kozak, M. (1991). *Nucleic Acids Res.*, **9**, 5233.
12. Chaudhary, V., Batra, J., Gallo, M., Willingham, M., Fitzgerald, D., and Pastan, I. (1990). *Proc. Natl. Acad. Sci. USA*, **87**, 1066.
13. Padlan, E. A. (1994). *Mol. Immunol.*, **31**, 169.
14. Carroll, W. L., Mendel, E., and Levy, S. (1988). *Mol. Immunol.*, **25**, 991.
15. Riechmann, L., Clark, M., Waldmann, H., and Winter, G. (1988). *Nature*, **332**, 323.
16. Kettleborough, C. A., Saldantia, J., Heath, V. J., Morrison, C. J., and Bendig, M. M. (1991). *Protein Eng.*, **4**, 773.

17. Adair, J. R., Athwal, D., Bodmer, M. W., Bright, S. M., Collins, A. M., Pulito, V. L., *et al.* (1994). *Hum. Antibod. Hybridomas*, **5**, 41.
18. Chothia, C. and Lesk, A. (1987). *J. Mol. Biol.*, **196**, 901.
19. Tramontano, A., Chothia, C., and Lesk, A. M. (1990). *J. Mol. Biol.*, **215**, 175.
20. Queen, C., Schneider, W. P., Selick, H. E., Payne, P. W., LAdolfi, N. F., Duncan, J. F., *et al.* (1989). *Proc. Natl. Acad. Sci. USA*, **86**, 10029.
21. Singer, I., Kawka, D., DeMartino, J., Daugherty, B., Elliston, K., Alves, K., *et al.* (1993). *J. Immunol.*, **150**, 2844.
22. Padlan, E. A. (1991). *Mol. Immunol.*, **28L**, 489.
23. Lewis, A. P. and Crowe, J. S. (1991). *Gene*, **101**, 297.
24. Daugherty, B. L., DeMartino, J. A., Law, M.-F., Kawka, D., Singer, I., and Mark, G. E. (1991). *Nucleic Acids Res.*, **19**, 2471.
25. Riechmann, L., Foote, J., and Winter, G. (1988). *J. Mol. Biol.*, **203**, 825.
26. King, D., Byron, O., Mountain, A., Weir, N., Harvey, A., Lawson, A., *et al.* (1993). *Biochem. J.*, **290**, 723.
27. Huston, J., Levinson, D., Mudgett-Hunter, M., Tai, M., Novotny, J., Margolies, M., *et al.* (1988). *Proc. Natl. Acad. Sci. USA*, **85**, 5879.
28. Bird, R., Hardman, K., Jacobson, J., Johnson, S., Kautman, B., Lee, S., *et al.* (1988). *Science*, **242**, 423.
29. Anand, N., Mondal, S., Mackenzie, C., Sadowska, J., Sigurskjold, B., Young, N. M., *et al.* (1991). *J. Biol. Chem.*, **266**, 21874.
30. Pantoliano, M. W., Bird, R. E., Johnson, S., Asel, E. D., Dodd, S. W., Wood, J. F., *et al.* (1991). *Biochemistry*, **30**, 10117.
31. Desplancq, D., King, D. J., Lawson, A. D., and Mountain, A. (1994). *Protein Eng.*, **7**, 1027.
32. Brinkmann, U., Reiter, Y., Jung, S., Byungkook, L., and Pastan, I. (1993). *Proc. Natl. Acad. Sci. USA*, **90**, 7538.
33. Rodgriques, M. L., Presta, L. G., Kotts, C. E., Wirth, C., Mordenti, J., Osaka, G., *et al.* (1995). *Cancer Res.*, **55**, 63.
34. Bebbington, C. R. (1991). *Methods: A companion to Methods in Enzymology*, **2**, 136.
35. Page, M. J. and Sydenham, M. A. (1991). *Bio/Technology*, **9**, 64.
36. King, D. J., Adair, J. A., Angal, S., Low, D. C., Proudfoot, K. A., Lloyd, J. C., *et al.* (1992). *Biochem. J.*, **281**, 317.
37. Stark, G. R. and Wahl, G. M. (1984). *Annu. Rev. Biochem.*, **53**, 447.
38. Urlaub, G. and Chasin, L. A. (1980). *Proc. Natl. Acad. Sci. USA*, **77**, 4216.
39. Subramani, S., Mulligan, R., and Berg, P. (1981). *Mol. Cell. Biol.*, **1**, 854.
40. Peakman, T. C., Worden, J., Harris, R. H., Cooper, H., Tite, J., Page, M. J., *et al.* (1994). *Hum. Antibod. Hybridomas*, **5**, 65.
41. Bebbington, C. R., Renner, G., Thomson, S., King, D., and Yarranton, G. T. (1992). *Bio/Technology,* **10**, 169.
42. Field, H., Yarranton, G. T., and Rees, A. R. (1989). *Protein Eng.*, **3**, 641.
43. Skerra, A., Pfitzinger, I., and Plückthun, A. (1991). *Bio/Technology*, **9**, 273.
44. Carter, P., Kelley, R., Rodrigues, M., Snedecor, B., Covarrubias, M., Velligan, M., *et al.* (1992). *Biotechnology*, **10**, 163.
45. Ghrayeb, J., Kimura, H., Takahara, M., Hsiung, H., Masui, Y., and Inouye, M. (1984). *EMBO J.*, **3**, 2437.
46. Better, M., Chang, C. P., Robinson, R. R., and Horwitz, A. H. (1988). *Science*, **240**, 1041.

47. Yarranton, G. T. and Mountain, A. (1992). In *Protein engineering* (ed. A. R. Rees), p. 303. Oxford University Press, Oxford.
48. Stark, M. J. (1987). *Gene*, **51**, 255.
49. Rose, R. E. (1988). *Nucleic Acids Res.*, **16**, 355.
50. MacKenzie, C. R., Sharma, V., Brummell, D., Bilous, D., Dubuc, G., Sadowska, J., *et al.* (1994). *Bio/Technology*, **12**, 390.

3

Production of human monoclonal antibodies from B lymphocytes

M. D. MELAMED and J. SUTHERLAND

1. Introduction

The discovery in 1975 by Kohler and Milstein (1) that murine antibodies of pre-defined specificity could be produced in monoclonal culture unleashed an explosion of activity in almost all fields of biomedical research. However, the transfer of this technology to the production of human monoclonal antibodies has proved to be a slow and difficult process. The essential event in the creation of a monoclonal antibody by the Kohler–Milstein approach is the fusion of a spleen B cell of appropriate specificity from an immunized mouse with a myeloma or plasmacytoma cell adapted for indefinite growth in culture, to create productive hybridomas. The additional requirements are:

(a) The proliferation of unfused myeloma cells must be selectively blocked.

(b) The hybridoma cells must be capable of cloning from a single cell.

(c) The resulting hybridoma should continue to express indefinitely the two essential properties of proliferation and antibody secretion.

The impetus to develop human Mabs came from two directions. First, it soon became apparent that murine antibodies are immunogenic in man and stimulate a significant immune response to the conserved regions of the murine antibody. Secondly, it was found with some antigens, e.g. the Rh(D) blood group determinant, that, although strongly immunogenic in man, they are not recognized by the rodent immune system. Thus for diagnostic or therapeutic use, either on their own or tagged with an indicator or immunotoxin, human Mabs would have clear advantages if they could be produced with a high avidity and specificity. Potential applications for human monoclonal antibodies include antibodies to many viral and bacterial diseases and tumour markers. The main diagnostic application has been in the field of blood group serology and, to a lesser extent, HLA typing, and the antibodies have proved to be valuable research tools.

For the production of human Mabs by the conventional approach, i.e. fusion of a committed B cell and a non-secreting myeloma cell to yield a productive hybridoma, the main limiting factor has proven to be the source of specific committed B cells. The other essential requirements of fusion partner, hybridoma selection and cloning, antibody secretion, and continuous hybridoma proliferation, have all proved to be attainable in various ways.

In most work on human Mabs the only source of B cells is the peripheral blood, where only about 5–10% of the lymphocytes are B cells and antigen-specific B cells are present at a frequency of 1 in 10^4 to 1 in 10^7 B cells. Sometimes it is possible to obtain lymph nodes removed surgically for medical purposes and post-mortem spleen cells have occasionally been used. None of these methods yield the B cell population equivalent to splenocytes from intentionally immunized and boosted mice.

Another approach to the production of human Mabs arose from the discovery that the Epstein–Barr virus (EBV) could infect human B cells and induce proliferation which is often accompanied by antibody secretion (2). This forms the basis of a great deal of work on human Mabs but it soon became apparent that EBV infection, on its own, had limited potential. In a few cases it did yield clones capable of permanent secretion of human Mabs but the frequency of success was very low and most clones had a limited life in culture.

A much more efficient method arose out of the combination of EBV transformation and fusion of the transformed cells with various myeloma partners (3). This approach takes advantage of the ability of the EBV virus to expand the population of specific B cells. These are then rescued by fusion with a myeloma partner before the common occurrence of clonal senescence occurs (4). A similar approach has been used with various mitogens (such as poke-weed mitogen) to induce proliferation of B cells followed by fusion. B cell stimulation with EBV followed by myeloma cell fusion has formed the basis of our work on the creation of human Mabs with useful blood group specificities. This has been particularly successful with volunteers who have agreed to be intentionally immunized in order to produce useful titres of blood group-specific antibody (5). Under these circumstances it has proved possible to attain success with Mab production from these donors. It has been shown that for a period of a few weeks after immunization there is a significantly increased chance of obtaining stable productive clones (6). The precursor frequency of specific B cells in circulation probably needs to be at least about 1 in 10^4. For other situations different strategies will need to be evolved.

2. Equipment and media

2.1 Basic requirements

Basic requirements for tissue culture are well known, so we shall mention them only briefly.

A laminar flow hood, an inverted microscope, and a CO_2 incubator are essential. The use of plastic disposables (e.g. pipettes) reduces the time required by the operator (i.e. washing, repackaging, and sterilizing), but does increase costs. Other pieces of equipment required include a phase-contrast compound microscope, a liquid nitrogen cell storage system, and some means of freezing cells at the required rate. We have found that the Nalgene 'Mr Frosty' freezing chamber (Cat. No. 5100–0001) placed in a –80°C freezer to be quite efficient. An autoclave will be required to dispose of used material as well as for sterile preparations.

2.2 Media

This is an area which encourages much discussion, which type to use, which formulation is best, which supplements to add, etc. The final selection is as much due to personal choice as anything else, although human cell lines tend to be a little more demanding than those of mouse origin. It is also a constantly evolving process in most laboratories, with small modifications being made to a basic media preparation. Powdered media provides the cheapest method, but does require more operator time than other formulations, and to cut down on the use of expensive filters a large batch must be made up each time. A ready-to-use preparation on the other hand needs minimal preparation time, but is expensive, and needs a large cold storage area for holding the material if large quantities are used. A 10 × concentrate bought as a batch can provide the solution. This needs little additional preparation (other than autoclaved water) and is very cost-effective. If a large batch is purchased it can provide a constant supply over a period of many months. A 1:1 mixture of DMEM/Hams F12 medium has provided consistent results in the culture of human hybridomas, although other media have been used successfully.

2.3 Supplements

Essential additions are glutamine which should be added at 2–4 mM, and serum or serum replacement. FCS is still the serum of choice although CPSR-3 from Sigma is also useful. Serum replacements tend to give variable results, and should be individually tested. Serum-free media with supplements tend to give better results than protein-free, with which we have obtained poor results.

Protocol 1. Preparation of culture medium

Materials

- Sterile distilled water
- 10 × concentrate DMEM/Hams F12 (Sigma)
- 7.5% (v/v) sodium bicarbonate (Sigma)
- 200 mM glutamine (Sigma)
- 1 M sodium hydroxide (Sigma)
- FCS (or alternative serum supplement) (Sigma)
- 1 M Hepes (Sigma)

Method

1. Make up stock solutions of the individual reagents. Hepes can be autoclaved, the sodium hydroxide is self-sterilizing, and others should be sterile filtered.

Protocol 1. *Continued*

2. To each 840 ml of sterile water, add 100 ml of 10 × medium, 20 ml of sodium bicarbonate, 20 ml of glutamine, and 15 ml of Hepes. Adjust the pH to 7.3 using sodium hydroxide. The amount needed will vary upon the supplier of the medium, however it is usually between 8–16 ml per litre.

3. After preparation, leave the medium overnight at 37°C to confirm sterility. Any colour change or turbidity in the medium indicates contamination and the bottle should be discarded. Store the clear medium at 4°C.

Other optional additives are antibiotics, anti-fungal agents, and some chemical additives such as pyruvate.

It is always best to avoid the use of antibiotics if possible. They are not needed with good technique and should be used sparingly to avoid generating antibiotic resistant strains of bacteria. On those occasions when they must be used penicillin/streptomycin is usually used at a working concentration of 100 U penicillin and 100 μg streptomycin/ml. It is usually bought as a concentrated liquid and stored at –20°C. Occasionally a broader spectrum antibiotic is required, in which case gentamycin is often used although it may affect growth, especially of LCLs (lymphoblastoid cell lines induced by EBV), which seem particularly sensitive. The working concentration of this is usually 200 μg/ml. This can be stored at 4°C and is quite stable.

Anti-fungal agents such as nystatin should also be used sparingly if possible. Nystatin is used at 250 U/ml and is also usually bought as a concentrate and stored at –20°C.

Selection media are also used as in the generation of hybridomas, with fusions plated out in HAT, then fed at the appropriate time with HT. These are bought as concentrates (usually 50 × or 100 ×) and diluted appropriately, and are discussed in greater detail in Section 6.

3. Assays for detection of human monoclonal antibodies

Choosing the correct assay to detect specific antibodies is a crucial part of monoclonal antibody production. The appropriate assay in terms of specificity and sensitivity needs to be established before antibody production is begun. The desired epitope of the antigen against which Mabs are to be made must be available for antibody binding, and any cross-reactivity should be eliminated. The assay must be sensitive enough to detect antibody in the levels found in tissue culture supernatant, which will be lower when using human hybridomas compared to their murine counterparts. To set-up an assay ideally one would use plasma from the donor as the source of known positive antibody. This may need to be diluted up to 1/1000 to give a concentration approximating that which will be found initially in hybridoma supernatant.

It is very important to include controls in any assay system, especially when developing a new system. Each step must have a separate control, both positive (containing the desired antibody) and negative (containing no antibody). Negative controls can be either normal plasma, assay buffer, or ideally both. Positive controls are most usually diluted sera or, if available, a purified preparation of the antibody. Once an in-house positive has been established this can be used in future assays.

The assay must be suitable for the screening of many hundreds of wells which will be produced by each fusion/transformation. Once the cells have reached the stage where an assay is appropriate the time-scale before something must be done with the positives (usually expansion to a larger volume) is quite short. It is usually impractical to expand all colonies; therefore a rapid but accurate and reliable assay system is essential.

3.1 ELISA

3.1.1 Indirect ELISA for human Ig

The usual assay of choice in work with monoclonal antibody production is ELISA. This does not need special skills and can be carried out in almost all laboratories, the only essential piece of specialized equipment needed is a plate reader. There are plate washers and computer-linked readers available, and these are useful if many assays are done each day. If the numbers are low and are likely to remain so this is not necessary. A wash bottle may be used for rinsing the trays, and results calculated manually.

There are two main types of ELISA, the direct assay where antigen is applied directly to the plate, and the indirect assay where antibody (or lectin) is bound to the plate, and then antigen is bound to the antibody (or lectin). ELISA may also be used for isotyping any monoclonal antibodies produced.

Protocol 2.   Indirect ELISA for human Ig

This is a basic protocol for the detection of antibody (non-specific) in cell culture supernatant. Here the culture supernatant is, in effect, the antigen to be measured.

Equipment and reagents

- ELISA trays (Greiner)
- PBS: 140 mM sodium chloride, 2.7 mM potassium chloride, 1.5 mM potassium dihydrogen orthophosphate, 0.1% sodium azide
- Anti-human IgG/IgA/IgM, as appropriate (Sigma)
- Wash buffer: PBS, 0.05% Tween 20 (Sigma)
- Plate washer or wash bottle

Protocol 2. *Continued*

- Blocking buffer: 5% serum or 3% casein in PBS
- Anti-human Ig conjugate[a] (Sigma)
- ELISA plate reader (Dynatec)

- Substrate in buffer (e.g. 1 mg/ml pNPP in diethanolamine buffer for alkaline phosphatase conjugates) (Sigma)

Method

1. Coat 96-well ELISA trays with anti-human IgG/IgM/IgA as required. The usual concentration is 10 µg/ml in PBS, and the usual volume is 100 µl/well. Leave the trays at 4°C until required, but at least overnight.

2. Remove the coating antibody and block the trays using either 5% serum or 3% casein in PBS. Incubate for 15 min at 37°C.

3. Wash three times in wash buffer.

4. Add supernatant samples, 100 µl/well, and incubate for 2 h at 37°C.

5. Wash three times as in step 3.

6. Add 100 µl/well of anti-human immunoglobulin conjugate diluted as specified by the manufacturer. Incubate for 1 h at 37°C.

7. Wash three times as in step 3.

8. Add appropriate substrate in buffer and leave until colour develops.

9. Read absorbance at the appropriate wavelength for the conjugate substrate used.

[a] The conjugates used are usually either alkaline phosphatase or horse-radish peroxidase. The decision on which to use will depend upon availability as well as sensitivity required.

3.1.2 Direct ELISA for specific antibody

In the direct ELISA, antigen is directly bound to the plate. This needs relatively large amounts of the antigen, but it is simpler to perform than the indirect ELISA as one less step is involved, and is often more reliable.

Protocol 3. Direct ELISA for specific human antibody

Equipment and reagents
- ELISA trays (Greiner)
- PBS (see *Protocol 2*)
- Wash buffer (see *Protocol 2*)
- Anti-human Ig conjugate (Sigma)

- Substrate (Sigma)
- Antigen
- Blocking buffer (see *Protocol 2*)

Method

1. Dissolve the antigen in PBS. The optimum concentration will vary, but 5–10 µg/ml should be suitable for most antigens.

2. Add 100 μl to each well, cover, and leave at least overnight at 4 °C.

3. Remove and discard excess antigen from the wells, then add 150 μl/well of blocking buffer. Incubate for at least 15 min at room temperature.

4. Wash the plate three times in wash buffer, then add 100 μl/well of cell culture supernatant. Incubate for 2 h at 37 °C.

5. Wash three times as in step 4 and add 100 μl/well of appropriately diluted conjugate. Incubate for 1 h at 37 °C.

6. Wash three times and add 100 μl/well of appropriate substrate.

7. Read the results using a plate reader at the appropriate wavelength.

3.1.3 Indirect ELISA for specific antibody

Indirect ELISA can also be used to detect specific antibody, by coating the trays with a specific antibody to the antigen (usually polyclonal, but occasionally monoclonal), blocking, and adding antigen. After a suitable incubation and wash, culture supernatant is added and any bound, specific antibody is detected, as above. Sometimes lectin can be used to bind the antigen to the tray. If this is done then the blocking step should be carried out after antigen has been added.

3.2 Haemagglutination assays

These are very quick and straightforward to use. They offer adequate sensitivity in detecting monoclonals, but have limited applications. If the antibodies produced are against a red cell surface antigen or if the antigen can be bound to red blood cells then the assay is very easy.

Protocol 4. Preparation of red blood cells

Equipment and reagents

- LISS: 180 mM glycine, 47 mM glucose, 30 mM sodium chloride, 1.5 mM disodium hydrogen phosphate, 1.5 mM sodium dihydrogen phosphate (dihydrate), 1.5 mM adenine, 1.5 mM inosine
- Anticoagulant (CPD, heparin, etc.) (Sigma)
- PBS (see *Protocol 2*)
- 0.22 μm filters (Millipore)

Method

1. Make up LISS by dissolving the glycine in 400 ml water and dissolving the remaining reagents in a further 400 ml of water. Mix the two solutions, adjust the pH to 6.65–6.75, and bring to a final volume of 1 litre. Sterile filter and store at 4 °C.

2. Collect venous blood into a suitable anticoagulant (e.g. CPD).

Protocol 4. *Continued*

3. Mix well and then aseptically remove 2.5 ml into a 15 ml tube.

4. Dilute to approx. 10 ml with PBS, and centrifuge for 10 min at 400 *g* (this should give approx. 1 ml of packed red cells).

5. Remove supernatant and repeat step 4 twice.

6. Resuspend the final pellet in LISS to a final volume of 10 ml to give a 10% suspension.

7. Store at 4°C. This should be stable for about two to three weeks.

8. To prepare a 1% suspension make a 1/10 dilution of the above. This is stable for three to four days.

This red cell suspension can then be used directly in the haemagglutination assay.

Protocol 5. Haemagglutination assay for antibody in hybridoma supernatants

Equipment and reagents

- Sterile 'V'-well trays (Greiner)
- Plate shaker, Titertek
- 1% red blood cells suspension (see *Protocol 4*)

Method

1. Transfer 25 μl of each supernatant to one well of a 96 V-well microtitre tray.

2. Add 25 μl/well of the red cell suspension to the V-well tray.

3. Mix well on a plate shaker and remove air bubbles (a hair-dryer is very useful for this). Allow to settle at room temperature for about 20–30 min.

4. Negative wells will show a button of cells which will 'streak' when the plate is tilted. Positive wells however will show either a tight irregular button of cells which will hold together, or alternatively a 'lawn' of cells spread over the base of the well.

5. This method can be used as a titration by serially diluting the neat supernatant in PBS/1% BSA. The end-point is the well in which the cells begin to streak (see step 4).

In conclusion, the assay system chosen will depend very much upon the antigen to be detected and the resources available in each laboratory. The assay needs to be well controlled and have sensitivity, reliability, and specificity, and the final choice will often be a compromise of all three. The most important point is that the assay is established prior to the initiation of mono-clonal production, as much work may be lost if this is not the case.

4. Donor selection

This is one of the most important steps in the production of human mono-clonal antibodies, as well as one of the most difficult. Unlike murine antibodies there are very few antigens with which donors can be immunized (notable exceptions being blood group antigens and vaccines). In general, therefore, we have:

- naive donors
- naturally-infected donors
- donors stimulated with vaccines or blood group antigens

Naive donors are obviously the most readily available. Hybrids secreting antibody can be isolated from them but are often of autoimmune origin or of unknown specificity, as well as being of low affinity. Large panels of such donors have been screened to isolate hybrids secreting autoimmune antibodies.

For antibodies to infective agents the use of cells from naturally-infected individuals has resulted in a large number of antibodies being generated (e.g. from patients with CMV) (7), but these tend to be of low affinity compared to murine antibodies. The antibodies which are easiest to obtain tend to be to immunodominant epitopes, which may not necessarily be those of interest. Hybridoma technology does however permit screening of large numbers of antibody-containing supernatants, which may allow isolation of the required specificity. Assays may also be designed to selectively reveal antibody bind-ing to an epitope of interest, for example by screening on peptides from the required epitope, or by treating or tethering the antigen in such a way that only certain sites are exposed.

Immunized and boosted donors are preferable for obtaining antibodies of highest affinity but there are only a few cases in which this is possible, for example, following deliberate vaccination to infectious diseases, e.g. tetanus (8) and hepatitis B (9). One successful experience with this method for the production of human Mabs has been with the Rh(D) specificity. It was shown that the success rate was particularly high within a few weeks after a booster immunization given to already anti-D positive volunteers (6).

Experience with other blood grouping antibodies has shown a decreasing spectrum of response from Rh(D) as the best, through the other Rh antigens which are more difficult to obtain (10,11), all the way down to some antibodies where success has not been achieved at all. For all blood group antigens the immunization schedule which yields successful Mabs seems to be the same. The structure of these antigens varies greatly, ranging from carbohydrate to proteins and polypeptides, and this difference does not seem to alter the immunization schedule required to give successful antibody production. This leads us to believe that mode of presentation may be at least as important as antigenic structure when determining immunization schedules for Mab production.

Cells obtained from individuals after vaccination have shown very good results, but detailed time course studies have not been conducted. The time course for other routes of immunization, e.g. natural infection, needs to be determined empirically in each case.

5. Isolation and proliferation of human B lymphocytes

5.1 Introduction

The most common source of naive human B cells for research purposes is tonsillar lymph nodes but these are not suitable for production of monoclonal antibodies of pre-defined specificity. For all practical purposes the only source of specifically-sensitized human lymphocytes is from peripheral blood. Direct fusion of peripheral blood lymphocytes has not generally yielded the desired specific hybridomas even after specific stimulation of the donors. Thus it is common practice to subject the human B cells to a non-specific cell proliferation stimulus before fusion. In all of our work we have used Epstein–Barr virus infection of the B cells to induce proliferation before fusion. Although only a small proportion of the peripheral blood B cells are induced into proliferation by the Epstein–Barr virus (12) the induced lymphoblastoid (LCL) cells fuse efficiently with mouse myeloma cells and give relatively high yields of secreting hybridomas. Where lymphatic material is available—e.g. from lymph nodes removed during surgery of metastatic tumours—it has also proved to be an excellent source of B lymphocytes for production of human monoclonal antibodies by this method (see Chapter 4 on anti-melanoma antibodies). A drawback to this way of expanding the pool of B lymphocytes for fusion lies in the fact that most donors have circulating cytotoxic T lymphocytes with a specificity for EBV-infected B lymphocytes. These cytotoxic T cells are induced into proliferation in a pooled EBV-treated culture and result in specific and highly effective killing of EBV-transformed B cells. This cytotoxic response can be eliminated by T cell depletion of the culture, or by inhibition of T cell proliferation by cyclosporin A at 1 µg/ml (13) or by the non-specific activation of the T cells with PHA at 5 µg/ml (14).

The EBV-infected B cell population expands fairly rapidly in the first 10–14 days after infection with a significant number of B cell lines proliferating. The lifespan of the lymphoblastoid cell lines is very variable and the number of different clones from the culture starts to decline after this point. Occasionally the desired clone will persist indefinitely and be amenable to limiting dilution cloning and expansion (15). Most clones have a limited lifespan and can be rescued at this stage by fusion with mouse myeloma (3). The hybridomas so produced are more stable and robust in culture than LCL but the efficiency of fusion is only about 1 in 10^4. Thus the timing of fusion is a matter of experience and judgement based on the rate of growth of the EBV-induced culture. We have found 10–14 days is usually about the best. At this point the EBV transformants are subjected to PEG fusion with a mouse myeloma cell line. The hybridomas are plated out at a cell density which gives a low number of clones/well on microplates over macrophage feeder layers. In this way any clonal competition, which may mask slow growing clones, is eliminated. Cultures from wells giving a positive specific antibody test are then recloned several times to clonality.

5.2 Isolation of PBMCs from human blood

Venous blood collected by routine procedure is anticoagulated by adding any of the common anticoagulants. For most purposes 10 ml of whole blood is the minimum required. At the other extreme, leucocyte-enriched buffy coat from about 450 ml of blood is about the maximum that can be handled. Leucocytes are isolated from the whole blood using centrifugation methods developed by Böyum (16) employing any one of a variety of proprietary products which have a density of 1.077 and containing an appropriate dextran to induce rouleaux formation by the red cell—e.g. Lymphoprep, Histopaque, Ficoll Hypaque, etc. This method isolates the peripheral blood mononuclear cells (PBMC) which are about 90% T lymphocytes with about 5%–10% B lympho-cytes. For some B cell transformation protocols the whole PBMC fraction can be used with no further purification. Specific T cell cytotoxicity can be avoided by addition of PHA or CSA (see Section 5.1).

When isolating B lymphocytes from buffy coats which are more than about 24 hours old leucocytes often become coagulated or cross-linked and are difficult (or impossible) to resuspend. This does not appear to affect the B cell transformation, but if required this problem can be reduced if all the centri-fugations of leucocytes are done over a small layer of 90% isotonic Percoll in PBS so that the cells are layered rather than pelleted. Once the cells have been washed a few times in this way the agglutination is no longer a problem and any residual Percoll can be removed by a slow speed centrifugation.

5.3 Production and use of Epstein–Barr virus

5.3.1 Handling of EBV in the laboratory

Epstein–Barr virus is a ubiquitous human herpes virus. About 90% of all adult populations are positive (17). The laboratory strains of the virus are of a very low infectivity and no authenticated case of laboratory contracted infection has ever been reported. Nevertheless it is reasonable to handle EB virus and EBV-transformed B cells in accordance with standard biological safety practice.

5.3.2 Laboratory growth of the virus

The cell line most frequently used for this is B95–8 which is a primate cell line derived from an *in vitro* infected cotton-topped marmoset with wild-type virus isolated from a patient with infectious mononucleosis (18). The line produces a relatively low but sufficient titre of the virus and the cell-free supernatant from this cell line is the most commonly used source of infective virus. Additional agents such as phorbol esters (17) significantly increase the viral production by the cell line but their presence in the supernatant is toxic to transformed lymphocytes and therefore EBV cultures which have been induced by phorbol esters then have to be purified before they can be used.

Protocol 6. Production of EB virus

Equipment and reagents
- Tissue culture flasks
- Cell culture medium containing 10% FCS
- B95–8 cell line (ECACC No. 85011419)
- 0.15 M sodium hydroxide

Method

1. Seed a cell culture flask with B95–8 cells using a standard cell culture medium.
2. Feed and expand the cell culture up to 250 ml in a 180 cm^2 cell culture flask.
3. Once the cells have become confluent replace the medium and allow the cells to continue to proliferate for seven to nine days without changing the medium.
4. Collect the contents of the flask, remove the cellular debris by centrifugation. If desired the supernatant can be further clarified by filtration through a 0.4 μm filter.
5. Finally adjust the filtrate back to neutrality with 0.15 M sodium hydroxide. Aliquot the virus stock into 5 ml or 10 ml aliquots and store at –70°C.

It is convenient to produce sufficient virus for several month's or even year's worth of work as the viral stock stored at –70°C appears to have a very long life.

5.4 T cell depletion

All the methods described in this chapter are directed at achieving productive proliferation of human B cells. The negative effect of leaving the T cells in the mixture during B cell activation by EBV has already been mentioned. This cytotoxicity can be prevented by addition of PHA or CSA. Whether

there are other beneficial or detrimental affects of T cells in the various protocols described in this chapter is not clear. Some authors take great pains to
eliminate T cells (19,20) but most of our work is done without T-depletion.
For situations where elimination of T cells is desirable *Protocols 7* and *8*
describe two ways of doing this (derived from refs 21 and 22).

Protocol 7. Depletion of T cells with AET

Equipment and reagents

- AET (Sigma)
- Sheep red blood cells (SRBC) (Serotec)
- Isotonic sodium bicarbonate (12 g/litre) sterile filtered (this is used in place of PBS for the reaction between AET and SRBC as a pH of about 8.5 is optimal)
- 0.2 μm filter
- 40% heat inactivated FCS in growth medium
- Lymphoprep or equivalent
- 0.8% NH_4Cl
- 0.5 M NaOH
- PBS (*Protocol 2*)

Method

The quantities in this protocol are given for 10^9 PBL, which is approximately the number of cells expected from about 450 ml of blood, i.e. one
transfusion unit (Section 5.2).

1. Ensure that the SRBC suspension (as supplied) is well and evenly dispersed and remove 3 ml into a 25 ml tube.

2. Wash the SRBC in isotonic sodium bicarbonate four times collecting the cells by centrifugation at 800 *g* for 5 min.

3. Dissolve 250 mg of AET in 6–7 ml of sterile water and bring the pH of the solution up to 8.5. This will require about 2 ml of 0.5 M sodium hydroxide. Make the volume to 10 ml with water and filter sterilize the solution through a 0.2 μm filter.

4. Add 1 ml of packed washed SRBC to 10 ml of the AET solution and mix at 37 °C for 15 min. 1 ml of the packed blood cells contains approx. 10^{10} cells and is sufficient to deplete the T cells from 10^9 peripheral blood lymphocytes.

5. Wash the AET-treated SRBC five times in PBS or HBSS by resuspen-sion, followed by centrifugation at 800 *g* for 5 min, and finally add the cells to 10 ml of 40% heat inactivated fetal calf serum (HIFCS) in medium.

6. Wash and count the PBL and resuspend the cells at 10^8/ml in medium containing 40% HIFCS.

7. Add an equal volume of the suspension of AET-treated SRBC from step 5 and mix the cell populations by rotation.

8. Centrifuge the mixture at 200 *g* for 15 min to form the rosettes and resuspend the mixture by rotation and gentle pipetting.

Protocol 7. *Continued*

9. Layer the mixture over 25 ml Lymphoprep (or equivalent), centrifuge at 800 *g* for 20 min, and collect the interface by pipetting. This layer should not be red as it contains only those leucocytes which do not form rosettes with the SRBC.

10. Dilute the cells with medium (10% FCS) and centrifuge at 800 *g* for 10 min. If there are residual SRBC at this stage they can be lysed by resuspending the pellet into 0.8% NH$_4$Cl. Agitate the suspension at 37 °C until lysis is complete (about 5 min), and remove the leucocytes by centrifugation.

11. Finally, wash the T cell depleted cells twice in medium.

An alternative method to obviate the problem of specific T cell cytotoxicity towards EBV-infected B cells is by depletion of cytotoxic cells with the peptide ester leucyl-leucine-methyl ester. This also removes natural killer cells and is a relatively simple procedure. The drawbacks are that if the cells are not thoroughly depleted of platelets coagulation can occur and yields can sometimes be disappointing.

Protocol 8. Depletion of cytotoxic cells with leucyl-leucine-methyl ester

Equipment and reagents

- Hanks balanced salt solution without calcium or magnesium but containing 0.01% bovine serum albumin, filter sterilized (H/B)
- Percoll (Sigma)
- Leucyl-leucine-methyl ester

Method

1. Wash the leucocytes several times in H/B. If there is a significant contamination with platelets they should be removed by resuspending the cells at 10^7–10^8/ml over two or three volumes of 40% isotonic Percoll in H/B and centrifuging at 300 *g* for 15 min. Wash the platelet-free cell suspensions once more to remove the Percoll.

2. Resuspend the leucocytes at 2–2.5 × 10^6/ml in H/B.

3. Weigh out 1 mg of Leu-leu-O-Me for each 10 ml of leucocyte suspension, dissolve in a small volume of H/B, filter sterilize, and add to the cell suspension. (A buffy coat from one unit of blood will yield about 10^9 leucocytes in 500 ml H/B and will require 50 mg of Leu-leu-O-Me dissolved in 10 ml H/B.) Or a 400 × stock solution of Leu-leu-O-Me can be prepared in methanol (0.1 M), sterile filtered, and stored at –20 °C. For use the appropriate amount of stock is dispensed, methanol is evaporated off, and the cells are added.

4. Incubate the mixture with continual rotation for 15 min.

5. Wash the suspension three times by centrifugation in H/B. The cytotoxic cells lyse over the following 1–3 h and should be rotated until
lysis is complete.

6. Finally, centrifuge at 300 *g* for 5 min to remove the cell debris.

6. Cell fusion

6.1 Introduction

The four essential components of the Milstein–Kohler method of producing
monoclonal antibodies are:

(a) An immortal cell line which, although a non-secretor, has the capacity to
produce and secrete antibodies (myeloma or plasmacytoma cell line).

(b) Primed B lymphocytes of appropriate specificity, in sufficient numbers to
allow success with a fusion frequency of about 1 in 10^4.

(c) The ability to bring about fusion of these two cell types in such a way as
to yield a hybrid cell secreting the desired antibody and continuing to
proliferate in culture.

(d) A method for selectively killing unfused myeloma/plasmacytoma cells.
That used by Kohler and Milstein (1) remains the standard method. The
fusion partner lacks HGPRT and therefore cannot use the salvage pathway, and can only make nucleotides *de novo*. If the *de novo* pathway is
blocked by the addition of aminopterin (A) the cells will die. However,
hybrids which have acquired a functional HGPRT enzyme will use the
salvage pathways making nucleotides from hypoxanthine (H) and thymidine (T). Thus the myeloma fusion partners are selectively killed by
HAT and only the hybrids survive.

Following the establishment of the murine system, a great deal of endeavour
was devoted to the search for a human myeloma fusion partner. A reliable,
efficient partner does not yet exist but either a HAT-sensitive mouse–human
hybridoma, or a mouse myeloma cell line, can serve as a fusion partner for
human B cells. For most purposes the only source of human B cells is
peripheral blood and only under very unusual circumstances can lymph
nodes or spleens be used. In addition intentional immunization, particularly
of a repeated nature, cannot normally be performed. A strategy which has
been adopted by several groups is the infection of peripheral B cells with
Epstein–Barr virus followed by fusion with mouse myeloma or mouse–human
hybridoma to yield the desired hybridoma secreting human antibody of
choice.

6.2 Fusion of LCLs

The method most commonly used is EBV transformation of all the PBMCs then allowing the pooled culture to begin to proliferate. After about 10–14 days the whole culture is fused with a mouse myeloma cell line. Fusion is performed essentially by a standard murine method as detailed in *Protocol 10*. One difference, however, lies in the fact that the EBV-transformed B cells are to some extent immortalized. Thus, to focus attention only on the true heterohybridomas after fusion, the growth of unfused EBV-transformed human B cells is inhibited by the addition of 10 μM ouabain (which is more toxic to human than to murine or heterohybridoma cells). The myeloma fusion partner we use is derived from U63 Ag8.653 by selecting for a highly ouabain-resistant variant. This facilitates the use of ouabain to eliminate unfused LCLs without impeding the growth of the hybridomas.

Protocol 9. PEG fusion

Equipment and reagents

- LCLs derived from expanded wells as described in *Protocol 11*; or bulk LCLs derived as described in *Protocol 12*
- Myeloma cells in log growth
- PBS without calcium and magnesium
- 45% PEG 4000 (Merck) in PBS
- Growth media containing 10 μM ouabain

Method

1. Count both LCL and myeloma cells and check viability. Mix together at a ratio of 1:1 to 1:2 and centrifuge at 400 *g* for 5 min.

2. Discard the supernatant taking care that the pellet is as dry as possible. Mix well preferably using a vortex mixer to ensure an even suspension of cells.

3. Add 1 ml of 45% PEG 4000 dropwise over 1 min whilst continuously mixing.

4. Transfer to a 37°C water-bath and mix well for 90 sec.

5. Add 2 ml of PBS again slowly over 2 min with constant agitation.

6. Add 2 ml of PBS over 1 min as in step 3.

7. Add 5 ml of PBS over 2 min as in step 3.

8. Make the volume up to 20 ml with PBS adding it slowly and mixing continuously.

9. Stand at room temperature for 15 min.

10. Centrifuge at 400 *g* for 5 min and remove supernatant. The cells are very delicate at this stage and great care must be taken with them. Add growth medium to the pellet and resuspend slowly and gently.

---

11. Resuspend the cells in the appropriate volume of growth medium containing ouabain HAT supplement. The cells should be seeded at 2×10^6 LCL per 96-well plate, and 100 μl per well.

12. Leave for seven days before replacing the medium with growth medium containing HT.

13. Approximately 14 days after fusion the plates should be ready for assay.

6.3 Electrofusion

Cell fusion using PEG, although widely used, requires relatively high numbers of lymphocytes for each fusion, and the fusion efficiency is only about 1 in 10^4. Electrofusion for production of hybridomas (i.e. the fusion of cells by the application of electrical fields), first reported in the early 1980s by Zimmerman *et al.* (23), can be performed on a smaller scale and with a higher fusion efficiency. The method is performed in a fusion chamber which consists of two metal electrodes connected to a voltage generator. First the cells are brought into close membrane contact (pearl chain formation) by subjecting them to low energy alternating current (A/C), and then a number of high energy direct current (D/C) pulses are applied to bring about cell fusion. As the electrical field is applied a critical voltage (the breakdown voltage) is achieved, this produces a build-up of potential difference which leads to reversible pore formation in adjacent cell membranes. The pores allow cytoplasmic exchange and therefore passage of electrolytes between the cells, which causes a release of potential difference and the membranes re-seal as a single structure.

The method consists of two main steps, each of which has several parameters which need to be optimized. These will also differ depending upon the machine used. One of the major problems with electrofusion is the optimization of these many parameters, which are different for each cell type. As well as optimizing cell concentration, ratio, and pre-treatment of the cells, the medium used during fusion must be optimized. It is usual to employ an iso-osmolar sugar-based medium containing trace amounts of membrane stabilizing agents such as magnesium and calcium acetate. Standard growth medium cannot be used immediately after fusion due to its phenol red content, which appears to be toxic if used at this stage. Phenol red-free medium is obtained in powder form, filter sterilized once, made up, and the pH adjusted. Routine growth medium can be used again at the first feed when the toxic effects are no longer seen.

The electrical parameters which must first be established for pearl chain formation are; A/C frequency, A/C voltage, and duration of field. Ideally the cells should be stable, non-revolving, have a high area of membrane contact, and yet not be under undue stress. The cells should align quickly, but the field should only be applied for as short a time as possible whilst still allowing efficient alignment.

Once the pearl chain parameters have been optimized it is necessary to repeat the optimization process for the fusion parameters. These are, pulse number, duration, and voltage, as well as time between pulses and the post-fusion ramp. The voltage is critical as too low a voltage will not allow pore formation whereas too high a voltage will cause irreversible pore formation and cell lysis. Short repeated pulses will increase the probability of fusion, but in excess may rupture the cell. The post-fusion ramp is the re-application of an A/C field immediately after fusion. This holds the cells together for enough time to allow efficient membrane re-formation.

Protocol 10. Electrofusion of LCLs

This protocol will need to be adjusted for each cell line used, and variation will occur between different machines.

Equipment and reagents

- Fusion medium: 280 mM sorbitol, 0.1 mM calcium acetate, 0.5 mM magnesium acetate, 1 mg/ml bovine serum albumin (Sigma)
- Culture medium without phenol red (see text)
- 96-well tray with feeder cells (see *Protocol 13*)
- HAT (Sigma)
- Electrofusion apparatus

Method

1. Mix myeloma and LCLs at a ratio of between 1:1 and 10:1.

2. Wash three times in fusion medium. Resuspend in 250 μl of fusion medium.

3. Transfer to electrofusion chamber and apply an A/C field of frequency 1500 kHz and voltage of 320 V/cm for 20 sec. Follow this by three D/C pulses each of 15 μsec, voltage 3.2 kV/cm, and a 1 sec delay between pulses.

4. Re-apply the A/C field for a further 20 sec and collect the cells into 5 ml of culture medium.

5. Incubate the cells for 1 h at 37°C and then plate over a tray of feeder cells at 50 μl/well after supplementing the medium with HAT.

6. Continue as for a PEG fusion, *Protocol 9*, step 12.

This technique has been successfully used to increase fusion efficiency in the production of both human and murine hybridomas. Up to 100-fold increases in efficiency have been reported compared with conventional PEG techniques. A great advantage of electrofusion is that once parameters have been optimized the technique is simple, quick, and easily replicated. Using the protocol above fusion efficiencies of 9.7×10^{-4} were achieved using just 2×10^5 LCL. This is approximately a 100-fold increase over that achieved by PEG fusion in the same laboratory.

7. Strategies for production of human Mabs

The general framework consists of isolation of appropriate lymphocytes, transformation of the B cells with EB virus, and establishment of LCL cultures. These can then be cloned out or, more often, fused with an immortal (HAT-sensitive) myeloma cell line.

7.1 EBV only

Cloning of EBV-transformed LCLs has been reported (15,24). The basis of this approach is that a few LCLs are capable of continuous growth and antibody secretion in culture. This is a rare event and production of large amounts of antibody from these human B cell transformants is relatively difficult. However, despite these drawbacks success is possible and has the advantage that the system is completely of human origin without the possibility of any murine components being present (25).

7.2 Fusion of selected LCLs

A more productive approach to the creation of human monoclonal antibodies from mouse–human heterohybridomas was achieved by the partial selection at the LCL stage followed by fusion and single cell cloning (6). Peripheral blood mononuclear cells collected from a density interface (see Section 5.2) are exposed to EBV and plated out at 2×10^5 cells/well in 96-well flat-bottomed cell culture microplates with 0.2 ml culture medium containing 5 μg/ml PHA (see *Protocol 11*). Wells are assayed weekly and those showing rising titres of the desired specificity are expanded into 24-well plates. After three to four weeks the well is virtually full. If the antibody titre has remained high the LCLs are subjected to PEG fusion (*Protocol 9*) with about twice as many myeloma cells in log phase. The hybrids are immediately plated out in 96-well trays over mouse peritoneal macrophages and approximately two weeks after fusion they are assayed. This method was particularly successful in yielding productive hybridomas from blood collected in the few weeks following a booster immunization. Many LCL wells show a transient secretion of antibody to the immunogen which decreased irreversibly after a few weeks.

Protocol 11. Infection with EB virus (for culture in 96-way trays)

Equipment and reagents
- Peripheral blood cells (Section 5.2) or cells derived from *Protocols 7* or *8*
- 96-well trays (Greiner)
- Frozen (–70°C) B95.8 supernatant containing EBV (see *Protocol 6*)
- PHA (Sigma)

Method

1. Suspend PBL to be EBV infected (derived as described in *Protocols 7* or *8* or Section 5.2) at a concentration of 10^6 to 10^7/ml in EBV-containing B95.8 culture supernatant.

2. Add 2 vol. of fresh medium.

3. Rock or rotate the cells gently at 37°C for 1–2 h.

4. Remove the cells from the culture supernatant by centrifugation and wash once in fresh medium.

5. Make the cell suspension to $1-2 \times 10^6$/ml and add 5 μg/ml of PHA (phytohaemagglutinin) and distribute into the wells of a 96-well tray to which has been added mouse peritoneal macrophages (*Protocol 13*). Proliferation will be at its maximum between 7–15 days after infection.

7.3 Fusion of bulk LCLs

The third approach to the derivation of human antibodies in this way is derived from the one above. Instead of plating out the transformants they are incubated together in a single tissue culture flask and the whole population of proliferating LCLs is fused with mouse myeloma cells and plated out in 96-well plates over peritoneal exudate feeder cells. Two weeks is the optimum time for fusion to maximize the yield of transient antibody-expressing cells as well as those capable of longer-term secretion.

Protocol 12. Infection with EBV (for bulk fusion of LCLs)

Equipment and reagents
- Peripheral blood cells (Section 5.2) or cells derived from *Protocols 7* or *8*
- B95.8 supernatants containing EBV
- 75 cm² flask
- PHA (Sigma)[a]

Method

1. To about 10^7 to 10^8 PBMNC add 5 ml B95.8 supernatant and incubate at 37°C for 1–2 h.

2. Dilute with 3 vol. of growth medium.

---

3. Add PHA at 5 μg/ml.

4. Incubate for about two weeks adding fresh medium as needed.

[a] Cyclosporin at 1 μg/ml can be used in place of PHA. This suppresses the T cells and hence stops proliferation.

The PHA stimulates a short-lived proliferation of the T cells which is usually followed by a slow but consistent outgrowth of LCLs. At about 10–15 days after EBV transformation all the cells are removed from the flask and subjected to fusion and cloning over mouse peritoneal cells.

This approach to the production of human monoclonal antibodies is the one which requires the least operator time to set-up and is the one which has been used in most of our work on production of human anti-blood group antibodies by our group (10,11).

Of these three approaches to human Mabs, the first (EBV only) is clearly the most difficult and has the lowest success rate, but has the advantage that the antibody is of completely human origin. The derivation of human Mabs by bulk EBV transformation and fusion followed by cloning is the most productive and efficient in terms of operator effort and time and several hundred Mabs have been produced by this method. Selection and expansion of EBV transformants before fusion requires more effort but has the theoretical advantage that slow growing EBV transformants would not be overwhelmed by clonal competition, but direct comparative studies have not been carried out.

7.4 Use of anti-CD40

A relatively new development in B cell proliferation studies has been derived from the discovery that the surface of resting B lymphocytes expresses the glycoprotein, CD40. Interaction of CD40 ligand (which is found on helper T cells) delivers a strong proliferation signal to the B cell (26). This signal can also be brought about by the binding and cross-linking of the CD40 with monoclonal anti-CD40 antibodies (27,28), particularly when the antibody is presented bound to Fc receptors (28,29). When EBV-infected B cells are stimulated in this way with anti-CD40 the effect is even stronger and more sustained (30,31). All reports with this system have used a mouse fibroblast cell line transfected with human CD32 Fc receptor (28).

B cells activated in this way can be fused to yield productive hybridomas (19,20). There is also one report of sustained growth of B cells in the CD40/EBV system without fusion (31).

8. Cloning

Cloning is an essential part of monoclonal antibody production, ensuring that the antibody detected is the product of a single clonally expanded B cell. This can be carried out either after immortilization of the B cells, or after fusion. If the cells have been fused the cloning should be carried out as soon as practical, but cloning after EBV immortilization can only be done after a stable line has been produced.

8.1 Limiting dilution

The most widely used technique in cloning is that of limiting dilution. Here the cells are diluted and plated out at concentrations down to theoretically one cell, or less, per well.

An important part of this technique is the use of feeder cells in the trays. These are essential in order to attain reasonable cloning efficiencies. The most productive system seems to be mouse peritoneal macrophages.

Protocol 13. Preparation of mouse peritoneal macrophages

Equipment and reagents

- Mouse
- Laminar flow hood
- Cold growth medium[a] (see *Protocol 1*)
- 10 ml syringe (Becton Dickinson)
- 25 G needle (Becton Dickinson)
- 16 G needle (Becton Dickinson)
- Forceps and scissors

Method

1. Kill the mouse by cervical dislocation or gassing.

2. Transfer the mouse to the laminar flow hood after surface sterilizing with 70% alcohol. Lay the mouse on its back and using forceps and scissors make a longitudinal incision into the outer skin of the abdomen. Pull the skin back to reveal the peritoneum. Carefully inject approximately 10 ml of cold medium into the peritoneum using a 25 gauge needle.

3. Gently massage the sides of the peritoneum to release macrophages from the tissues.

4. Using a larger gauge needle (e.g. 16 G) remove the media containing the macrophages and transfer to a sterile container.

5. Count the cells on a haemocytometer and adjust to 2–4 × 10⁵/ml. Plate out at 50 μl/well in 96-well trays and incubate at least overnight before use.

[a]The medium should be cold (0–4°C) as macrophages are very sticky. The procedure should also be carried out in as short a time as possible to minimize losses in this way.

The dilutions which are to be used in cloning will be determined by a number of factors, such as the cloning efficiency and the growth characteristics of the cells. The first cloning should be carried out using relatively high cell numbers which can then be reduced on subsequent re-clonings. In general human cell lines will not show as much growth as their murine counterparts and initially dilutions of ten, five, and two cells per well have been successfully used with human heterohybrids. If a LCL line is being cloned this will need to be higher, especially for the initial cloning.

Protocol 14. Cloning by limiting dilution

Equipment and reagents
- Three trays containing feeder cells (see *Protocol 13*)
- Multichannel micropipette (Titertek) sterile tips (Micronics)
- Sterile media trough (Micronics)
- Haemocytometer (BDH)
- Culture medium (see *Protocol 1*)
- Cells to be cloned

Method

1. Label the trays containing macrophages as 10, 5, and 2 cells/well.

2. Resuspend and determine the viable count of the cells which are to be cloned.

3. Add 2000 cells to 20 ml of complete medium. Mix this well to give an even suspension of the cells.

4. Add 10 ml of this suspension to the plate labelled 10 cells/well at 100 μl/well.

5. To the remaining 10 ml add a further 10 ml of medium and mix well. Take 10 ml of this and add 100 μl/well to the plate labelled 5 cells/well.

6. Repeat step 5, and label the final tray 2 cells/well.

7. Discard the final 10 ml of suspension.

8. Incubate the trays for approximately two weeks by which time colonies should develop. They should obviously be checked during this time and the media topped up as required.

Other systems such as using mixed thymocyte medium (MTM) do not seem to be as efficient in the cloning of human hybridomas.

To be certain that monoclonality has been achieved clonings should be carried out several times. The number of clonings required can be calculated statistically using the Poisson distribution. This is based upon the equation:

$$P(1) = y + x(1 - y)$$

where $P(1)$ = the probability that a well contains a single clone; $\times$ = the fraction of wells from the first cloning showing growth that contain only a single cell/well; y = the fraction of wells from the second cloning showing growth that contain only a single cell/well.

If more subclonings are carried out then the value of y is taken from the last subcloning, and $\times$ is replaced by the $P(1)$ value from the previous clonings (32).

A rough guide for this is that three clonings with no more than 37% growth in the plate from which the re-cloning is carried out will give statistical mono-clonality. Obviously if the number of wells with growth is lower then less cloning are needed, but conversely if it is higher then more clonings should be carried out. Another valuable indicator of clonality is the microscopic observation that there is only one growing cluster of cells in the well.

8.2 Other techniques

Another technique which has not proved as popular is that of using soft agar. Here the cells are diluted and spread over a dish containing a low concentration of agar. Individual colonies can be picked off using a sterile pipette and a microscope.

Protocol 15. Cloning in soft agar

Equipment and reagents

- 10 cm Petri dishes (Greiner)
- Sterile Pasteur pipettes(BDH)
- 1.2% agar (keep at 45 °C)

- Cloning medium: as for *Protocol 1*, but make up to 500 ml at double the concentration

Method

1. Add 7 ml of agar to 7 ml of cloning medium and pour into a 10 cm Petri dish. Allow to solidify at room temperature.

2. Prepare the relevant dilution of cells in culture medium and mix 1 ml of this with 1 ml of cloning medium. Overlay this onto the agar plate and leave for 7–14 days.

3. When colonies are visible pick single colonies (in a laminar flow hood) using a microscope and sterile Pasteur pipette.

4. Transfer these to a 96-well plate with growth medium and assay when appropriate.

8.3 Points to note

Attention should be paid to the selection of the cells for expansion. These should always be taken from the tray with the lowest cell density, and from wells or plates showing single colonies whenever possible. If there are only few of such wells at the lowest cell density then, when the line is re-cloned, this should be taken to a higher dilution than that used previously. Hybrids producing human monoclonal antibodies tend to grow at quite a slow rate, so care should be taken in expanding them to larger volumes at this critical stage. Wells should be left until they are confluent and expanded gradually for best results. The first cloning tends to be the most difficult with human hybrids, and any genetic instability will show up at this stage, usually resulting in loss of antibody production. For this reason the first cloning should be carried out as soon as possible in order to 'rescue' antibody-producing cells. Thereafter the re-clonings can be carried out when convenient, and usually after expansion for the production of supernatant, so that the antibody can be investigated before proceeding to the next stage.

There are various supplements available for use in cloning which are variable in their usefulness. This tends to differ from line to line and will depend to a certain extent upon the stringency of the conditions under which the cells are grown.

A novel extension to the use of soft agar cloning has been described (33). Clones are plated in low melting point agar and antibody-positive foci are detected by a blot and lift technique. If a suitable blot detection can be devised this technique is capable of detecting Mabs with great efficiency.

9. Scale up of production

Once an antibody-secreting line has been established it is usually desirable to grow larger quantities of it. Human monoclonal antibody-producing cell lines will not produce ascites under normal circumstances, so alternatives must be found.

The volume of cells grown will depend upon the intended use of the antibody, as well as the level of productivity of the hybridoma. The methods used for human antibodies will be the same as for murine monoclonals with the exception of ascites. Small volumes (i.e. up to about 200 ml) can be grown in static flask cultures, but for larger amounts a different method of production is needed. The easiest is probably roller culture. The vessels can be either plastic disposables available from most of the manufacturers of tissue culture ware, or glass reusable. The protocol for production of rollers is given below and can be used for up to 1.5 litre per roller (using a 2 litre vessel).

Protocol 16. Roller bottle culture

Equipment and reagents
- Roller bottle (sterile) (Corning)
- Growth medium (see *Protocol 1*)
- Minimum of 10^7 cells at > 85% viability
- 1 M sodium hydroxide
- Sodium azide

Method
1. Resuspend the cells and add to the roller bottle.
2. Add approximately 50 ml of growth medium and leave overnight in a CO_2 incubator on its side with the lid ajar.

Protocol 16. *Continued*

3. Check microscopically that cells are viable, and add medium to double the volume. Place on a rolling apparatus.

4. Check daily and double the volume when the medium becomes acidified, until the vessel is about half full.

5. If cells are required, harvest the culture when it appears confluent. The roller should be centrifuged at 400 *g* for 10 min and the supernatant discarded. The cells can then be frozen or processed as required.

6. If supernatant is needed, harvest the roller when the culture viability has reduced to approximately 30%. The contents should be centrifuged at 400 *g* for 10 min and the supernatant retained. The pH of this should be adjusted with sodium hydroxide and sodium azide added to 0.1% (w/v).

Above this level one can go to spinner vessels or hollow fibre bioreactors. Spinners are a little more specialized requiring spinner vessels, and either a stirrer base or an overhead motor. The vessels are glass and range in capacity from 100 ml to about 32 litres. They require 5% CO_2 in air which is sparged into the larger vessels, or into the airspace of the smaller ones. A hotbox or hot room will be required to house the vessels.

Hollow fibre bioreactors are often used as an alternative to spinners and to an extent in place of small fermenters. These are usually self-contained units, again linked directly to a CO_2 line, but they usually have temperature control built in. They produce a more concentrated product than the spinner vessel, but are sometimes more difficult to keep sterile.

The Milstein bottle (34) also allows the production of concentrated antibody. This bottle holds the cells, a large air bubble, and the secreted antibody in dialysis tubing suspended in growth media, and is placed on a roller apparatus to ensure circulation of nutrients.

The Heraeus Miniperm system works on a similar system producing concentrated antibody on a laboratory scale (35). It uses a disposable cell chamber and a reusable (autoclavable) media compartment. The membrane separating the two compartments allows low molecular weight nutrients and waste products to pass but retains the hybridomas and concentrates the antibody produced. By replacing the medium in the media compartment waste products are removed and fresh nutrients supplied. The cells are seeded into the gas-permeable cell chamber with 30–40 ml of growth medium, and the media chamber also filled with 300 ml of medium. This is placed on a roller apparatus in a CO_2 incubator, therefore not needing a dedicated gas supply or temperature control. Once cell growth is established the medium in the media chamber must be changed daily, but that in the cell chamber (containing the antibody) remains until the viability begins to fall. The cells can then be harvested and supernatant containing antibody collected. The cells can be used to re-seed the same cell chamber and the Miniperm run again. This can be done three or four times before the number of residual non-viable cells builds up to a critical level. This system was designed to be used with mouse hybridomas, but has been used with some success with human lines too. The results with these types of lines are very variable. Some lines work very well indeed and give antibody levels of ten times that found in static cultures whilst growing rapidly, but others do not adapt to this system very well.

If very large volumes are required (i.e. on a commercial scale) large hollow fibre bioreactors, or fermenters of various sizes are needed. The decision on which system to use will depend upon the purity of antibody required, and the available space. Air lift fermenters start at about 5 litres, to a maximum of about 10000 litres. The limiting factor is the pressure differential which builds up inside of the vessel. The lower part of the vessel will be under much higher pressure than the top and cells may be damaged during circulation.

A recent innovation in this area is a compact gravitational settling device (36). By recirculating the cell suspension through an inclined series of settling surfaces the cells can be recirculated and the culture supernatant harvested in an efficient and continuous manner.

Acknowledgements

The authors would like to acknowledge the collaboration of S. O'Reilly on the section on electrofusion and the hard work of Mrs Eileen Layton in the preparation of this manuscript. The Monosera laboratory is wholly supported by Serologicals Corporation, Atlanta, GA, USA.

References

1. Köhler, G. and Milstein, C. (1975). *Nature*, **256**, 495.
2. Rosen, A., Gergely, P., Jondal, M., Klein, G., and Britton, S. (1977). *Nature*, **267**, 52.
3. Thompson, K. M., Hough, D. W., Maddison, P. J., Melamed, M. D., and Hughes-Jones, N. (1986). *J. Immunol. Methods*, **94**, 7.
4. Melamed, M. D., Gordon, J., Ley, S. J., Edgar, D., and Hughes-Jones, N. C. (1985). *Eur. J. Immunol.*, **15**, 742.
5. Thompson, K. M., Melamed, M. D., Eagle, K., Gorick, B. D., Gibson, T., Holburn, A. M., *et al.* (1986). *Immunology*, **58**, 157.
6. Melamed, M. D., Thompson, K. M., Gibson, T., and Hughes-Jones, N. C. (1987). *J. Immunol. Methods*, **104**, 245.
7. Yoshiaki, K., Mikihiro, Y., Tomokuni, T., Yukio, S., Shoji, I., Etsuko, M., *et al.* (1992). *Hybrdioma*, **11**, 594.
8. Kozbor, D. and Roder, J. C. (1981). *J. Immunol.*, **127**, 1275.

9. Andris, J. S., Ehrlich, P. H., Ostberg, L., and Dapra, J. D. (1992). *J. Immunol.*, **149**, 4053.

10. Thompson, K., Barden, G., Sutherland, J., Beldon, I., and Melamed, M. (1991). *Transfusion Med.*, **1**, 96.

11. Thompson, K., Barden, G., Sutherland, J., Beldon, I., and Melamed, M. (1990). *Immunology*, **71**, 323.

12. Mevissen, M. L. C. M., Kwekkeboom, J., Goormachtig, E., Lindhout, E., and de Groot, C. (1993). *Hum. Antibod. Hybrdomas*, **4**, 66.

13. Bird, A. G., McLachlan, S. M., and Britton, S. (1981). *Nature*, **289**, 300.

14. Moss, D. J., Rickinson, A. B., and Pope, J. H. (1979). *Int. J. Cancer*, **23**, 618.

15. Doyle, A., Jones, T. J., Bidwell, J. L., and Bradley, B. A. (1985). *Hum. Immunol.*, **13**, 199.

16. Böyum, A. (1968). *Scand. J. Clin. Lab. Invest.*, **21**, Supp. 97: 77.

17. Walls, E. V. and Crawford D. H. (1987). In *Lymphocytes: a practical approach* (ed. G. G. B. Klaus), p. 149. Oxford University Press.

18. Miller, G. and Lipman, M. (1973). *Proc. Natl. Acad. Sci. USA*, **1**, 190.

19. Darveau, A., Chevrier, M.-C., Nêron Delage, R., and Lemieux, R. (1992). *J. Immunol. Methods*, **159**, 139.

20. Thompson, J. M., Lowe, J., and McDonald, D. F. (1994). *J. Immunol. Methods*, **175**, 137.

21. Madsen, M., Johnson, H. E., Hansen, P. W., and Christiansen, S. E. (1980). *J. Immunol. Methods*, **33**, 323.

22. Ohlin, M., Danielsson, L., Carlsson, R., and Borrebaeck, C. A. K. (1989). *Immunology*, **66**, 485.

23. Zimmerman, V. and Pilwat, G. (1982). *J. Biol. Phys.*, **10**, 43.

24. Kumpel, B. M., Poole, G. D., and Bradley, B. A. (1989). *Br. J. Haematol.*, **71**, 125.

25. Kumpel, B. M., Rademacher, T. W., Rook, G. A. W., Williams, P. J., and Wilson, I. B. H. (1994). *Hum. Antibod. Hybridomas*, **5**, 143.

26. Clark, E. A. and Ledbetter, J. A. (1986). *Proc. Natl. Acad. Sci. USA*, **833**, 4494.

27. Gordon, J., Millsum, M. J., Guy, G. R., and Ledbetter, J. A. (1988). *J. Immunol.*, **140**, 1425.

28. Banchereau, J., de Paoli, P., Vallé Garcia, E., and Rousset, F. (1991). *Science*, **251**, 70.

29. Wheeler, K., Pound, J. D., Gordon, J., and Jefferis, R. (1993). *Eur. J. Immunol.*, **23**, 1165.

30. Rousset, F. (1992). Personal communication.

31. Peyron, E., Nicolas, J.-F., Réano, A., Roche, P., Thivolet, J., Haftek, M., *et al.* (1994). *J. Immunol.*, **153**, 1333.

32. Coller, H. A. and Coller, B. S. (1986). In *Methods in enzymology* Vol. 131, p. 412.

33. Gherardi, E., Pannell, R., and Milstein, C. (1990). *J. Immunol.*, **126**, 61.

34. Pannell, R. and Milstein, C. (1992). *J. Immunol. Methods*, **146**, 43.

35. Falkenberg, F. W., Weichert, H., Krane, M., Bartels, I., Palme, M., Nagels, H.-O., *et al.* (1994). *J. Immunol. Methods*, **179**, 13.

36. Thompson, K. J. and Wilson, J. S. (1995). In *Animal cell technology* (ed. R. E. Speir, J. B. Griffiths, and W. Berthold), p. 227.

4

Generation and selection of human monoclonal antibodies against melanoma-associated antigens: a model for production of anti-tumour antibodies

M. D. THOMAS, M. D. MELAMED, and J. NEWTON

1. Introduction

Melanomas are generally resistant to chemotherapy and radiotherapy, and often give rise to metastases (1). However, the neoplastic transformation of melanocytes induces qualitative and quantitative modulation of cell membrane antigens (2,3). These tumour-associated immunogenic markers suggest the possibility of a strategy for selective diagnosis and even therapy for melanoma (4–8), e.g. by the production of tumour-specific human monoclonal antibodies. The melanoma tumour can thus be used as a favourable case for the establishment of techniques for the production of anti-tumour Mabs.

2. Production of hybridomas from lymph node lymphocytes

One of the major limitations of human monoclonal antibody production is the difficulty in obtaining specifically sensitized B lymphocytes. In many cases during block dissection for surgical treatment of melanoma, lymph nodes draining the tumour site are removed (9). This routine procedure, therefore, can provide a source of lymphatic material containing lymphocytes sensitized to tumour-associated antigens. The successful employment of this material as a source of lymphocytes for immortalization is critically dependent upon a short transition time (< 25 h) between surgical removal and lymphocyte processing.

Protocol 1. Generation of antibody-secreting hybridomas from lymph node lymphocytes

Equipment and reagents

- 70 μm cell strainers (Falcon)
- Lymphoprep (Nygard)
- Preservative-free heparin (C. P. Pharmaceuticals)

- HBSS, penicillin, streptomycin, nystatin, phytohaemagglutinin (Sigma)
- Culture medium DMEM/F12 containing 10% CPSR-3 (Sigma)

Method

1. Lymph nodes surgically removed by block dissection are obtained in culture medium supplemented with 100 U/ml penicillin, 100 μg/ml streptomycin, and 10 U/ml preservative-free heparin.

2. Aseptically dissect the lymph nodes into HBSS supplemented with heparin/antibiotics, and prepare a single cell suspension by repetitive pipetting. Remove tissue debris by filtration through 70 μm cell strainers, and wash the cells three times by centrifugation at 300 g for 5 min with HBSS. Resuspend the cells in HBSS, centrifuge over Lymphoprep at 1500 g for 20 min, and collect mononuclear cells at the interface. Wash the cells twice more with HBSS and process either directly for hybridoma production or resuspend the cell pellet in freezing medium and store aliquots under liquid nitrogen.

3. The methods for exposure of the cells to EBV and PHA, the fusion of the lymphoblastoid cells to mouse myeloma cells are as described in Chapter 3, *Protocols 9* and *11*.

4. From an average size lymph node this method will yield several hundred wells with proliferating heterohybridomas. A preliminary selection for wells secreting human Ig can be performed by a human Ig assay as described in Chapter 3, *Protocol 2*.

---

3. Identification of tumour-specific Mabs

The method for identification of tumour-specific Mabs has to be suitable for use with several hundred human Ig-secreting culture supernatants. For this purpose a method was devised using cultured melanoma cells lightly fixed in 96-well trays.

The 96-well format of the fixed cell assay allows for the screening of large numbers of wells. Once this selection has been made a further selection with higher sensitivity and specificity can be used to identify those antibodies which react well with native unfixed cultured melanoma cell lines which are known to express tumour-specific melanoma antigens.

Up to this stage selection was performed on proliferating malignant cells in culture. Although the cell lines were selected on the basis of their ability to secrete tumour-associated antigens, nevertheless the final selection of candidate anti-tumour antibodies can only be done on malignant human tissue. Fresh frozen cryostat sections of malignant and normal skin tissues is used as the final selection step.

Protocol 2. Selection of antibodies against melanoma-associated antigens using fixed melanoma cells (see *Figure 1*)

Equipment and reagents

- Half area 96-well tissue culture trays (Costar)
- Cell scrapers (Falcon)
- Culture medium—DMEM/F12 containing 5% CPSR-3 (Sigma)
- Cultured melanoma cell lines: WM266/4 (ATCC CRL 1676)
- Gluteraldehyde
- Gelatin
- Mouse antibody to human Ig
- Rabbit antibody to mouse Ig
- Mouse alkaline phosphatase anti-alkaline phosphatase (APAAP) (Dako)
- Antibody diluent: PBS containing 3% goat serum (Sigma)
- Nitrophenyl phosphate (Sigma)
- Solution of diethanolamine (10%)
- Levamisol (Sigma)
- ELISA plate reader (Dynatech)

Method

1. Harvest cell lines from confluent culture using cell scrapers and wash the cells twice by centrifugation at 300 *g* for 5 min with PBS.

2. Resuspend cells at 4×10^4 cells/ml in culture medium and dispense 50 μl aliquots to wells of half area 96-well tissue culture trays.

3. Incubate trays in standard cell culture conditions until the monolayers are established in approximately two days.

4. Wash the monolayers twice with PBS and fix with 0.02% gluteraldehyde in PBS for 15 min at room temperature.

5. Wash the monolayers with PBS and block with 100 μl/well 0.1% gelatin in PBS for 1 h at room temperature. Wash again as above.

6. Incubate the monolayers with 50 μl/well spent hybridoma supernatant for 4 h at 4°C, and wash again with PBS.

7. Incubate the wells with 50 μl (suitably diluted) mouse anti-human immunoglobulin (Ig) per well for 1 h at 37°C.

8. After washing incubate the wells with 50 μl (suitably diluted) rabbit anti-mouse Ig and incubate for 1 h at 37°C.

9. Wash the wells again and incubate with 50 μl diluted mouse APAAP for 1 h at 4°C.

10. Finally, wash the wells four times with PBS and develop with 1 mg/ml *p*-nitrophenyl phosphate in 10% diethanolamine pH 9.6 containing 5 mM levamisol. Read the plates using a standard ELISA plate reader at 405 nm.

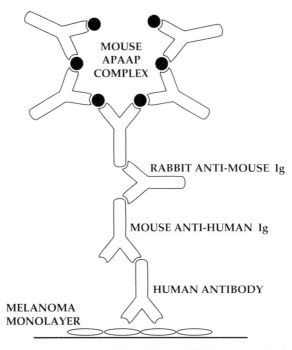

Figure 1. Schematic representation of the cell-based ELISA system for the detection of human antibody reactive against melanoma cell antigens.

Protocol 3. Selection of antibodies reacting with native melanoma cells using fluorescence microscopy

Equipment and reagents

- Melanoma cells in culture (*Protocol 2*)
- Sterile 30-well format multiwell microscope slides (Hendley)
- Goat anti-human Ig FITC conjugated (Sigma)
- Microscope with ultraviolet illumination

Method

1. Harvest the melanoma cells from confluent culture using cell scrapers and wash the cells twice by centrifugation at 3000 *g* for 5 min in PBS.

2. Resuspend the cells at 4 × 10⁵/ml in culture medium and dispense 50 µl/well to the multiwell slides. Allow the cells to adhere by overnight incubation under standard culture conditions.

3. Wash the cells with PBS and incubate for 1 h at room temperature with 20 µl per well heterohybridoma supernatant.

4. Wash the cells with PBS, and incubate with 20 µl per well goat anti-

human Ig FITC conjugate, suitably diluted, for 40 min at room temperature.

5. The cells are washed four times with PBS, mounted in PBS, and
   membrane fluorescence detected using ultraviolet microscopy.

A major problem associated with screening for exogenous human antibody
in human tissue, using anti-human Ig reagents, is the high background of
endogenous tissue Ig. The methodologies employed for screening are largely
dependent on the isotype of antibody being screened. Usually endogenous
tissue IgM levels are low, and binding of IgM test antibody can be performed
using anti-human IgM conjugates. But IgG levels are high and, apart from
directly conjugating the test antibody with the signal, alternative approaches
have to be employed.

Protocol 4. Test for specificity of anti-melanoma antibodies in
tissue sections

Equipment and reagents

- Cryopreserved melanoma tissue
- Cryopreserved normal skin
- Antibody diluent: 5% goat serum in PBS
- Wash buffer: PBS
- Fab fraction of goat anti-human Ig (Jackson)

- Goat anti-human IgM FITC conjugate (Sigma)
- Goat anti-human IgG FITC conjugate (Sigma)
- Microscope with ultraviolet illumination

A. *Preparation of slides*

1. Cut 2 μm sections of cryopreserved malignant and non-malignant
   skin and air dry onto microscope slides. Store sections at –70°C until
   required.

2. When needed equilibrate slides to room temperature and air dry.

3. Flood the slides with PBS/5% BSA and incubate for 1 h at 37°C.

B. *Detection of human IgM monoclonal antibody*

1. Wash the sections three times and incubate with test IgM monoclonal
   Ab for 1 h at 37°C.

2. Wash the section three times with PBS, incubate for 1 h at 37°C with
   suitably diluted goat anti-human IgM FITC conjugate, and examine in
   the fluorescence microscope.

C. *Detection of human IgG monoclonal antibody*

1. Wash the section three times and incubate for 1 h at 37°C with excess
   Fab fraction of goat anti-human immunoglobulin.

2. Wash the section three times and incubate with test IgG monoclonal
   antibody for 1 h at 37°C.

Protocol 4. *Continued*

3. Wash the sections three times and incubate for 1 h at 37 °C with suitably diluted goat anti-human IgG FITC conjugate.

4. Wash the section four times with PBS, mount in Citiflor, and detect fluorescence microscopically.

References

1. Riley, P. A. (1974). *The physiology and pathology of the skin* (ed. A. Jarret), Vol. 3.
2. Herlyn, M. and Koprowski, H. (1988). *Annu. Rev. Immunol.*, **6**, 283.
3. Lloyd, K. O. and Old, L. J. (1989). *Cancer Res.*, **49**, 3445.
4. Cheung, N. V., Lazarus, H., Miraldi, F. D., Abramowski, C. R., Lallick, S., Saarinen, U. M., *et al.* (1987). *J. Clin. Oncol.*, **5**, 1430.
5. Spitler, L. E. (1990). *Front. Radiat. Ther. Oncol.*, **24**, 186.
6. Beverley, P. C. L. and Riethmuller, G. (1987). *Immunol. Today*, **8**, 101.
7. Reisfeld, R. A. (1989). *Immunity to Cancer* II, 183.
8. Thomas, A., Steffens, M. D., Dean, F., Bajorin, M. D., and Houghton, M. D. (1992). *World J. Surg.*, **16**, 269.
9. Binder, M., Pehamberger, H., Steiner, A., and Wolff, K. (1990). *Eur. J. Cancer*, **26**, 871.

5

Antibodies packaged in eggs

JENS CHRISTIAN JENSENIUS and CLAUS KOCH

1. Chicken antibodies

Like mammals, chickens make antibodies in response to challenge with anti-gen. Numerous publications have described the humoral antibody response in chickens to various antigens given by different routes and with different adjuvants (reviewed in refs 1 and 2). It is well established that chickens possess at least three immunoglobulin isotypes: a high molecular weight IgM type immunoglobulin, and two 7–8S immunoglobulins, an IgG type constituting the major plasma immunoglobulin, and an IgA type in external secretions, e.g. in the gall bladder and oviduct (3). The term 'IgY' is often used for the chicken IgG type immunoglobulin. This nomenclature was suggested by Leslie and Clem (4) to reflect some unusual features of chicken IgG heavy chains. Later data, especially recent results from molecular cloning, have shown that IgY has a fourth constant heavy chain domain, which shows homology to the hinge region of mammalian IgG (5,6). These data also suggest that avian IgY may be related to a common ancestor for mammalian IgG and IgE. However, we find that the available information justifies the use of 'IgG' to denote the chicken 7S immunoglobulin and we shall do so throughout this text.

Compared with mammalian antibodies, chicken antibodies are formed by a distinctive genetic process which involves the presence of only one re-arranged V(D)J gene. This primordial gene is modified during ontogeny by a process of gene conversion using homologous, non-functional pseudogenes as donors (7). This process serves as a means of creating the chicken antibody repertoire available for selection by antigens which have entered the organ-ism. The distinct mechanism in chickens may in theory generate antibodies of different characteristics from those of mammalian antibodies, but this has yet to be demonstrated experimentally.

While the available data do not show any general differences in affinities between chicken and mammalian antibodies, the extent and mechanism of affinity maturation of chicken antibodies remain unresolved. Not only are the genetic mechanisms involved in antibody production in the chicken different in detail from those of mammals, but the whole immune system of birds is somewhat differently organized. Chickens have no lymph nodes, but instead have the bursa of Fabricius, which historically gave its name to the B (bursa-derived) lymphocytes. For this reason, the chicken (as the most commonly studied avian species) was used as an experimental model in which Ig-secreting B lymphocytes could be depleted by bursectomy. Most of the earlier immunological literature treats the chicken immune system from this point of view. Only relatively recently has attention focused more on the special features of chicken antibodies that may be put to practical use in the laboratory. Experience with chicken antibodies is considerably less than with mammalian antibodies, and their practical advantages have yet to be widely realized.

Chickens diverged about 300 million years ago from the reptilian branch which led to mammals. As a consequence of this phylogenetic separation, the chicken has some attractive features as a source of antibodies compared with mammalian sources. Antigenic molecules (e.g. proteins) in mammals resemble each other more than the homologous molecules in chickens, and the chicken immune system might therefore be able to recognize mammalian structures that are not recognized as foreign by other mammals (8). Chicken antibodies may be advantageous in various solid phase immunoassays because they do not interact with mammalian rheumatoid factors, which might otherwise pose a problem (9). Furthermore, chicken antibodies do not activate mammalian complement systems, and chicken IgG can thus be used in solid phase immunometric assays to reduce interference from complement components (10). Chicken IgG does not react with protein A (11) or protein G (12), which may again be useful in some applications (e.g. in a sandwich ELISA developed with protein A-coupled enzyme, or assays for estimating protein A or G). On the other hand one should be aware that some applications require different conditions from those optimized for mammalian antibodies. This applies particularly to precipitation techniques, including precipitation in gel.

2. Antibodies in eggs

Mammals provide their offspring with passive immunity, either through transplacental transfer of antibodies or via antibodies in the colostrum. The newly hatched chicken is likewise provided with protective antibodies from the hen. These antibodies have been acquired *in ovo* through absorption from the yolk (13). The possibility of using eggs as a source of antibodies was realized around 1980 independently by us (14) and by a South African group (15), and chicken antibodies harvested from yolk have by now been used in a large number of publications, and are also available from many commercial sources.

The main advantage of chicken antibodies for most researchers lies in the convenient acquisition of abundant amounts without venepuncture. The concentration of IgG in the yolk is roughly the same as in serum, i.e. 10–15 mg/ml (14). The volume of one yolk is about 15 ml, and yolk is thus an excellent and easily accessible source of antibodies. In one week a chicken can supply as much antibody as can be obtained from a rabbit in three months. Chickens are traditional domestic animals around the world, and easily and cheaply kept in laboratory animal houses. Fewer resources are required for collecting eggs than for bleeding rabbits. The only source of antibody that can compete in several of these respects is cow's milk which contains about 0.6 mg of IgG per ml (16), and can thus provide about 20 g of IgG per cow per day, but we will refrain from economic considerations in this case, since keeping cows in a laboratory animal house is usually impractical and researchers will rarely need antibodies in such vast quantities.

It is interesting to note that our early suggestion (14) of exploiting the cheap and abundant source of antibody provided by eggs for the treatment of gastro-intestinal infections has been exploited in a number of studies (17–19). The results suggest that oral administration of yolk antibodies may indeed find a place in the treatment of certain bacterial and viral infections.

Diluted yolk may be used directly in some techniques such as Western blotting with a secondary, enzyme-labelled anti-chicken IgG antibody, but for most purposes it is convenient or necessary to purify the IgG from the yolk. This is easily done by salt precipitation, but the abundant lipids first have to be removed. It is the necessity of removing the lipids from egg yolk (and from chicken plasma or serum as well) that seems to be the major obstacle inhibiting many researchers from using chicken antibodies. The lipids also need to be removed prior to purification of the antibody by affinity or ion exchange chromatography.

3. Chicken husbandry

Chickens are easily kept in conventional animal facilities. They can be kept in cages identical to rabbit cages, with one chicken per cage when the eggs have to be identified as laid by individual immunized chickens. Alternatively one may have one brown egg layer (e.g. New Hampshire) and one white egg layer (White Leghorn) sharing a cage. Another possibility, which we prefer, is to reserve one animal room for chickens, where they can move about freely and rest on a chicken ladder. The chicken coop is provided with a rack of specially constructed nesting boxes (*Figure 1*), each of which has a trap-door which closes when the chicken enters (available from America A/S). At suitable intervals during the day the coop is inspected, the chicken freed, and the egg numbered.

4. Immunization

Chickens are purchased from the supplier at about the time when they start laying eggs, i.e. at about four months of age. About two weeks later they may

Figure 1. Chicken coop with nesting boxes.

be immunized (*Protocol 1*). Immunizations are usually carried out by repeated intramuscular injection in each of the breast muscles with the antigen emulsified in complete or incomplete Freund's adjuvant or adsorbed onto aluminium hydroxide gel (20,21). Intraperitoneal or intravenous immunizations are also feasible.

Protocol 1. Immunization procedures

Equipment and reagents

- Freund's adjuvant, complete and incomplete (Difco, Sigma)
- Phosphate-buffered saline (PBS): 0.14 M NaCl, 10 mM sodium phosphate buffer pH 7.4
- Vortex mixer
- Syringes (1 ml or 2 ml) with 21 gauge needles

A. *Emulsifying antigen in Freund's adjuvant*

1. Use 0.5 ml of antigen solution containing 1 μg to 100 μg of antigen[a,b] and 0.55 ml of complete Freund's adjuvant per chicken.

2. Stir a 10 ml glass or polypropylene tube containing the adjuvant vigorously on a vortex mixer and add the antigen solution SLOWLY

---

dropwise. For an optimal antibody response it is essential to produce a viscous water-in-oil emulsion.

3. For an alternative method to achieve this goal use reciprocating glass syringes connected via their tips (for this and other methods consult manuals of immunological techniques; a thorough description is given in ref. 22).

4. Check the stability of the emulsion by placing two drops on the surface of some water. The second drop should remain intact for some time. Adequate antibody responses may be achieved even without too much attention to the above, but we have experienced occasions when useful responses were not obtained until a proper emulsion was used.

B. *Injection*

1. Withdraw the emulsion into a 1 ml or 2 ml syringe equipped with a 21 gauge needle (this is the time when the emulsion is conveniently tested on water).

2. Inject about 0.5 ml into each breast muscle while you or a helper holds the chicken breast up.

3. Repeat the injection at intervals[c] with the antigen in incomplete Freund's adjuvant.

---

[a] There is no way of telling the minimum amount of antigen needed to obtain a useful antibody response. If plenty of material is available, 50 μg or 100 μg is suggested as a suitable dose. Even when only a few micrograms or less is available we recommend dividing it between two or three chickens.

[b] Antigen available as a stained protein band on nitrocellulose after Western blotting (e.g. a band across half the width of the gel) is cut out and finely ground in a scrupulously cleaned mortar and suspended in buffer, or the nitrocellulose is solubilized in DMSO (23). 1–2 cm^2 of nitrocellulose is dissolved in 0.3 ml of DMSO and slowly added dropwise to 3 ml 15 mM sodium carbonate buffer pH 9.6, under vortexing. The nitrocellulose precipitates as fine particles, which are sedimented at 10 000 g, washed twice with PBS, and resuspended in 1 ml PBS.

[c] We have found that weekly immunizations result in a fast and strong antibody response after only three immunizations. This is different from the rabbit where experience shows that prolonged intervals often result in better responses. High titres of antibodies can then be maintained by re-immunizing every three or four weeks.

5. The antibody response

The IgG antibody appears in the yolk about three to four days later than in serum, and it may occasionally be expedient to test the response on serum, which is recovered from blood withdrawn with a syringe from a vein under the wing or in the neck. Otherwise, just collect the eggs and test the antibody in the yolk diluted with four volumes of buffered saline. This is conveniently done by ELISA (*Protocol 2*).

Protocol 2. Test of antibody response in chickens by ELISA

Equipment and reagents

- ELISA plates (e.g. Nunc or Costar)
- ELISA plate washer—a manual device (Nunc) works as well or better than many expensive automatic plate washers
- ELISA reader (many suppliers)
- Alkaline phosphatase-labelled anti-chicken IgG (Dako, Sigma, and others)
- PBS (see *Protocol 1*)
- Tris-buffered saline (TBS): 0.14 M NaCl in 10 mM Tris–HCl pH 7.4, containing 15 mM NaN$_3$

- Coating buffer: TBS or PBS with 15 mM NaN$_3$ is suitable for most antigens
- Carbonate buffer (0.05 M, pH 9.6) is traditionally used for IgGs but may denature some antigen epitopes
- TBS–Tween washing buffer: TBS with 0.05% (v/v) Tween 20
- Enzyme substrate: 0.12 mM bromochloroindolyl phosphate, 0.12 mM nitroblue tetrazolium, 4 mM MgCl$_2$, 0.1 M ethanolamine pH 9.0

Method

1. Dilute the antigen to 1 μg/ml (or less) of coating buffer and add 100 μl to each well of an ELISA plate.

2. Keep the plate in a humidified box overnight at 4°C and washed three times with TBS–Tween.[a,b]

3. Add 100 μl of yolk or serum dilutions in duplicates to the wells. Start at a 1/100 dilution in TBS–Tween and serially dilute (five- or tenfold dilution steps) to 1/10^6.

4. Incubate for 2 h at room temperature (or longer) and wash four times with TBS–Tween.

5. Add 100 μl alkaline phosphatase-labelled anti-chicken IgG diluted 1/2000 in TBS–Tween (the suitable dilution or highest useful dilution may vary with the source).

6. Incubate for 2 h at room temperature and wash three times with TBS–Tween.

7. Add 100 μl of substrate and incubate at 37°C for 0.5 h or longer to develop a suitable colour intensity. Development may be continued overnight at room temperature.

8. Read absorbance at 415 nm in an ELISA reader.

9. Plot the results on three or five decade semi-log graph paper. Read dilutions at 50% of maximal response. The reciprocal of this value is the titre and is used for comparing individual chickens and the course of the antibody response.

[a] Denaturation of the antigen through adsorption to the polystyrene surface may impede the assay. This is circumvented by using biotinylated antigen added to wells coated with avidin (1 μg/ml TBS). We also find that less antigen is needed in this modification (1/10 to 1/100 of the amount used for direct coating). Biotinylation is usually effectively carried out as described below (*Protocol 7*).
[b] Some protocols call for treatment with a special blocking solution (e.g. 1% BSA). In our experience this is rarely needed.

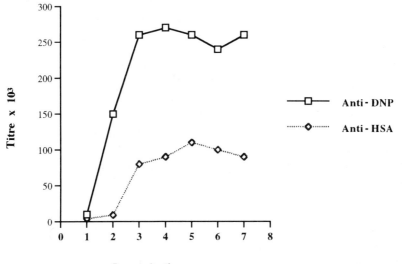

Figure 2. The antibody response in chickens immunized with DNP–HSA. Chickens were injected intraperitoneally with 50 μg of DNP–human serum albumin (DNP–HSA) adsorbed onto 1 mg Al(OH)$_3$ gel. For 20 doses, 1 mg DNP–HSA was diluted in 6.7 ml PBS containing 0.01% merthiolate and mixed with 3.3 ml Al(OH)$_3$ gel (6 mg Al per ml). Injections were repeated every 14 days and the chickens were bled from the neck vein 10 days after each immunization. Antibody titres were measured as the reciprocal of the dilution giving 50% of maximal signal in ELISA. Anti-DNP titres were measured on ELISA plates coated with DNP–ovalbumin, while anti-HSA titres were measured on HSA coated plates.

As for other animals, no chickens specially bred for a high antibody response are available, and it is thus advisable to immunize three or four chickens with the same antigen and not to persist with those that give a poor antibody response after three injections. Only rarely is it possible to improve much on a poor initial response. Typical responses are illustrated in *Figure 2*.

6. Purification of IgG from yolk

As mentioned above, the purification of IgG from yolk poses no special problems once the lipids have been removed. It is a common experience, even with chicken serum, that salt precipitation of IgG is ineffective unless steps have been taken to reduce the lipid content, either by withdrawing food (but maintaining free access to water) for 12–24 h before withdrawing the blood, or by dextran sulfate precipitation of serum lipids (24). Several methods for removing lipid from yolk have been published: extraction with organic solvents (25,26), precipitation with polyethylene glycol (PEG) (15,27), precipitation with dextran sulfate (14), or precipitation at low ionic strength (euglobulin precipitation) (14). Unfortunately, no careful and systematic comparison of the various methods is available in the literature. Judged by IgG quantification and antibody titration, recoveries of about 75% or 60% were achieved with salt precipitation after removing lipids with dextran sulfate or by euglobulin precipitation, respectively (14). We cannot refrain from mentioning that the euglobulin method has been rediscovered a couple of times (28,29) since it was originally reported (14).

PEG precipitation was reported to yield a recovery of only about half that obtained by dextran sulfate or euglobulin precipitation (30), and Polson (25) reported a more than doubled recovery when chloroform was used instead of PEG for removing lipids. The propanol method yields only about half as much IgG as the PEG method (31). PEG precipitation is reported to require high speed centrifugation to recover the IgG in the water phase between the precipitate and an upper lipid layer, whereas the dextran sulfate and euglobulin methods only need conventional low speed centrifugation. Recently, the use of caprylic acid for precipitating non-IgG proteins (32) was exploited for use on yolk IgG (33). While the principle was found to work, the reported recovery of IgG (about 35%) and the purity of the final product (about 65%) indicate that further methodological development is needed.

Whichever method is chosen, one will immediately note that the volume of precipitate is prodigious. The loss of IgG in the lipid fraction will thus be inversely proportional to the volume of buffer used, and one may wish to wash the precipitate one or more times to increase the yield. However, one will also sooner or later realize that chickens lay a lot of eggs, and in a research laboratory the vast amounts of antibodies, i.e. eggs, suddenly available may pose hitherto unknown storage problems. Eggs can, of course, be kept for extended periods of time at 4°C, but in our experience the yolk membrane becomes progressively weaker with storage, thus making the separation of yolk from egg white more tricky. Alternatively, one can separate yolk from egg white soon after collecting the eggs and store the yolk mixed with buffer and preservative, or one can keep the eggs in the freezer (see below).

Separation of yolk from egg white should be carried out carefully in order to reduce contamination with egg-white proteases, the abundant ovalbumin, and avidin. Conventionally, the initial step of this separation is carried out as in the kitchen, after which the yolk is placed on a piece of heavy absorbent paper in which a hole (approx. 5 mm diameter) has been cut. The yolk is placed above the hole and the yolk membrane is pierced with a needle (e.g. from a syringe). The ovalbumin and the yolk membrane will stick to the paper and the yolk is collected in a container (e.g. a 50 ml tube).

We have found that the separation can be made much easier if the egg is first frozen and thawed. When the shell is subsequently broken the yolk will have coagulated and remains semi-solid. The ball of yolk is now easily separated from the egg white and the yolk membrane by rolling it on a piece of absorbent paper. The use of a simple mechanical device will facilitate the subsequent suspension of the yolk in water or buffer, e.g. passage through a sieve (a tea strainer will do for a few yolks) or mixing in a kitchen blender. The freezing of the entire egg does not impede the recovery of IgG. On the contrary, it improves the subsequent removal of lipids in the first step of the dextran sulfate method (i.e. precipitation in TBS, *Protocol 4*).

Protocol 3. Removal of yolk lipids by euglobulin precipitation

Equipment and reagents
- Refrigerated centrifuge
- Freezer, –20°C
- TBS (see *Protocol 2*)
- 0.1 M NaOH
- 0.4 M sodium phosphate buffer pH 7.6

Method

1. Dilute the yolk with 9 vol. of distilled water and mix thoroughly.

2. Adjust the pH to 7.0 with 0.1 M NaOH.

3. Freeze the suspension at –20°C.

4. Thaw the suspension and centrifuge for 20 min at 2000–3000 g at 4°C with the brake off.

5. Harvest the supernatant.[a]

6. Add 1 ml 0.4 M phosphate buffer pH 7.6 per 100 ml supernatant (and preservative if the supernatant is to be left for some time before proceeding with step 7).

7. Purify the IgG by salt precipitation (*Protocol 5* and *6*).

[a] Chickens and eggs vary and occasionally one will encounter turbid supernatants, indicating residual lipids. These must then be removed before salt precipitation, which is easily accomplished with dextran sulfate and $CaCl_2$, using one-tenth of the volumes described in *Protocol 4*.

Protocol 4. Removal of lipids by dextran sulfate precipitation

Equipment and reagents
- Refrigerated centrifuge
- TBS (see *Protocol 2*)
- 1 M $CaCl_2$
- Dextran sulfate (e.g. Pharmacia, Sigma), 10% (w/v) in TBS

Method

1. Dilute the yolk with 4 vol. of TBS.

2. Centrifuge at 2000–3000 g for 20 min at room temperature[a] with brake off.

Protocol 4. *Continued*

3. Add 120 μl dextran sulfate solution per ml of supernatant. Mix well and incubate at room temperature for 30 min or longer.

4. Add 50 μl 1 M $CaCl_2$ per ml and mix. Incubate for a further 30 min.

5. Centrifuge for 30 min at 2000–3000 *g* and collect the supernatant. This supernatant should be clear—if not, repeat steps 3 and 4 with half the amount of dextran sulfate and $CaCl_2$.[b]

[a] The amount of lipid precipitated varies and tends to increase upon storage of the diluted yolk at 4°C. Smaller amounts of dextran sulfate/$CaCl_2$ than those stated will then be required to clear the suspension of lipids.
[b] The precipitate is voluminous—about 10 ml per yolk. The loss of IgG in the lipid fraction will thus be inversely proportional to the volume of buffer used. The buffer volume may be increased, or, for greater efficiency, the precipitate may be washed one or more times with buffer to increase the yield. We routinely wash once with 50 ml of TBS per yolk processed and combine this supernatant with the first dextran sulfate supernatant.

After removal of the lipid the proteinaceous solution can be subjected to salt precipitation for concentration and purification of the IgG. Alternatively, the antibody can be purified by affinity chromatography. An IgG preparation of very high purity is obtained by sodium sulfate precipitation (*Protocol 5*).

Protocol 5. Purification of yolk IgG by precipitation with sodium sulfate

Equipment and reagents

- Delipidated yolk (*Protocol 3* or *4*), 100 ml
- Centrifuge
- Powdered Na_2SO_4
- 36% (w/v) Na_2SO_4 in water[a]
- TBS (see *Protocol 2*)

Method

The entire procedure including centrifugation is carried out at 20–25°C—otherwise the sodium sulfate will precipitate.

1. Stir the delipidated yolk and slowly add 20 g Na_2SO_4.[b]

2. When all of the Na_2SO_4 has dissolved, let the solution stand for 30 min.

3. Centrifuge for 20 min at 2000–3000 *g* with the brake off. Discard the supernatant.

4. Redissolve the sediment in 10 ml TBS.

5. Centrifuge as in step 3. Discard the pellet.

6. Stir the supernatant and add 8 ml 36% (w/v) Na_2SO_4.[c] Let the solution stand for 30 min.

7. Centrifuge as in step 3.

8. Discard the supernatant.

9. Redissolve the pellet in 5 ml TBS (or PBS, if coupling to amino groups is to be carried out) and dialyse against the same buffer.

10. Remove any precipitate by centrifugation or filtration.[d]

[a]This is a supersaturated solution. The sodium sulfate is dissolved by boiling, and the solution cooled to 30–40°C before adding it to the protein solution, which has been warmed to room temperature.
[b]The resulting sodium sulfate concentration will be 18% (34).
[c]The resulting sodium sulfate concentration will be 16%.
[d]Purified chicken IgG shows a tendency to aggregate on freezing. For frequent use it is recommended to store solutions at 4°C with a preservative.

The recovery of IgG should be around 10 mg per ml of yolk. The IgG purified by this procedure shows a high degree of purity (*Figure 3*).

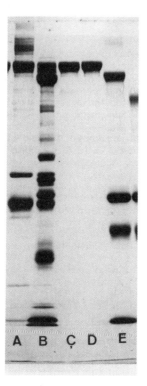

Figure 3. SDS–PAGE in the absence of reducing agents of (A) 1 µl chicken serum; (B) 1 µl yolk; (C) 10 µg IgG purified from clarified yolk by sodium sulfate precipitation; (D) purified chicken serum IgG. (E) Molecular weight markers: upper band, human IgG; then bovine serum albumin (slightly large M_r than chicken serum albumin) (see lane A), ovalbumin, and cytochrome *c*. Note the absence of ovalbumin in the yolk preparation, demonstrating the efficiency of separating yolk from the egg white, in which ovalbumin dominates. (Reproduced from ref. 14 with permission.)

For purposes not requiring high purity, the process may be simplified by only carrying out one precipitation step with 18% sodium sulfate (*Protocol 5*, steps 1 to 5).

Precipitation of IgG from yolk can also be carried out with ammonium sulfate. Sodium sulfate has been the first choice of immunochemists, but it is easier to work with ammonium sulfate since there is no problem of handling supersaturated solutions and no precipitation of the salt in the cold. The final IgG preparation may, however, be less pure.

Protocol 6. Purification of IgG by ammonium sulfate precipitation

Equipment and reagents

- Delipidated yolk (*Protocol 3* or *4*), 100 ml
- Refrigerated centrifuge cooled to 4°C
- Powdered ammonium sulfate
- 3.6 M ammonium sulfate in 4 mM sodium phosphate buffer pH 7.0
- TBS (see *Protocol 2*)

Method

1. Stir the delipidated yolk and slowly add 18.5 g powdered ammonium sulfate.[a] Incubate at 4°C for 30 min or longer.[b]

2. Centrifuge at 2000–3000 g for 20 min at 4°C with the brake off.

3. Discard the supernatant, and dissolve the precipitate in 10 ml TBS.

4. Centrifuge as in step 2.

5. Collect the supernatant. For many purposes the purity at this stage will suffice, and the product is dialysed against the desired buffer. If greater purity is desired continue with steps 6 and 7.

6. Add an equal volume of 3.6 M ammonium sulfate under stirring.

7. Repeat steps 2 to 5.

[a] The resulting ammonium sulfate concentration will be 2.0 M.
[b] If incubated overnight the precipitate will have settled, allowing the removal of part of the supernatant by suction and thus reducing the volume to be centrifuged.

It will be apparent from the previous discussion that we have some reservations about the PEG method. However, this method is preferred by some workers, and it may be felt that the simplicity of the method offsets its disadvantages. We therefore offer a suitable protocol (slightly modified from refs 15 and 27).

The PEG purification method is easy to perform but gives a lower recovery and a less pure IgG preparation than the other methods described. Ready-made reagents for purifying yolk IgG by PEG precipitation are available from Promega.

Protocol 7. Purification of yolk IgG by PEG precipitation

Equipment and reagents
- Refrigerated centrifuge
- Polyethylene glycol 6000 (PEG)
- 4.4% (w/v) PEG 6000 in TBS (see *Protocol 2*)

Method

1. Dilute the yolk with 4 vol. of 4.4% PEG in TBS and incubate for 20 min at 4°C.

2. Centrifuge at 3000 *g* for 20 min at 4°C with brake off.

3. Collect the supernatant and add PEG to a final concentration of 12% PEG (w/v).

4. Incubate for 20 min at 4°C.

5. Centrifuge at 3000 *g* for 20 min at 4°C.

6. Discard the supernatant and dissolve the precipitate in a volume of TBS similar to the volume of the yolk.[a]

[a]The remaining PEG does not interfere in many applications. It can be removed by ion exchange chromatography. It has also been reported that it can be removed by precipitation with ethanol below 0°C (27). It is not easily removed by dialysis.

---

7. Use of chicken antibodies

Chicken antibodies can be used in most techniques in which mammalian antibodies have traditionally been employed, and a few examples will be given below. As mentioned above, chicken antibodies have certain advantages over mammalian antibodies due to differences caused by their phylogenetic separation. However, the use of chicken antibodies has not been thoroughly optimized for the multitude of immunological techniques in which rabbit antibodies are generally used. There are no specific problems with the use of chicken antibodies in techniques relying on direct binding to target antigen (such as solid phase assays or affinity chromatography), but precipitation with chicken antibodies is less efficient because the complexes formed are more soluble, and the overall lower pI of chicken IgG affects its electrophoretic mobility.

With mammalian antibodies it has been found desirable to use Fab or F(ab')$_2$ fragments for many applications, primarily in order to avoid the interactions that can take place via the Fc region. In most cases such interactions simply do not occur with chicken IgG. In this context it should, however, be pointed out that no enzyme has been found that will produce divalent chicken F(ab')$_2$ fragments. Both papain and pepsin digest chicken IgG to monovalent Fab (35).

The poor precipitation seen with chicken antibodies in conventional buffers can be remedied by slight changes in conditions. For gel diffusion techniques the presence of high salt concentrations (1.5 M NaCl) or polyethylene glycol (e.g. 2–6% PEG 6000) in the gel buffer, or a low pH (pH 5.5–6) will dramatically improve the formation of precipitin lines (36,37).

For immunoelectrophoretic techniques (crossed immunoelectrophoresis, rocket immunoelectrophoresis) the low pI of chicken IgG requires the use of a lower pH for it to remain stationary in the gel (or move in the direction opposite to the direction of movement of the antigen). It is thus possible at pH 5.5 or 6 to obtain rockets of a quality equal to those seen with rabbit antibody at pH 8.6. The procedure by Altschuh *et al.* (37) for quantifying human IgG by rocket immunoelectrophoresis (RIE) is given in *Protocol 8*. In the same paper the authors show that chicken anti-human IgG can also be used in RIE for quantifying goat, rabbit, and mouse IgG, thus illustrating the marked species cross-reactivity obtainable when chickens are used to generate antibodies against mammalian proteins.

Protocol 8. Quantification of human IgG by rocket immunoelectrophoresis using chicken antibody

Equipment and reagents

- Glass plates (7 cm × 10 cm)
- Gel puncher (2 mm diameter)
- Horizontal gel electrophoresis apparatus (e.g. Bio-Rad)
- Thermostated water-bath
- Thermostated oven
- Voltmeter with electrodes fixed 4 cm apart
- Antigen: human IgG (e.g. Sigma) solutions of known concentrations (1 mg/ml to 10 μg/ml) for standard curve, and test samples of unknown IgG concentration (if sera, then dilute 1/50)
- Levelling table
- Agarose (HSA®, Litex A/S; or A4679, Sigma)
- Electrophoresis buffer: 50 mM 2-(N-morpholino) ethane sulfonic acid–NaOH pH 6.0
- Chicken antibody against human IgG
- Electrophoresis wicks (7 cm × 10 cm, cut from Whatman 3MM filter paper)
- TBS (see *Protocol 2*)
- Destaining solution: acetic acid/water/ethanol (1:6:7)
- Staining solution: 0.2% (w/v) Coomassie Brilliant Blue R-250 in destaining solution

Method

1. Dissolve the agarose in electrophoresis buffer (0.7 g/100 ml) in a boiling water-bath.

2. Pipette 10 ml into a tube in a 56°C thermostated water-bath (use pipettes taken from an oven set at 56°C).

3. Add a suitable amount of chicken antibody to the agarose solution. (The amount depends of the antibody titre and the concentration range of the antigen in the samples. 1–20 mg may be required if measuring IgG in undiluted human serum, less if the analysis is carried out on pre-diluted serum samples.)

4. Mix and pour onto a glass plate on a levelled table.

5. When the agarose has solidified, punch 2 mm holes in the agarose (with the gel puncher attached to a vacuum hose) in a row across the plate, 1 cm from the edge that is to be at the anode end (see *Figure 4*). This is conveniently done with the plate placed on a template drawn on graph paper.

6. Place the agarose plate on the cooling surface in the electrophoresis apparatus and fill the buffer reservoirs with 1 litre of electrophoresis buffer.

7. Wet the filter paper wicks in the buffer and place them so that they cover about 0.5 cm of the anode and cathode ends of the gel.

8. Apply 2 μl samples to the sample wells.[a]

9. Turn on the power with the cathode opposite to the row of holes and electrophorese at 10 V/cm overnight.

10. Turn off the current, remove the gel plate, and dry the gel. Fill the sample holes with water and overlay the gel with a pre-cut piece of wetted filter paper avoiding any trapping of air, add several layers of absorbent paper, a glass plate, and a weight of about 1 kg. Leave for a minimum of 20 min.

11. Remove the absorbent paper and filter paper carefully, so that the gel stays attached to the glass plate. Pouring water on the filter paper may facilitate this.

12. Place the plate in a container (Petri dish) with TBS for about 20 min to allow the gel to swell.

13. Repeat steps 10 and 11 and allow the gel to dry completely (this may be speeded up by using a hot air blower).

14. Stain the gel by submerging in staining solution for 5–10 min, destain, and dry in a fume-hood.[b]

15. Measure the rocket heights. Plot the standard curve on linear graph paper and interpolate to estimate the concentration of the antigen (human IgG) in the unknown samples.

[a] Larger holes and larger sample volumes may be used if samples with lower antigen concentrations are to be analysed.
[b] Staining is required only for permanent records or when the precipitates are too weak to be easily measured under indirect illumination of the untreated agarose plate.

Chicken IgG shows a greater tendency to aggregate on freezing and thawing than mammalian IgG, and it is therefore especially important to keep the purified preparations at 4 °C rather than in the freezer. This is, of course, a convenience rather than a problem, except that the preparations either have to be kept sterile or must have a preservative added. We recommend 15 mM

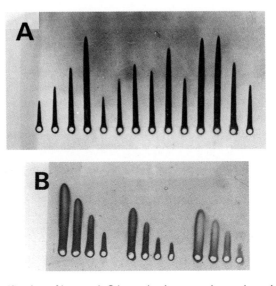

Figure 4. (A) Quantification of human IgG by rocket immunoelectrophoresis (RIE). Samples (2 μl) of serial doubling dilutions of a standard human serum were applied to the first four wells, while the rest of the wells were loaded with other human serum samples. (B) RIE of dilutions of goat, rabbit, and mouse sera in agarose containing chicken anti-human IgG antibody. (Reproduced from ref. 37 with the permission of the publishers.)

sodium azide or 0.01% (w/v) merthiolate (thimerosal). Sodium azide should not be used in acidic solutions because of the possible generation of carcinogenic substances and the alleged risk of generating explosive products in the drains. Merthiolate, an organic mercurial compound, is thought to be safe in the amounts used in a research laboratory, and it is used as a preservative in vaccines. It has one practical advantage over azide in that it does not have to be removed (by dialysis) before labelling the IgG via amino groups.

For use in a number of applications, e.g. ELISA, Western blotting, or immunohistochemical techniques, it is convenient to label the chicken antibody with biotin or directly with enzyme, and the protocols used in our laboratories are given below (*Protocol 9* and *10*).

Protocol 9.   Labelling of chicken IgG with biotin

Equipment and reagents

- High speed refrigerated centrifuge
- IgG purified from eggs of immunized chickens
- BNHS solution: *N*-hydroxysuccinimidobiotin (e.g. Sigma) 1 mg/ml in dry DMSO—this solution may be kept for up to three months at −20°C

- PBS (see *Protocol 1*)
- TBS (see *Protocol 2*)
- PBS/carbonate: PBS adjusted to pH 8.5 with 5% Na_2CO_3

Method

1. Dialyse 1 ml of purified chicken IgG at 1 mg/ml[a] extensively at 4°C against PBS and PBS/carbonate. We routinely dialyse overnight and the following day against two changes of 1 litre of PBS, and then overnight against 1 litre of PBS/carbonate.

2. Open the dialysis bag (it is convenient to use 1/8 inch diameter tubing closed with a clamp). Add 0.166 ml of the BNHS solution, close the bag, and incubate for 4 h at room temperature in a humid chamber.

3. Dialyse at 4°C against four changes of TBS.

4. Centrifuge at 10 000 *g* for 20 min and keep the supernatant with the biotin-labelled IgG at 4°C.

[a]Occasionally one may wish to biotinylate small amounts of IgG, e.g. affinity purified antibody. The whole procedure should then be carried out in the presence of 0.1% (w/v) Tween 20 in order to minimize loss of protein through adsorption and denaturation. The labelling is not very sensitive to changes in ratio of BNHS to IgG, and the experience in our respective laboratories is that adequate labelling is obtained at low protein concentration if the stated concentration of BNHS is maintained, as well as when maintaining the proportion between BNHS and protein.

Protocol 10. Labelling of chicken IgG with alkaline phosphatase

Equipment and reagents

- High speed refrigerated centrifuge
- Alkaline phosphatase (e.g. Sigma Type VII, P5521) 2000–3000 U/mg protein
- Glutaraldehyde, 25% (w/v)—aged reagents (after extended storage) may be more efficient than fresh reagents, and use of a high grade is not recommended
- Purified chicken antibody
- 0.1 M phosphate buffer pH 6.8
- 1 M ethanolamine–HCl pH 8.0
- 0.1 M $ZnCl_2$ and 0.1 M $MgCl_2$
- PBS (see *Protocol 1*)
- TBS (see *Protocol 2*)

Method

1. Mix 1 mg of antibody in 1 ml with 0.5 mg alkaline phosphatase (50 μl of the Sigma preparation), dialyse twice against 1 litre of PBS, and once against 1 litre of phosphate buffer at 4°C.

2. Add, with mixing, 50 μl 1% (w/v) glutaraldehyde (stock diluted 1/25 in phosphate buffer) and incubate for 4 h at room temperature.

3. Add 100 μl 1 M ethanolamine pH 8.0, and incubate for 1 h at room temperature.

4. Dialyse at 4°C against TBS.

5. Centrifuge at 10 000 *g* for 20 min and collect the supernatant.

6. Add $ZnCl_2$ and $MgCl_2$ to 1 mM each. Store at 4°C.[a]

[a]Some investigators prefer to add an equal volume of glycerol and store the enzyme conjugate at –20°C. There is no information on the suitability of this method of storage for conjugated chicken antibody.

References

1. Kubo, R. T., Zimmerman, B., and Grey, H. M. (1973). In *The antigens* (ed. M. Sela), p. 417. Academic Press, New York and London.
2. Benedict, A. A. and Yamaga, K. (1976). In *Comparative immunology* (ed. J. J. Marchalonis), p. 335. Blackwell Scientific Publications, Oxford.
3. Lebacq-Verheyden, A. M., Vaerman, J.-P., and Heremans, J. F. (1972). *Immunology*, **22**, 165.
4. Leslie, G. A. and Clem, L. W. (1969). *J. Exp. Med.*, **130**, 1337.
5. Magor, K. E., Warr, G. W., Middleton, D., Wilson, M. R., and Higgins, D. A. (1992). *J. Immunol.*, **149**, 2627.
6. Warr, G. W., Magor, K. E., and Higgins, D. A. (1995). *Immunol. Today*, **16**, 392.
7. Reynaud, C. A., Anquez, V., Grimal, H., and Weill, J. C. (1987). *Cell*, **48**, 369.
8. Gassmann, M., Thommes, P., Weiser, T., and Hubscher, U. (1990). *FASEB J.*, **4**, 2528.
9. Larsson, A., Karlsson-Parra, A., and Sjöquist, J. (1991). *Clin. Chem.*, **37**, 411.
10. Larsson, A., Wejåker, P.-E., Forsberg, P.-O., and Lindahl, T. (1992). *J. Immunol. Methods*, **156**, 79.
11. Kronvall, G., Seal, U. S., Finstad, J., and Williams, R. C. (1990). *J. Immunol.*, **104**, 140.
12. Larsson, A. and Lindahl, T. (1993). *Hybridoma*, **12**, 143.
13. Ramon, G. (1928). *C. R. Soc. Biol. (Paris)*, **99**, 1476.
14. Jensenius, J. C., Andersen, I., Hau, J., Crone, M., and Koch, C. (1981). *J. Immunol. Methods*, **46**, 63.
15. Polson, A., von Wechmar, M. B., and van Regenmortel, M. H. V. (1980). *Immunol. Commun.*, **9**, 475.
16. Butler, J. E. (1983). *Vet. Immunol. Immunopathol.*, **4**, 43.
17. Kuroki, M., Ohta, M., Ikemori, Y., Peralta, R. C., Yokoyama, H., and Kodama, Y. (1994). *Arch. Virol.*, **138**, 143.
18. Ebina, T., Tsukada, K., Umezu, K., Nose, M., Tsuda, K., Hatta, H., *et al.* (1990). *Microbiol. Immunol.*, **34**, 617.
19. O'Farrelly, C., Branton, D., and Wanke, C. A. (1992). *Infect. Immunol.*, **60**, 2593.
20. Steinberg, S. V., Munro, J. A., Fleming, W. A., French, V. I., Stark, J. M., and White, R. G. (1970). *Immunology*, **18**, 635.
21. French, V. I., Stark, J. M., and White, R. G. (1970). *Immunology*, **18**, 645.
22. Herbert, W. J. (1978). In *Handbook of experimental immunology* (ed. D. M. Weir), Vol. 3, A3. Blackwell Scientific Publications, Oxford.
23. Abou-Zeid, C., Filley, E., Steele, J., and Rook, G. A. W. (1987). *J. Immunol. Methods*, **98**, 5.
24. Masseyeff, R., Gombert, J., and Josselin, J. (1965). *Immunochemistry*, **2**, 177.
25. Polson, A. (1990). *Immunol. Invest.*, **19**, 253.
26. Akita, E. M. and Nakai, S. (1993). *J. Immunol. Methods*, **162**, 155.
27. Polson, A., Coetzer, T., Kruger, J., von Maltzahn, E., and van der Merwe, K. J. (1985). *Immunol. Invest.*, **14**, 323.
28. Akita, E. M. and Nakai, S. (1992). *J. Food Sci.*, **57**, 629.
29. Svendsen, L., Crowley, A., Østergaard, L. H., Stodulski, G., and Hau, J. (1995). *Lab. Animal Sci.*, **45**, 89.
30. Akita, E. M. and Nakai, S. (1993). *J. Immunol. Methods*, **160**, 207.

31. Bade, H. and Stegemann, H. (1995). *J. Immunol. Methods*, **72**, 421.
32. Steinbuch, M. and Audran, R. (1969). *Arch. Biochem. Biophys.*, **134**, 279.
33. McLaren, R. D., Prosser, C. G., Grieve, R. C. J., and Borissenko, M. (1994). *J. Immunol. Methods*, **177**, 175.
34. Heide, K. and Schwick, H. G. (1978). In *Handbook of experimental immunology* (ed. D. M. Weir), Vol. 1, p. 7.1. Blackwell Scientific Publications, Oxford.
35. Dreesman, G. and Benedict, A. A. (1965). *J. Immunol.*, **95**, 855.
36. McKinney, K. L. and Rimler, R. B. (1981). *J. Immunol. Methods*, **43**, 1.
37. Altschuh, D., Hennache, G., and van Regenmortel, M. H. V. (1984). *J. Immunol. Methods*, **69**, 1.

6

Standardization of immunochemical procedures and reagents

ROBIN THORPE, MEENU WADHWA, and TONY MIRE-SLUIS

1. Introduction

A prerequisite for producing valid, correctly calibrated immunochemical results is that the procedures and reagents are appropriately standardized. This is also necessary if results obtained on different occasions and particularly in different laboratories are to be compared.

Qualitative and even semi-quantitative procedures are generally less prone to standardization problems than fully quantitative methods. However, even these can provide incorrect results if standardization of procedures and reagents is ignored. Although much of this is simply a matter of ensuring that assay procedures are appropriately designed for the intended purposes, and have been validated for the types of samples to be processed, some important issues may be less obvious.

2. Quality of immunochemical reagents

Use of the appropriate quality of reagents is clearly important if immuno-chemical procedures are to be optimized. This can affect the sensitivity and specificity of most immunochemical methods. In some cases, e.g. immuno-assays, quite different results can be obtained if reagents are not of comparable quality and purity. The quality of immunoglobulin reagents can be especially important. For some procedures it is necessary to purify the IgG fraction from antisera, ascitic fluid, or hybridoma culture fluid (e.g. for labelling with enzymes, radioisotopes, or fluorochromes) whereas unpurified antibodies can be used equally successfully for other methods (e.g. immunohistochemistry or immunodiffusion techniques). In many cases, it is advantageous to purify the antigen-binding subpopulation from antibody-containing fluids using affinity chromatography (1). For procedures which are particularly sensitive

Robin Thorpe et al.

to antibody quality such as immunoassays, it is best to prepare a large batch
of material which will last for a considerable time and preserve this in
aliquots either lyophilized or at −70 °C.

Many immunochemical methods involve the use of 'inert' carrier proteins
for incubation and washing steps. These need to be chosen carefully to
ensure that they are efficient and do not adversely influence the assay results.
Serum albumin is often used for this, but particular preparations of this can
be appreciably contaminated with IgG which can compromise some methods.
Bovine haemoglobin or dried milk powder are often better than serum
albumin. In some cases, use of whole serum (e.g. from pig, sheep, or cow) can
produce low background signals in immunoassay techniques, but this needs
to be established experimentally.

3. Standardization of immunochemical procedures

The indirect nature of immunochemical procedures necessitates the inclusion
of a preparation of defined antigen content in all determinations if quantification of a substance in 'unknown' samples is to be achieved. This is particularly important for immunoassays and other quantitative immunological
assay methods, where concentration of analyte(s) is calculated by comparison
with a dose–response curve ('standard curve') generated using a preparation
of known antigen concentration. Such preparations are known as 'standards',
'reference preparations', 'reference reagents', or 'calibrators'.

3.1 Use of standards

For standardization and calibration of most quantitative immunochemical
procedures, a series of dilutions of the standard are prepared and assayed
using the adopted assay procedure. Unknowns or preparations being calibrated as 'secondary' or working standards (see Section 3.2) are included in
the same assay run. Results are usually plotted as assay readout (e.g.
absorbance for enzyme immunoassays or c.p.m. for radioimmunoassays or
immunoradiometric assays) against concentration of analyte (see Section 3.3).
Only the linear part of the plot can be used to derive comparative data, and
in some instances mathematical transformation of the data improves this (2).

Assessing the quantity of analytes in a preparation involves comparison of
a dilution series of this preparation with the linear part of the response produced using the primary standard. The displacement of the two lines represents the difference in antigen content (see *Figure 1*). The linear portions of
the dose–response curves for all preparations should be parallel if the analytes are similar. If not, the molecular species responsible for the assay
response are not the same and the results must be treated with care and even
suspicion.

In order to obtain a statistically valid estimation of antigen content it is

110

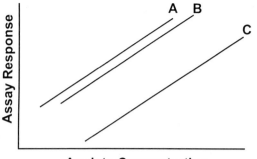

Figure 1. Typical parallel line analysis obtained using an immunoassay. Results obtained for a dilution series of (A) primary standard, (B) working standard, (C) unknown sample. All three samples produce parallel dose–response curves, showing that the assay is detecting the same molecular species in the different preparations. Analyte concentration can be calculated by measuring the displacement of the dose–response curves relative to the primary standard.

best to compare a number of dilutions of each sample with the dose–response curve. This not only ensures accuracy, as the data derived from the different dilutions essentially represent multiple determinations of the sample, but also allows a check for parallelism with the standard preparation, which assesses the comparability of the antigens being measured. However, this is not always feasible, particularly for routine measurement of large numbers of samples. In such cases, it is essential that the methods are carefully validated for the types of sample being measured. In particular, it is necessary to ensure that 'matrix effects', i.e. artefacts produced by the non-analyte content of the sample are not adversely affecting the assay and consequently the results obtained (see Section 4.1).

3.2 Nature and availability of standard preparations

The validity and calibration of quantitative immunochemical procedures depends on the quality and characteristics of the standard preparation used. It is therefore essential that the standard adopted is correct for the intended purpose.

A general principle of comparing 'like-with-like', i.e. that the standard and unknown samples are of similar non-analyte composition is well established for the assay of biological materials. Whilst matrix composition may be less important for immunochemical procedures than for bioassays, it is vital to en-sure that the molecular species in the standard is known, well characterized, and is of appropriate structure.

Use of a single standard or reference preparation world-wide allows universal comparison of results obtained in different laboratories. Such materials, often referred to as 'primary' or 'gold' standards need to be prepared with particular care and normally should be capable of being used in an assay-independent manner. They are normally used to calibrate 'in-house' or working standards (often called secondary standards).

It is imperative that all standards are stable and are stored appropriately; primary standards are normally lyophilized. Assessment of stability can be problematical as deterioration may only be apparent after storage for some time. In some cases, accelerated degradation studies in which potential candidate standards are subjected to a range of temperatures, some of which are relatively high, can allow prediction of instability. However, these conditions may not necessarily reflect conditions or processes which lead to instability under 'normal' storage conditions and can provide misleading information.

Production of primary standards for biological materials has mainly been carried out by, or under the auspices of, the World Health Organization (WHO). WHO International Standards and reference preparations and reagents are available for cytokines, immunoglobulins, hormones, allergens, and numerous other analytes; some of these are listed in *Tables 1* and *2*. Assessment and selection of International Standards almost always includes an International Collaborative Study involving a range of expert laboratories world-wide (3,4). Such standards are intended to last for a considerable length of time and so distribution is normally restricted to a few ampoules/laboratory/year.

It is often necessary to produce 'secondary' standards (also known as 'in-house' or 'working' standards) which are included every time an assay is carried out, or even on every microtitre plate for well controlled immunoassays using this format. Such standards are normally calibrated directly against the primary standard, e.g. using the parallel line approach (see *Figure 1*). It can be useful in some instances to include 'performance indicators' in immunoassays. These materials are not standards or calibrators, but are used to ensure that the assay performance characteristics are appropriate, e.g. sensitivity is as expected.

3.3 Unitage of standard preparations

Standard preparations do not necessarily need to be pure if they are to be used for immunoassay calibration and validation. However, the amount of relevant analyte(s) present in such materials must obviously be known. Biological potency standards, produced primarily for use in bioassays, are usually calibrated in terms of units of biological activity such as WHO International Units. These may reflect an assay related parameter or more usually (and better) are arbitrarily defined and therefore directly relate to the ampoule contents of the standard. In some cases this approach can be adopted for standardizing immunochemical procedures, but problems can arise where results obtained using immunoassays do not necessarily reflect biological

Table 1. Cytokine standards

Preparation	Product code	Status[a]	Depository[b]
Interleukin 1 alpha rDNA	86/632	IS	NIBSC
Interleukin 1 beta rDNA	86/680	IS	NIBSC
Interleukin 2 cell line derived	86/504	IS	NIBSC
Interleukin 2 rDNA	86/564		NIBSC
Interleukin 3 rDNA	91/510	IS	NIBSC
Interleukin 4 rDNA	88/656	IS	NIBSC
Interleukin 5 rDNA	90/586	RR	NIBSC
Interleukin 6 rDNA	89/548	IS	NIBSC
Interleukin 7 rDNA	90/530	RR	NIBSC
Interleukin 8 rDNA	89/520	RR	NIBSC
Interleukin 9 rDNA	91/678	RR	NIBSC
Interleukin 10 rDNA	92/516	IR	NIBSC
Interleukin 11 rDNA	92/788	RR	NIBSC
Interleukin 12 rDNA	95/544	RR	NIBSC
Interleukin 13 rDNA	94/622	RR	NIBSC
Interleukin 15 rDNA	99/554	RR	NIBSC
M-CSF rDNA	89/512	IS	NIBSC
G-CSF rDNA	88/502	IS	NIBSC
GM-CSF rDNA	88/646	IS	NIBSC
Leukaemia inhibitory factor rDNA	93/562	RR	NIBSC
Oncostatin M	93/564	RR	NIBSC
Stem cell factor/MGF rDNA	91/682	IR	NIBSC
flt 3 LIGAND rDNA	96/682	IR	NIBSC
Bone morphogenetic protein-2	93/574	IR	NIBSC
RANTES rDNA	92/520	IR	NIBSC
GRO alpha rDNA	92/722	IR	NIBSC
MCP-1 rDNA	92/794	IR	NIBSC
MCP-2 rDNA	96/594	IR	NIBSC
MIP-1 alpha rDNA	92/518	IR	NIBSC
MIP-1 Beta rDNA	96/588	IR	NIBSC
IFN alpha leucocyte	69/19	IS	NIBSC
IFN beta fibroblast	Gb23-902-531	IS	NIAID
IFN gamma rDNA	Gxg01-902-535	IS	NIAID
TGF beta 1 rDNA	89/514	IR	NIBSC
TGF beta 1 (Nat bovine)	89/516		NIBSC
TGF beta 2 rDNA	90/696	IR	NIBSC
TNF alpha rDNA	87/650	IS	NIBSC
TNF beta rDNA	87/640	RR	NIBSC
Interleukin-1 Soluble Receptor Type 1 rDNA	96/616	IR	NIBSC

[a] IS, International Standard; RR, WHO Reference Reagent; IR, Interim Reference Reagent.
[b] NIBSC: The National Institute for Biological Standards and Control, Blanche Lane, South Mimms, Potters Bar, Hertfordshire EN6 3QG, UK. NIAID: The National Institute for Allergy and Infectious Diseases, Solar Building, 6003 Executive Drive, Maryland, USA.

Table 2. Other immunological standards

Category	Preparation	Product Code	Status[a]	Depository[b]
Serum immunoglobulins	Human serum IgE	75/502	IRP	NIBSC
	Human serum IgG, A, and M	67/86	IRP	NIBSC
	Human serum IgD	67/37	MRC-RS	NIBSC
Complement Components	Human serum complement components C1q, C4, C5, factor B and whole functional complement CH50	W1032	IS	CLB
	Anti-dsDNA	W1065	IS	CLB
Autoantibodies	Anti-nuclear factor serum, human	W1064	IS	CLB
	Rheumatoid arthritis serum, human[c]	W1066	IS	CLB
	Anti-nuclear ribonucleoprotein serum, human	W1063	IRP	CLB
	Anti-smooth muscle serum, human	W1062	IRP	CLB
Other human serum proteins	Alphafetoprotein	AFP	IS	SSI
	Human serum proteins[c], for immunoassays; albumin; alpha-1-antitrypsin; alpha-2-macroglobulin; ceruloplasmin; complement C3; transferrin	W1031	IS	CLB
	Pregnancy specific $\beta 1$ glycoprotein	SPG1	IS	SSI
	$\beta 2$ microglobulin	$\beta 2m$	IS	SSI
	Ferritin	80/578	IS	NIBSC
	Primary biliary cirrhosis serum	67/183	MRC-RS	NIBSC

Alloantibodies
(a) For quantitation of anti-D

• in prophylactic Ig preparations	Anti-D immunoglobulin, human	68/419	IRP	NIBSC
• in plasma	Anti-D antibodies	73/515	BS	NIBSC
(b) For quantitation of anti-C in plasma	Anti-C antibody	84/628	BS	NIBSC

Other antibody reagents

(a) For blood grouping (intended for standardization of haemagglutinin assays)	Anti-A[d]	88/722	Brit RR	NIBSC
	Anti-B[d]	88/724	Brit RR	NIBSC
	Anti-D[d]	91/592	Brit RR	NIBSC
(b) For detection of platelet antigen	Monoclonal antibody against platelet IIb/IIIa complex	85/661	RR	NIBSC

[a] BS, British Standard; Brit RR, British Reference Reagent; IS, International Standard; IRP, International Reference Preparation; MRC-RS, MRC Research Standard; RR, Reference Reagent.
[b] CLB: Central Laboratory, Netherlands Red Cross Blood Transfusion Service, Plesmanlaan 125, Amsterdam, The Netherlands. SSI: Statens Seruminstitut, 80 Amager Blvd., 2300 Copenhagen S, Denmark, NIBSC: The National Institute for Biological Standards and Control, Blanche Lane, South Mimms, Potters Bar, Hertfordshire, UK.
[c] The 1st British Standard for Rheumatoid Arthritis Serum (code 64/2) and human serum proteins (code 74/520) are available from NIBSC.
[d] International Standards for these and other blood typing serum preparations (e.g. anti-C and anti-E) are available from CLB.

activity (5,6). In such cases, use of biological units may provide incorrect or at least misleading information, and this has resulted in alternative approaches to standard calibration.

It is tempting to calibrate immunoassays in terms of mass units and therefore use standards in which the weight of analyte in ampoules is accurately known. However, this can be problematical for biological substances where 'mass' determination has been carried out indirectly and therefore can be only approximate. Immunoassays can also produce incorrect mass estimates for samples containing an analyte which differs even slightly from the structure of the molecular species in standard preparations due to the epitope specificity properties of the antibodies used. This can result in different 'mass' estimates when different immunoassays are employed, especially if different antibodies are used (see Section 4.3). Use of mass calibration in cases where the analyte exists in different molecular forms (aggregates, precursor molecules, clipped and truncated forms, fragments, etc.) can lead to incorrect mass estimates as the assay may not discriminate between the different structures (4). In such cases, calibration of assays and standards in terms of molarity of analyte can partially overcome the problem, or at least avoid reporting incorrect results.

4. Problems with the standardization of immunochemical procedures

Several standardization problems have been clearly identified with the use of immunochemical procedures, some of the most common are discussed in this section.

4.1 Matrix effects

Matrix effects, i.e. phenomena influencing assays which are unrelated to the analyte and are primarily due to the sample composition, can considerably affect some immunoassays. This can appear as an artefactual reduction or increase in observed assay readout, or shift in parallelism or shape of the response curve relative to the standard curve. It is essential that assays are validated to ensure that such effects are absent; this normally involves estimating different types of sample containing the analyte and 'spiking' a pure preparation of known antigen content with analyte-free biological fluid (7), e.g. serum, plasma, tissue homogenate, etc. (see *Figure 2*).

Matrix effects can limit the sensitivity of immunoassays and this can be appreciable if sensitivity is calculated using purified recombinant DNA-derived analyte in a simple buffer. Sensitivity will often be less if complex biological samples are assayed.

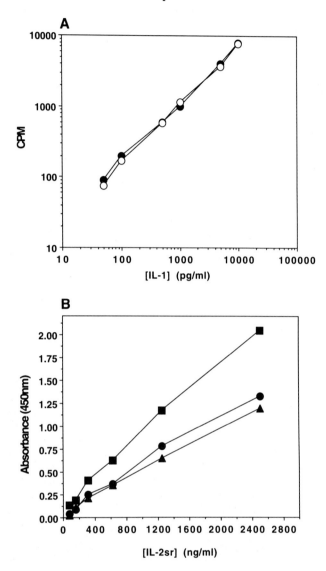

Figure 2. (A) An IRMA specific for human IL-1α (7) which does not show matrix effects. Open circles, rDNA derived IL-1α prepared in buffer alone; solid circles, the same preparation of IL-1α spiked with human serum. The superimposable dose–response curves show that the presence of serum does not affect the dose response to analyte. (B) An ELISA for IL-2 soluble receptor which shows matrix effects. Squares, rDNA derived IL-2 soluble receptor in buffer alone; circles and triangles, the same preparation 'spiked' with human plasma or human serum respectively. In this study, the plasma and serum clearly influence the dose–response curves.

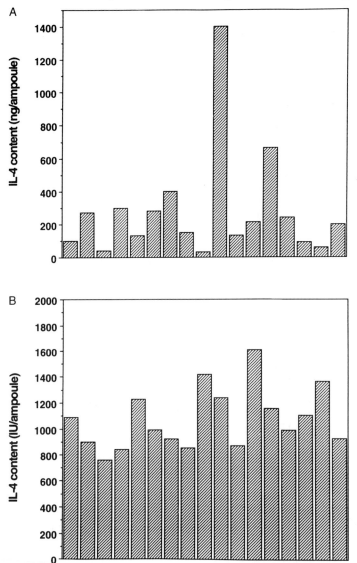

Figure 3. Estimates of IL-4 levels using immunoassays (8). A result from the WHO International Collaborative Study for the standardization of IL-4 in which 17 laboratories contributed immunoassay data to estimate levels of IL-4 in ampouled preparations. (A) The estimates of the levels of IL-4 by the different laboratories of a single ampouled preparation of IL-4 estimated to contain 100 ng of material. The laboratories used their own in-house standards calibrated in 'mass units' to estimate these levels. (B) The same laboratories estimating the level of IL-4 in the same ampouled preparation estimated to contain approximately 1000 IU of IL-4. The laboratories were using the international standard for IL-4 as a reference preparation for all of these assays.

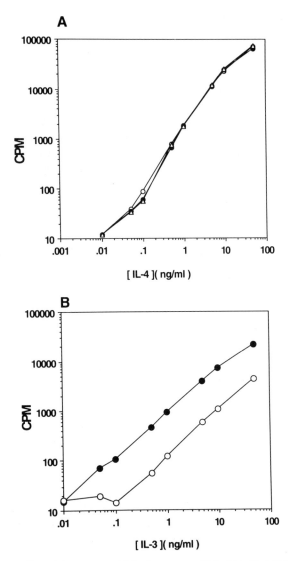

Figure 4. Immunoradiometric assays which show essentially identical (A) or preferential (B) reactivity with different molecular forms of analyte (9). (A) An IRMA for human IL-4 which equally detects three different subspecies of the cytokine which differ in primary sequence and/or glycosylation. (B) An IRMA for human IL-3 which reacts differently to subspecies which differ slightly in primary structure. Note that considerably different (incorrect) estimates of analyte content for the two preparations would be obtained if the assay was standardized using the 'incorrect' species.

4.2 Use of multiple standards

As mentioned in Section 3.3, calibration of standards can be more difficult than is immediately obvious and this is especially the case if calibration in terms of mass units (or what are assumed to be mass units) is attempted. Experience has shown that vastly different estimates for analyte content of a single sample can be produced if different immunoassays are used, in different laboratories, using different indirectly calibrated standards (8). Much of this can be due to the use of multiple in-house standards since results obtained are significantly less variable if a single standard is adopted (see *Figure 3*). This does not solve the problem of accurately calibrating standards in terms of mass, but does allow a degree of comparability of results between laboratories and assays and the use of a common arbitrary unitage for analyte.

4.3 Problems due to the epitope specificity of antibodies

The considerable specificity shown by antibodies can be of great advantage; however, it can cause problems with some immunochemical procedures. Many antibodies can discriminate between molecular species which differ very slightly from each other in amino acid sequence or conformation. This can lead to detection of analyte from different sources or expression systems to differing extents. This effect can be total (i.e. an assay can completely fail to detect a particular form of antigen) or partial (9). The latter effect is particularly misleading as the assay will respond to antigen, but the estimate of concentration will be incorrect (see *Figure 4*).

Post-translational modification of antigen may also affect immunoassay results. Glycosylation, proteolytic cleavage, aggregation and complexing with binding proteins, soluble receptors, etc. can all cause significant deviation for comparability with the standard and lead to appreciable errors in estimation of concentration of analyte. In some cases, this can be addressed by the production of standard preparations intended for estimation of particular forms of antigen (this can also reduce matrix effects in some cases). This approach needs great caution as it may result in the preparation of multiple standards with all the problems associated with this (see Section 4.2).

Acknowledgements

We thank Chris Bird for data and Deborah Richards for processing this manuscript.

References

1. Johnstone, A. and Thorpe, R. (1996). *Immunochemistry in practice* (3rd edn). Blackwell Science, Oxford.

2. Finney, D. J. (1978). *Statistical methods in biological assay* (3rd edn). Charles Griffin & Co. Ltd., London.
3. World Health Organization. (1990). *WHO, Technical Report Series*, **800**, 181.
4. Mire-Sluis, A. R., Gaines-Das, R., Thorpe, R., and Participants of the Study. (1995). *J. Immunol. Methods*, **179**, 141.
5. Thorpe, R., Wadhwa, M., Bird, C. R., and Mire-Sluis, A. R. (1992). *Blood Rev.*, **6**, 133.
6. Wadhwa, M., Seghatchian, M., Lubenko, A., Contreras, M., Dilger, P., Bird, C., and Thorpe, R. (1996). *Br. J. Haematol.*, **93**, 225.
7. Thorpe, R., Wadhwa, M., Gearing, A. J. H., Mahon, B., and Poole, S. (1988). *Lymphokine Res.*, **7**, 119.
8. Mire-Sluis, A. R., Gaines-Das, R., and Thorpe, R. (1995). *J. Immunol. Methods*, **186**, 157.
9. Bird, C., Wadhwa, M., and Thorpe, R. (1991). *Cytokine*, **3**, 562.

7

Synthetic peptides in epitope mapping

STUART J. RODDA, GORDON TRIBBICK, and N. JOE MAEJI

1. Introduction

Synthetic peptides can be used as a tool for probing the nature of antibody-defined (B cell) epitopes or T cell receptor (TCR)-defined (T cell) epitopes. Linear synthetic peptides homologous with the sequence of a protein antigen are suitable for mapping the 'linear' (continuous) type of B cell epitope, and all T cell epitopes. The power of the synthetic peptide approach to epitope mapping has been greatly expanded by the development of techniques for the parallel synthesis of hundreds to thousands of peptides, and by improvements in the chemistry of peptide synthesis which make the synthesis of peptides beyond 20mers in length feasible on a mass scale.

Of the several methods currently available for parallel synthesis of peptides, the Multipin method was the first to give practical access to large numbers of peptides (1). The 'T bag' method followed shortly thereafter (2), followed by methods for synthesis of peptides on paper and cotton (3) which permit low cost qualitative screening of synthetic peptides for binding by antibodies. For the well-funded researcher there is now a range of automated parallel synthesizers, many of which can make up to 96 peptides simultaneously (3). The Multipin method was adapted to synthesis of solution phase peptides (4–6) which allowed it to be applied not only to T cell epitope mapping but also to a greater variety of methods for study of B cell epitopes (7,8).

This chapter will focus on the Multipin method of peptide synthesis and the assays in which large numbers of synthetic peptides can be used. Many of the protocols are also applicable to peptides made by a number of alternative synthetic methods. This chapter does not give practical methods for the synthesis of peptides on resin (in T bags or multisynthesizers), or on paper or cotton.

2. Multiple peptide synthesis on pins

The Multipin method uses plastic pins, arranged in the format of a 96-well microtitre plate, to make sets of up to 96 peptides (or multiples thereof) simultaneously (9,10). Depending on the chemical nature of the linker between the peptide and the plastic, the peptides may be permanently bound to the plastic surface, or may be cleavable under more or less harsh conditions. The scale of synthesis for cleaved peptides is set by the type of pin—the two commercially available scales (types) are 1 μmol and 5 μmol (approx. 1 mg or 5 mg of a decamer respectively).

2.1 Cleaved or non-cleavable?

The original pin design was based on peptides permanently bound to the pin surface, and such pins are still being used for applications which require simple binding on a solid phase. They have the advantage, over some alternative methods, that the peptide density is relatively high, giving high sensitivity for detection of binding. The sensitivity (avidity) can be enhanced by interaction of more than one peptide molecule with each antibody molecule. 'Non-cleavable' peptides can be reused many times by disrupting the antibody–peptide interaction, using sonication in a warm detergent bath. Synthesis protocols are the same for non-cleavable as for cleavable peptides up to the point of side chain deprotection, after which the non-cleaved peptides are simply washed and are ready for use in a binding assay, while cleaved peptides require further processing before use.

2.2 Choice of ending

If cleaved peptides are desired, a choice must be made as to the type of C-terminal ending. This choice determines the type of linker used on the pins on which the peptides are synthesized, and also the type of cleavage method to be employed. *Table 1* lists the choices available and the cleavage method appropriate to each.

The choice of ending may be dictated by the biology of the application, such as the need for peptides with a C-terminal carboxyl (free acid) group when minimal cytotoxic T cell epitopes are being mapped; or it may be dictated by convenience, such as the easy handling of DKP-ended peptides when T helper epitopes are being mapped (11).

3. Peptide synthesis

Synthesis can be performed by following the recipes provided in the kits but it would be advisable to consult a reference work on peptide synthesis (12,13) for supplementary information. *Protocol 1* deals with the process of assembling the protected peptide sequences on pins, and *Protocol 2* describes the side chain deprotection and cleavage of peptides from pins as it applies to each of the pin types. There is insufficient space here to deal with the issue of peptide purification and characterization, but it is a major issue when ascribing the biological activity of a synthetic peptide to the target sequence synthesized.

Table 1. Types of cleavable pin kit

Desired ending	Pin kit name[a]	Scale	Linker	Cleavage method[b]
Free carboxyl (acid)	MPS Kit	5 μmol	Ester	TFA + scavengers
Amide	MPS Kit	5 μmol	Rink amide	TFA + scavengers
Glycine free acid	GAP Kit	1 μmol	Glycine ester	Base (NaOH)
Diketopiperazine	DKP Kit	1 μmol	Lysine–proline ester	Neutral buffer
Multiple choice	5-in-1 Kit	1 μmol × 5	Multiple types	Various

[a] MPS, multiple peptide synthesis; GAP, glycine acid peptide; DKP, diketopiperazine; 5-in-1, five component parts to each pin, giving multiple choices as to endings.
[b] TFA, trifluoracetic acid.

Presence of a minor contaminant with unexpectedly high specific activity can mislead the researcher into incorrect conclusions. Results obtained with unpurified peptides should therefore always be confirmed with peptides of higher purity. If any doubts still exist, peptides can also be prepared using an alternative chemistry as a way of reducing the likelihood that bioactivity is due to trace amounts of an unexpected contaminant.

Protocol 1. Synthesis of peptides on pins

Equipment and reagents

- Pin synthesis kit (can be non-cleavable or cleavable) (*Table 1*): kits include pins, plastic baths and reaction trays, software, manual, pre-synthesized control pins, and monoclonal antibody against the control peptide (Chiron Mimotopes)
- Computer (IBM compatible or Macintosh)—optionally, use a computer-controlled display unit, PinAID (Chiron Mimotopes)
- Solvent-resistant pipettors (e.g. Labsystems and Brand) and tips
- Activating agent (diisopropylcarbodiimide, DIC) and catalyst 1-hydroxybenzotriazole (HOBt) (Sigma, Fluka, Merck, Aldrich)
- Solvents: dimethylformamide (DMF), methanol (MeOH) (BDH, Merck, Fluka, Aldrich)

- Fluorenylmethyloxycarbonyl (Fmoc)-protected amino acids with appropriate side chain protecting groups (Chiron Mimotopes, Bachem (Switzerland), Novabiochem, Sigma)
- Deprotection reagent: piperidine (Merck, Fluka, Aldrich)
- Indicator: bromphenol blue (Merck) 0.7% in DMF
- Miscellaneous reagents: acetic anhydride, diisopropylethylamine (Merck, Fluka, Sigma)
- All reagents should be AR or the best quality available—DMF should only have low levels of amine (tested with fluorodinitrobenzene, FDNB)

Method

1. Read the manual supplied with the pin synthesis kit.

2. Use the software to generate a set of peptide sequences to be synthesized.[a]

3. Use the software to generate a list of solutions to be made each day (amino acid solutions, activation reagent) and the plate positions at which each amino acid will be added on each day.

Protocol 1. *Continued*

4. Following the instructions given in the manual, and the printout of the positions of the first day's amino acid additions OR the positions for placement of the appropriate amino acid pre-derivatized pins for the first day, put the required pins onto the 96-pin holder.

5. Fmoc deprotect the pins with 20% piperidine to prepare them for acceptance of the first amino acid. Wash them in DMF and methanol, and allow them to air dry.

6. Dissolve and activate the stated amount of each amino acid.

7. Pipette each activated amino acid into the target wells of the reaction tray, guided by the computer printout of the required amino acid locations. Alternatively, use the PinAID to light up the target wells for each amino acid.

8. Place the pins into the filled wells of the reaction tray and allow coupling to occur for 2 h or longer at 20–25 °C.

9. Check the pins for completeness of coupling by disappearance of blue staining from the reactive pin surface.

10. Wash the pins in DMF and methanol, then air dry.

11. Repeat the Fmoc deprotection with 20% piperidine in DMF (step 5).

12. Repeat steps 6–10 for the second day's amino acid additions.

13. Repeat steps 11 and 12 for each subsequent day's amino acid additions until all peptides have been made.

14. If the N-terminal end of any peptide is to be blocked by acetylation, then after the final Fmoc deprotection place the pin into a 193:6:1 mixture of DMF/acetic anhydride/diisopropylethylamine for 90 min, then wash in methanol and DMF, before proceeding to side chain deprotection.

[a] Note that peptides are conventionally written left to right from amino to carboxy end, but are synthesized from carboxy (closest to the pin surface) to amino end. The software simplifies design of systematic peptide sets such as overlapping sets. Alternatively, sets of unrelated peptide sequences are simply typed in at the keyboard.

Protocol 2. Side chain deprotection and cleavage of peptides

Equipment and reagents

- TFA (Merck, Aldrich, Fluka) (bulk TFA can be obtained from Halocarbon, USA)
- Scavengers and reducing agents: ethanedithiol (EDT), anisole, and mercaptoethanol (Fluka, Aldrich, Merck)

- Additionally, depending on the type of cleavage, solvent for the extraction of peptide from the pins or for washing of the dried peptide may be required—appropriate solvents include acetonitrile (HPLC grade), ether, petroleum ether

Method

1. Side chain deprotect the peptides by immersion of the pins in TFA/EDT/anisole (38:1:1) for 2.5 h at room temperature.[a] The volume of side chain deprotection solution used depends on the scale of synthesis: $\geq$ 0.3 ml per pin for the 1 μmol scale and 1.5 ml per pin for the 5 μmol scale.

2. Remove the pins from the side chain deprotection solution. For the MPS kit, the peptides are now in the TFA solution but for the other kits the peptides are deprotected but still on the pin.

3. After side chain deprotection of pins from non-cleavable, DKP and GAP kits, wash the pins in methanol for 10 min, then in 0.5% acetic acid in 1:1 methanol/water for 60 min, and finally twice in methanol again. They are now ready for the first test (non-cleavable) or for cleavage (DKP and GAP kits).

4. For the MPS kit, dry down the TFA solution containing the peptide, using a gentle stream of dry nitrogen. Each peptide is now extracted with 8 ml of a cold (4°C) mixture of ether/petroleum ether/mercapto-ethanol 1:2:0.003 for 30 min. Centrifuge the tubes for 6 min at approx. 3200 g in a flameproof centrifuge. If there is a substantial pellet (the peptide), decant the supernatant fluid and repeat the extraction with 4 ml cold ether/petroleum ether, 1:2, and centrifuge as above. Decant the ether wash and dry down the peptide pellet. If at any stage the peptide dissolves in the ether washes, do not discard the washes but dry them down to recover the peptide.

5. For the GAP kit, cleave the peptides from each pin by incubation for 0.5–1 h in 0.73 ml of 0.1 M sodium hydroxide made up in 40% (v/v) acetonitrile/water. The cleavage time can be kept to 30 min if sonication is used to assist in cleavage. Immediately after cleavage, remove the pins from their cleavage solution and neutralize with 0.07 ml of an appropriate acid, e.g. 2 M NaH_2PO_4. The peptide solutions are now ready for use in an assay.

6. For the DKP kit, the peptide is self-cleaving at neutral or alkaline pH. Cleave the peptide by immersing each pin into 0.8 ml of buffer, e.g. 0.1 M sodium phosphate pH 7.6 or 0.05 M Hepes pH 7.6, made up in 40% (v/v) acetonitrile/water, at room temperature for 16 h. This time can be decreased to 1 h if the solutions containing the pins are sonicated during cleavage. The peptides are now ready for use in an assay.

[a] Non-cleavable peptides and peptides in DKP and GAP kits remain on the pins; peptides in MPS kits are simultaneously side chain deprotected and cleaved from the pins.

3.1 Storage and handling of peptides

Peptides are liable to chemical and biological degradation. Peptides containing methionine or cysteine are readily oxidized, and some peptides are light-sensitive. Solutions of peptides can quickly degrade if exposed to proteases or if unsterile solutions of peptide are kept under conditions where micro-organisms can grow. Follow *Protocol 3* to handle and protect your peptides from damage up until the point where you use them in a biological assay.

Protocol 3. Redissolving, storing, and handling peptides

Equipment and reagents

- Synthetic peptides, either non-cleavable or cleaved, made on pins as described above or obtained by other means
- Solvent-resistant tubes, e.g. Bio-Rad racked polypropylene tubes Cat. No. 223–9395
- Solvents: acetonitrile (HPLC grade), dimethyl sulfoxide (DMSO), DMF, etc. (Merck, Fluka, Aldrich)

Method

1. If the peptides are in the form of a dry powder, store in the dark at 4°C or colder. If they are in solution, either in a solvent such as DMSO or in aqueous solution, store frozen at –20°C or colder.

2. When working with peptide powders which have been stored in the cold, allow them to equilibrate to room temperature before opening the container, so that atmospheric moisture does not rapidly condense on the peptide.

3. For redissolving peptides, choose solvent-resistant tubes which allow you a clear view of the solution so you can detect turbidity or the presence of undissolved solid. Glass is preferable but polypropylene Bio-Rad tubes have some advantages for mass handling: compact storage, strip lids, break resistance.

4. As a general strategy, dissolve peptide sets in 1 ml of 40% (v/v) acetonitrile/water or 0.5 ml DMSO per peptide.[a] Sonicate to assist dissolution if possible.

5. Prepare dilutions from these stock solutions into water or aqueous buffer and use immediately or re-freeze.

6. For oxidation-sensitive peptides, work with degassed solvent, keep the pH below 7 if possible, and store under a nitrogen atmosphere if practical to do so.[b]

7. For non-cleavable (pin-bound) peptides, store the dry pins (dried after each test, see *Protocol 4*) at 4°C or colder in a plastic bag, with a satchel of desiccant.

8. Avoid contamination of peptides with organic matter (dust, hair, skin particles, etc.). Avoid possible sources of proteolytic enzymes, and extremes of pH.

[a] These levels of solvent will generally not compromise the assay when diluted by a factor of 1/500 or greater. For example, cell proliferation assays will tolerate 2% of residual acetonitrile or 0.3% DMSO.
[b] Peptides prepared using the reducing agents/scavengers EDT and mercaptoethanol are initially in the reduced form and can be kept largely that way if precautions are taken against oxidation. Re-reduction may be necessary after a period of storage.

4. Antibody (B cell) epitope mapping with synthetic peptides

It is feasible to synthesize and test all linear peptides of a given length homologous with the known sequence of a protein antigen (9,10). This process has become known as a Pepscan (14,15). Testing of such a set of peptides has the potential to identify all the linear epitopes of an antigen and to give some insight into the identity of parts of discontinuous epitopes. This is most striking when working with a monoclonal antibody, where binding to two or more peptides of an antigen is presumptive evidence for the existence of a discontinuous epitope. Decisions which have to be made when proceeding to map epitopes in this way include: whether to use non-cleavable or cleaved peptides such as biotinylated peptides (*Table 2*); the most appropriate length of peptide; the offset and overlap between successive peptides in the set; whether to acetylate the N-terminus of pin-bound peptides or to leave them as the free amine (1); in the case of pin-bound peptides, the number of 'copies' of each peptide which are to be made. The 5-in-1 synthesis kit (*Table 1*) gives the option of both non-cleavable and cleavable peptides from the one synthesis. As a starting point, we would suggest that a set of overlapping

Table 2. Non-cleavable or cleaved (biotinylated) peptides for antibody (Ab) epitope ELISA?

Aspect	Non-cleavable[a]	Cleaved (biotinylated)[b]
Amount of peptide made	< 0.1 μmol/pin	1 μmol/pin
Detection of Ab binding	Highest sensitivity	Lower sensitivity
Number of tests per pin	Up to 50	Thousands
Verification of peptide quality	Difficult (AA analysis)	Simple (standard methods)
Reproducibility of ELISA	Reactivity declines	Reproducible
Variety of assays	Binding only	Binding, competition, bioassay
Speed of assays	One assay per day	Many assays simultaneously

[a] Pins are reused by disrupting the antibody–peptide interaction (see *Protocol 4*).
[b] A fresh aliquot of peptide is taken for each assay and coated on avidin or streptavidin plates (see *Protocol 5*).

peptides should not drop below 8mers as the longest peptide length for which all possible sequences are present in the scan. This can be achieved by making (for example) all 10mers offset along the sequence by three residues, or all 11mers offset along the sequence by four residues, and so on.

Protocol 4 presents the method for screening with a set of non-cleavable pin peptides and *Protocol 5* with a set of biotinylated peptides.

4.1 Direct binding of antibodies on pins

Peptides synthesized on pins can be reacted with either a monoclonal or polyclonal antibody preparation, and the presence of bound antibody can then be detected by a direct ELISA, if the primary antibody has been labelled, or by an indirect ELISA (9). The assay should be conducted in a way which minimizes the possibility of damage to the peptide on the pin, although as the peptide is present in many hundredfold molar excess above the level of bound antibody, it may therefore degrade substantially before any effect on antibody binding becomes detectable. We recommend making pin-bound peptides in duplicate at least, to provide a test of reproducibility, and to allow test and control sera to be run in parallel if necessary.

Protocol 4. ELISA on pin-bound (non-cleavable) peptides

Equipment and reagents

- Pin-bound peptides
- Anti-species horse-radish peroxidase (HRP) conjugate (KPL, Dako, Sigma)
- PBS: 0.15 M sodium chloride buffered with 0.01 M sodium phosphate pH 7.2
- PBST: PBS with 0.1% (v/v) Tween 20
- PBSTA: PBST with 0.1% (w/v) sodium azide
- Conjugate diluent: PBST with 1% (v/v) sheep serum and 0.1% (w/v) sodium caseinate (United States Biochemical)
- Substrate: 0.01% (w/v) hydrogen peroxide and 0.5 mg/ml ABTS (Boehringer Mannheim Cat. No. 756407) in substrate buffer: 0.1 M sodium phosphate and 0.08 M sodium citrate pH 4.0
- Shaking table, operating at 80–100 r.p.m.
- Flat-bottom microtitre plates, or shallow plastic baths, e.g. the upturned lids of pre-packaged pipette tip trays (ICN Cat. No. 77–987-H2)
- Specific antibody (monoclonal or polyclonal)
- ELISA reader, e.g. Labsystems Multiskan
- Computer and software to receive and store the ELISA data stream from the ELISA reader
- Disruption buffer: 0.1% 2-mercaptoethanol in 1% sodium dodecyl sulfate/0.1 M sodium phosphate pH 7.2, pre-heated to 60°C
- Sonication bath with power input of approx. 7 kW/m^2

Method

1. Titrate the anti-species HRP conjugate in a checkerboard titration with the target immunoglobulin coated on a plate, to establish a dilution of the conjugate which saturates the plate (excess conjugate[a]) without being wasteful (10).

2. Carry out the conjugate-only test on the pin-bound peptides, following step 3, and steps 8 to 16 below, to show that the conjugate does not bind directly to the pins.

3. Pre-coat the pin peptides by immersing them in 0.2 ml PBST per pin in a microtitre plate[b] for 1 h at room temperature to reduce non-specific binding of the primary antibody to the pins.

4. Dispense 0.2 ml/well of suitably diluted[c] primary antibody solution in PBSTA into each well of a microtitre plate.

5. Flick excess pre-coat buffer from the pins and place them into the primary antibody solution.

6. Incubate at 4°C overnight[d] with mechanical agitation (shake table at 100 r.p.m.).

7. Wash four times in PBS at room temperature.

8. Dilute the conjugate in conjugate diluent[e] to its working strength determined in step 1 above.

9. Dispense 0.2 ml/well of diluted conjugate into a microtitre plate.

10. Place pins in diluted conjugate and incubate for 1 h at room temperature with agitation.

11. Wash four times in PBS.

12. Dispense 0.2 ml/well of substrate into the required number of high optical quality[f] microtitre plates.

13. Place pins in substrate pre-equilibrated at a temperature between 15°C and 20°C, and agitate for 30–45 min.

14. Remove pins from wells and read the absorbance of the wells on a microtitre plate reader, such as a Multiskan, in dual wavelength mode.

15. Regenerate the pins by sonication in disruption buffer for 10 min.

16. Rinse the pins in warm water, then in warm methanol, and air dry for storage or return to PBST for pre-coating prior to commencing a further ELISA test.

[a] Excess conjugate is necessary to give linearity of absorbances if a comparison between pins is to be meaningful.
[b] This and other steps where all pins are treated equally can be carried out in a bath rather than in wells of a microtitre plate, with a gain in sensitivity of the assay.
[c] Initial tests should be over- rather than under-diluted to avoid loading pins with excessive bound antibody. Suggested starting dilution for convalescent sera is 1/1000 and for hyper-immune sera 1/5000.
[d] The low temperature may be of benefit in reducing the potential damage to the peptides from proteases present in sera. The long incubation time (overnight) compensates for the slower binding kinetics and allows a convenient one day cycle time for each ELISA test on the pins.
[e] The conjugate diluent described contains sheep serum to help block non-specific binding of the conjugate to the pins. It is suitable for antibody conjugates made in sheep or goats but is totally unsuitable if the primary antibody is sheep or goat antibody. In that case, a conjugate diluent containing rabbit serum may be substituted.
[f] The highest quality should be used to prevent artefacts in the absorbance readings due to optical imperfections: scratches, bubbles, dirty marks. Use of a dual wavelength mode of reading can also reduce the effect of such imperfections.

If an indirect ELISA is being used, it is theoretically possible that the anti-species conjugate used in developing the ELISA will bind directly to the peptides. A test of direct binding of the conjugate should thus be carried out, preferably before the peptides are used for any binding assay on a primary (unlabelled) antibody. *Protocol 4* describes the indirect ELISA, including regeneration of the pins ready for a further ELISA test.

4.2 Direct binding on biotinylated peptides

Biotinylated peptides are widely used because of the convenience of the strong biotin–avidin interaction for capture of the peptide. To carry out an ELISA on a biotinylated peptide, the peptide is usually captured onto avidin or streptavidin which has been previously coated onto the surface of a polystyrene plate. This method has the potential to create a surface with a predictable amount of captured peptide on it, in contrast to the method of passive coating of peptide onto plastic (16). Because of the defined limit to the amount of peptide which can be bound by the coated avidin, and the efficient capture method, the amount of biotinylated peptide required is small (< 1 µg/ml) by comparison with the amounts usually used in passive coating (1–10 µg/ml). Thus, the amount of biotinylated peptide obtained from one pin is sufficient to coat thousands of wells. We have found that even when this method is used to capture the peptide in a defined way, additional peptide will in many cases coat passively on the plate surface, even when Tween 20 or bovine albumin are used as blockers of non-specific binding. We therefore recommend the use of sodium caseinate for effective blocking of the avidin-coated plates prior to addition of the biotinylated peptide (*Protocol 5*).

Protocol 5. ELISA with biotinylated peptides

Equipment and reagents

- Sets of biotinylated peptides
- Microtitre plates, e.g. Nunc Immuno-Plate MaxiSorb F96 (Cat. No. 4–42404)
- Streptavidin (Sigma Cat. No. S-4762)[a]
- PBS/BSA/azide: 0.1% BSA and 0.1% sodium azide in PBS
- Bovine serum albumin (BSA)
- 1% Na caseinate/PBS/Tween: PBS containing 1% sodium caseinate (USB) and 0.1% Tween 20
- Conjugate, substrate, ELISA plate reader (see *Protocol 4*)

Method

1. Coat Nunc Immuno-Plate MaxiSorb F96 flat-bottomed plates with 100 µl of 5 µg/ml streptavidin diluted in purified water. Leave plates exposed to the air at 37 °C overnight to allow the solution to evaporate to dryness.

2. Wash plates four times with PBST. After the washings, remove excess buffer from the wells by vigorously 'slapping' the plates,

---

wells down, on a bench-top covered with an absorbent material (paper towels).

3. To block non-specific absorption, add 200 μl/well 1% Na caseinate/PBS/Tween and incubate the plate at 20 °C for 1 h.

4. Repeat washes as described in step 2.

5. Reconstitute the biotinylated peptides in 200 μl of either a pure solvent (e.g. DMSO or DMF) or solvent/water mixture. Each peptide should be diluted just before use to 1 μg/ml in PBS/BSA/azide.[b]

6. Transfer 100 μl of each of the diluted peptide solutions into the corresponding well positions of the plate, place the plate on a shaker for 1h at 20 °C.

7. Repeat washes as described in step 2.[c]

8. Dilute the serum to be tested in PBS/Tween containing 0.1% sodium azide.[d] Add 100 μl to each of the wells of the plate containing captured peptides. Incubate with shaking for 1 h at 20 °C or overnight at 4 °C.

9. Repeat washes as described in step 2.

10. React for 1 h at 20 °C with 100 μl conjugate solution, made in 1% Na caseinate/PBS/Tween. Note: do not use a diluent containing azide for HRP conjugates.

11. Repeat washes as described in step 2.

12. Wash the plate twice with PBS only (no Tween) to remove traces of Tween remaining from the washing buffer.

13. Add 100 μl/well of a freshly prepared enzyme substrate solution and incubate for up to 45 min at 20 °C.

14. Read the absorbance in each well using a microtitre plate reader in the dual wavelength mode at 405 nm against a reference wavelength of 492 nm.

[a] Other brands of streptavidin, or avidin, may be used to coat plates; however the coating conditions may need to be varied. Avidin is cheaper than streptavidin but avidin-coated plates tend to result in higher background absorbance readings.
[b] The peptide stock solution may be diluted further (down to 0.2 μg/ml), however some loss in ELISA sensitivity may occur if used too dilute. The diluted peptide solutions may be stored for several days at 4 °C. For longer storage the diluted peptide solutions should be stored frozen, preferably at −70 °C.
[c] For convenience, several sets of immobilized peptides may be prepared simultaneously. The plates should be dried at 37 °C before storage at 4 °C in the dry state if they are not to be used immediately.
[d] The dilution factor to be used with the serum will depend to some extent on the source and the level of antibodies present in the sample. The recommended dilutions are 1/1000 for hyperimmune serum from experimental animals and ascites fluid from hybridoma-bearing mice, and 1/500 for human serum.

4.3 Confirmation of relevance of binding

Binding of an antibody to a peptide from the homologous protein antigen is not, on its own, proof that the peptide is a linear epitope of the antigen. There are several approaches to confirming the relevance of observed binding. One line of evidence is to use sets of control peptides and sera to establish absolute correlation between specific peptide binding and the presence of the specific antibodies under investigation. A second test which can be applied is to see if a putative peptide epitope will compete for binding to the native antigen. A third line of evidence arises when peptide-binding antibodies are eluted from the peptides and re-tested in another antigen-specific assay to demonstrate identity between the peptide- and antigen-binding antibodies.

4.3.1 Choice of controls

When hyperimmune animal sera are to be scanned, a good control for each serum is the pre-immune serum from the same animal. Certain pre-immune sera demonstrate a tendency to bind any peptide with a minimal recognition property, such as a particular N-terminal residue or sequence. Any such binding in the pre-immune serum compromises the reliability of epitopes identified in those peptides with immune serum. With human sera, or convalescent animal sera, it is not usually possible to obtain pre-immune sera from the same individuals who supplied the immune sera, so it is usual to compare the findings for a panel of sera which have specific antibodies for the antigen under test, with a control panel of negative sera.

A second set of control reagents is a set of random or unrelated peptides. It is not always feasible to obtain or synthesize a set of peptides just for this purpose, so we have chosen to treat as controls many of the 'non-binding' peptides from a scan. It is rare that more than 75% of the peptides in a scan are able to bind antibodies from a polyclonal serum. We therefore take the mean of the lowest 25% ELISA absorbance values in a scan, plus three standard deviations above the mean, to give a cut-off above which a peptide is considered to be a potential epitope (9).

4.3.2 Competition test

If an antibody binds to a linear epitope of a protein antigen, it should be possible, by use of a suitably high molar ratio, to use the peptide as an effective competitor in the antigen–antibody binding test. Likewise, an homologous peptide competition test can be used to confirm that the actual sequence being bound on a solid phase is the same as the intended (target) sequence. A purified solution phase form of the target peptide should compete effectively with the solid phase peptide for the antibody binding site, with the proviso that the molar ratio of the solution phase peptide may have to be high to overcome the avidity effect of the high peptide density on the surface of a pin. In this and other competition tests, controls should be included to show that the competitor peptide does not inhibit the binding of an irrelevant antibody to its epitope. *Protocol 6* outlines a method for the competition test as applied to the homologous pin peptide/solution phase peptide situation.

4.3.3 Elution test

A more definitive way to test for the significance of peptide-binding antibodies in polyclonal sera is to see if the same antibodies are active in a second, antigen-specific assay. This can be done by elution of the peptide-bound antibodies in a way which does not destroy their integrity, followed by re-assay in the antigen-specific assay (17).

Protocol 6. Homologous competition test with peptide

Equipment and reagents

- Replicate pins[a] with test and control[b] peptide synthesized on them
- Purified test peptide (> 95% purity) [d] of the target sequence
- Specific test and control[c] antibodies
- Phosphate-buffered saline/Tween (PBST) (see *Protocol 4*)

Method

1. Carry out an ELISA using each of the antibodies ('test' and 'control') on their respective pins, to establish a baseline reactivity (for comparison with the competition result), and to allow calculation of the amount of antibody and peptide to use in the competition assay. Include in the assay the test serum on the control peptide and vice versa.

2. Prepare antibody at a concentration designed to give an ELISA absorbance of 1.5–2.0.

3. Prepare dilutions of the competitor peptide in PBS, covering the range 0.1–100 μg/ml.

4. Add 1 vol. of each peptide dilution, or buffer alone, to 9 vol. of the diluted test or control antibody. This gives final peptide concentrations covering the range 0.01–10 μg/ml.

5. Incubate the mixtures at 4°C for 1 h.

6. Carry out a standard pin ELISA with each mixture (*Protocol 4*) or buffer control, on both the test and control pin peptides.

7. Examine the absorbance readings of the controls to see if they conform with the requirements for a valid test. Uncompeted (buffer) test and control antibodies assayed on the heterologous peptide (i.e. test serum on control peptide and vice versa) should show negligible binding (close to substrate blank value). Competed control antibody, assayed on control peptide, should not show significant inhibition by the soluble peptide.

Protocol 6. *Continued*

8. If these conditions have been met, and the soluble test peptide shows a clear competition curve against the test antibody/test pin peptide, the specificity for the putative peptide epitope sequence is proven.

[a] The solid phase could also consist of avidin-captured biotinylated peptides, but if purified biotinylated peptides are used, it is unnecessary to carry out a competition test to establish absolute peptide specificity.

[b] See Section 4.3.1.

[c] An antibody preparation similar to the test antibody should be chosen, e.g. for polyclonal antibodies, the control serum can be the test serum itself if an unrelated peptide within the scan is chosen as the control peptide; or for a monoclonal antibody, the control can be another well-characterized MAb–peptide epitope system, preferably where the peptide has some (amino acid residue) similarity with the epitope under test.

[d] The higher the purity, the better. The identity of the purified peptide should also be checked, e.g. by mass spectrometry, as it is possible that a 'purified' fraction from a peptide synthesis run may not be the target peptide at all!

4.4 Analoguing for detailed epitope analysis

Having located a peptide containing a linear epitope for a particular antibody population, one can narrow down the active sequence and examine the role of each amino acid within that sequence for the strength and specificity of binding. This is dealt with later in the context of the T cell epitope mapping (*Protocol 8*).

5. T cell epitope mapping with synthetic peptides

Due to the linear peptide nature of the 'foreign' portion of the MHC–peptide complex recognized by the TCR, linear synthetic peptides homologous with the protein antigen under study are highly suitable for T cell epitope mapping. The major difficulty in applying such peptides to T cell epitope mapping is the fact that peptides are normally produced and loaded onto MHC molecules intracellularly for export to the cell surface and presentation to T cells. Methods which use synthetic peptides for epitope mapping need to recognize this and either direct the peptide to the natural pathway or enhance the entry of peptide to the cleft on MHC molecules by an alternative pathway. The simplest of the latter methods is to drive the filling of empty MHC molecules with synthetic peptide by use of a high, non-physiological concentration of peptide. Alternative methods include low pH pre-treatment of cells, use of APC with defective peptide processing mechanisms, enhancement of peptide uptake by chemical agents or liposomes, or osmotic shock. In its simplest form, the mapping of a T cell epitope with synthetic peptides involves the demonstration that one, and only one, peptide sequence homologous with the protein antigen sequence will substitute for that antigen in activating a specific T cell population (monoclonal or polyclonal). Further enhancement of the mapping process can involve establishing the minimal active sequence; determination of the recognition motif for the MHC molecule and the contact residues with the TCR; and identification of analogue peptide sequences which are antagonistic for activation of the specific T cells (19). For each of these purposes, the design of an appropriate peptide set is critical to the success of the experiment.

5.1 Design of peptide sets

For the primary purpose of mapping the sites of T cell epitopes within a protein sequence, the most thorough and reliable way of using synthetic peptides is to make a set of homologous overlapping peptides covering the entire sequence, without making assumptions or predictions as to where the epitopes might be. The design of such sets must encompass choices as to peptide length, degree of overlap, and quantity and purity of each peptide. Practicalities of peptide synthesis and handling are recognized in making further choices such as N- and C-terminal endings of the peptides. The sequences selected by the mechanical process of stepping through the sequence may have to be modified by other criteria, such as hydrophobicity. For example, it may be possible, by adjusting some peptide lengths by one or two residues, to make peptides which are more water soluble or pose fewer synthesis problems. This can be done without losing the presence (in the synthesized peptide set) of all possible sequences of a given length from within that protein. *Protocol 7* describes the process of designing a suitable peptide set. The design process takes into account the need to ensure that no potential epitopes are 'overlooked', and that peptide synthesis problems are recognized at the peptide design stage. Affordability is also a major factor in design of complete screening strategies. *Protocol 8* describes the process of designing a truncation set and analogues of an epitope.

Protocol 7. Designing a peptide set to scan a protein sequence

Equipment and reagents

- Computer
- Computer file of the protein sequence, in single letter amino acid code

- Software: word processor OR Pepmaker (commercially available from Chiron Mimotopes) and Pinsoft (available free from Chiron Mimotopes)

Method

1. Choose the maximum continuous sequence length (M) for which **no** sequence within the protein is to be missing from the set. For T cell epitopes, this should be 9 residues or more.[a]

2. Choose a target peptide length (L) for the peptides to be made. Peptides for Th mapping, or for preliminary Tc location, can be from 12 residues upward; peptides for accurate Tc mapping should be around 9mers,

---

Protocol 7. *Continued*

with the making of all 8mers, 9mers, and 10mers being closer to the ideal set.[b]

3. The 'offset' (F), or step from one peptide to the next, is now automatically determined, being $F = L - M + 1$.

4. Estimate the number of peptides which must be made to cover the entire protein sequence. An approximate figure (slightly overestimated) can be calculated by dividing the number of residues in the protein (S) by the offset (F) and rounding down.[c]

5. Choose the quantity and quality of peptides to be made, based on the estimated number of peptides and their affordability.[d] If the peptides are not affordable, try modifying the choice of peptide length (in step 2). Increasing the length of the peptides will make the project more affordable up to a peptide length of about 20 residues.[e]

6. Prepare the list of peptides, either manually by editing the protein sequence file, or automatically by using the Pepmaker software and entering the length (L) and offset (F) parameters decided above.

7. Screen the peptides for hydrophobicity and synthesis problems using Pinsoft,[f] or for hydrophobicity alone using one of the published algorithms.

8. Modify any peptides which are too hydrophobic, or too difficult to synthesize, by lengthening or shortening the peptides as appropriate. This may also involve decreasing or increasing the number of peptides in the set[g] while ensuring the coverage does not fall below the maximum continuous length limit, M (see step 1).

[a] Minimal recognition sequences of both Tc and Th epitopes are around 9 residues.
[b] Naturally processed Tc peptides are usually 8, 9, or 10 residues; natural Th peptides are approximately 13–18 residues. Peptides for preliminary Tc epitope location can be found using longer peptides if the peptides are processed, e.g. by proteases from serum, cell surface, or intracellular locations.
[c] An accurate value for the number of peptides can be calculated as: No. of peptides needed = $2 + \text{Ord} \{(S-L-1)/F\}$. (Note: Ord=Ordinate, the integer value obtained by rounding down.)
[d] Unpurified, small scale (approx. 1 mg scale) peptides are the most affordable; purified and larger scale (20+ mg) peptides are the least affordable.
[e] Increasing length will decrease the number of peptides required but decreases the resolution of the method; beyond 20 residues, the proportional decrease in the number of peptides becomes minimal and the cumulative effect of incomplete amino acid coupling during each cycle of synthesis has a rapidly accelerating adverse effect on the cost and quality aspects of synthesis.
[f] Pinsoft is a peptide assessment and ordering utility for IBM or Macintosh computers, available free from Chiron Mimotopes.
[g] If the peptide(s) containing a 'problem' sequence are increased in length to overcome the problem, it may be possible to omit a peptide from the set while still maintaining the desired value of M; if the peptides at the problem area are decreased in length, it will be necessary to insert at least one peptide to ensure that M is maintained.

Protocol 8. Design of truncations and analogues of an epitope

Equipment and reagents

- Results of mapping of a protein with long overlapping peptides
- Computer

- Software: word processor OR Pepmaker (commercially available from Chiron Mimotopes) and Pinsoft (available free from Chiron Mimotopes)

Method

1. From the sequence of the peptide found to contain the T cell epitope, prepare a list of peptides truncated from each end independently, e.g. by one residue at a time, down to an 8mer peptide.[a]

2. Screen the peptides for hydrophobicity and synthesis problems using Pinsoft, or for hydrophobicity alone using one of the published algorithms.

3. Modify any peptides which are too hydrophobic, or too difficult to synthesize, by lengthening or shortening the peptides as appropriate.[b]

4. After synthesis and testing, infer the N- and C-terminal residues of the minimal epitope, based on the shortest peptide in each set of truncations which retained full activity.

5. Synthesize and test the minimal epitope in highly purified form to confirm that it is fully active as an epitope.[c]

6. Using a word processor or Pepmaker, prepare a list of analogue peptides of the epitope with defined single amino acid substitutions.[d] Assess and select from this list of peptide sequences as per step 2.

7. After synthesis, test both for T cell activation, and for competitive inhibition of activation by the parent epitope. If possible, test for direct binding to the MHC restriction element, e.g. with biotinylated peptides (18).

8. Infer the pattern of MHC binding and of TCR recognition of individual residues, and the relative potency of any antagonistic peptides found.

[a] No activity is expected from a 7mer or shorter peptide.
[b] See *Protocol 7*.
[c] The inferred minimal epitope is not always fully active due to peptide size considerations. It may be necessary to make one or more peptides which are slightly longer than the inferred minimal epitope.
[d] As a minimum, each residue could be substituted with glycine; a very thorough study would use most of the alternative genetically coded amino acids in each position; or a diverse set such as Gly, Lys, Glu, Leu, and Phe could be used to test the dependence of the epitope for activity on the side chain characteristics of each residue.

5.1.1 T helper epitopes

A typical peptide set for mapping Th epitopes would be all 16mers homologous with the sequence, offset along the sequence by four residues. This gives a maximum continuous sequence length (M) of 13 residues; thus, some 14mers are absent, many 15mers are absent, most 16mers are absent, and no 17mers are present. The number of peptides would be approximately one-fourth of the number of amino acid residues in the protein. The peptides could be N-terminally acetylated and/or C-terminally blocked or extended, e.g. with a diketopiperazine group (4,11), which has the potential to enhance the effectiveness of the peptides.

5.1.2 Cytotoxic T cell epitopes

A typical peptide set for mapping Tc epitopes would be all possible 9mers homologous with the sequence, i.e. offset along the sequence by one residue. The number of peptides is nearly equal to the number of amino acid residues in the protein. The peptides should have free amine and free acid endings, as occurs in the native epitope bound on MHC Class I molecules. The effectiveness of synthetic peptides in loading Class I molecules, and the high sensitivity of assays, means that epitope activity in preparations of longer peptides can be due to contaminating peptides such as truncated or cleaved versions of the nominal peptide (20).

5.2 Epitope mapping with T cell clones

Results of epitope mapping with Th clones are usually very clear-cut: an all-or-nothing effect, although dose effects make it important to test the clone with a range of peptide levels. Anyone using a T cell clone is assumed to know the best conditions for culture of their own clone, including choice of an arrest method for the antigen-presenting cells (APC), choice of medium and supplements, whole antigen controls, and the T cell response readout assay.

5.2.1 T helper cell clones

Epitopes for Th are most readily identified by showing that a peptide containing the epitope, when added to an *in vitro* culture with the clone and APC expressing high amounts of MHC Class II, will stimulate the clone into proliferation or into release of a cytokine such as IL-2. Naturally, if the stimulatory effect of the peptide is to be observed, the Th cells should ideally be in a resting state. The peptide can be pre-loaded onto the APC and the excess washed away, or can be added to a mixture of APC and Th cells and the assay carried out without washing away excess peptide. *Protocol 9* describes one type of Th assay in which peptides may be used. Due to the large number of variations possible, only a general method is given.

Protocol 9. Response of helper T cell clone to peptide

Equipment and reagents

- Th clone cells in resting (non-activated) state, and arrested APC
- Test and control peptides (10 mg/ml) dissolved in DMSO
- Culture medium including serum
- Tritiated thymidine (Amersham), cell harvester, beta counter, OR metabolic labelling method (MTT or XTT assay)

Method

1. Prepare a series of tenfold or semi-log dilutions of the peptide in water or medium without serum.[a]

2. Dispense 10 000 Th clone cells and 10 000 arrested APC[b] to each replicate well of a flat-bottom microtitre plate.

3. Add peptide at final concentrations from 10–0.001 µg/ml to triplicate wells.[c] Include the following controls: Th clone plus APC plus whole antigen (positive control); Th clone plus APC with no antigen (negative control); Th cells alone (this checks on the 'resting' state of the Th cells); APC alone (checks on arrested state of APC); Th clone plus APC with an irrelevant peptide made up exactly as per the peptide under examination (control for non-specific effects of peptide[d] or medium in which peptide is made up).

4. Incubate for 24–48 h at 37°C, 5% CO_2.

5. Sample the supernatant of each well for subsequent titration of cytokine (IL-2) level; OR pulse label the cultures with tritiated thymidine (see *Protocol 11*).

6. Calculate the mean and SD of the response of triplicate cultures in each control and treatment group.

7. If the five types of control have given the expected results, observing a statistically significant (T-test; $p < 0.05$ when compared to negative controls) dose-dependent response to a peptide is conclusive evidence that the peptide contains the specific epitope for that Th clone.

---

[a] Adding the peptide to medium containing serum can result in extensive degradation of the peptide before it is added to the APC, reducing the chances of successful presentation by the APC.

[b] The APC need to be incapable of DNA synthesis or of otherwise responding to the favourable culture conditions in the assay. This is usually achieved by γ-irradiation or drug-induced (mitomycin C) inhibition of the cells.

[c] Testing a concentration range allows an estimate of the optimum dose and avoids the possibility of toxicity effects at high peptide concentration masking an otherwise strong response.

[d] While no single peptide is a 'sufficient' control, choices would include a peptide having the reverse sequence to the putative epitope under test, or a peptide having no sequence homology with the putative epitope under test.

5.2.2 Cytotoxic T cell clones

Epitopes recognized by cytotoxic T cells (CTL, Tc) are likewise readily identified by showing that a peptide containing the epitope, when added to APC expressing high amounts of MHC Class I, will sensitize the APC (target cells) to specific attack by the Tc or will stimulate release of cytokines by the Tc. Because the killing assay is rapid, and the effect is produced entirely upon the APC, no special treatment (arrest) of the APC is necessary for a cell killing assay and the Tc do not need to be in a resting state. If the assay contains too many bystander cells it may be necessary to use a high ratio of Tc to APC to get efficient killing. It may also be difficult to get good sensitization of the APC by added peptide unless the peptide has exactly the right sequence and length, because the peptide may not be able to enter the natural processing pathway and may not be readily processed to the correct epitope by other proteases. *Protocol 10* is a general guide to the use of peptides in sensitizing APC for a target cell killing Tc assay.

Protocol 10. Assay of peptide in a CTL killing assay

Equipment and reagents

- CTL clone
- APC expressing high levels of MHC Class I
- Sodium [^{51}Cr]chromate, 200–600 mCi/mg (Amersham) for labelling the APC (target) cells

- Test and control peptides as a concentrated stock solution dissolved in DMSO
- Cell culture media (e.g. RPMI 1640), serum, PBS, flat-bottom microtitre plates, plate-spinning centrifuge, gamma counter

Method

1. Label 4×10^6 APC with ^{51}Cr by incubation for 1 h with 250 μCi sodium [^{51}Cr]chromate in a small volume of cell culture medium at 37°C, then wash to remove excess [^{51}Cr]chromate.[a]

2. Add the test peptide to aliquots of washed, labelled APC at two or more final concentrations between 10–0.001 μg/ml.[b] Prepare similar cell suspensions with the control peptides, and keep one with no added peptide. Incubate for 1 h at 37°C, then wash to remove excess peptide.[c] Adjust the APC concentration to 100 000 cells/ml.

3. Dispense 0.1 ml APC suspension to wells of a flat-bottom microtitre tray.

4. Add 0.1 ml of Tc suspension to triplicate wells for a final Tc : APC ratio of 10, 1, and 0.1. To one set of replicate of APC of each type, add 0.1 ml detergent instead of Tc suspension (100% lysis control). To one set of triplicate APC of each type, add medium alone (no Tc) (spontaneous lysis control).

5. Incubate for 4 h at 37°C, 5% CO_2, with gentle agitation if possible.

6. Centrifuge the tray at 200 *g* for 5 min and harvest 0.1 ml of the super-natant fluid from each well for gamma counting, taking care not to disturb the cells.

7. Take the mean and SD of the c.p.m. for each set of triplicate wells.

8. Calculate the per cent specific chromium release for each test group as

$$\frac{\begin{array}{c}\text{(mean c.p.m for test group)-}\\\text{(mean c.p.m. for spontaneous release control)}\end{array}}{\begin{array}{c}\text{(mean c.p.m for detergent release control)-}\\\text{(mean c.p.m. for spontaneous release control)}\end{array}} \times 100$$

9. Look for significant differences[d] in per cent specific chromium release (killing) between the unsensitized APC and the sensitized APC, and between the putative epitope peptide and the control peptide.

[a] The chromate constantly leaks from the cells and thus labelling should be done just before the assay.
[b] Use of a dose range will help to avoid misleading conclusions due to artefacts arising from peptide toxicity or inefficient processing. However, the need for controls on each APC preparation limits the number which can be readily included in the experiment.
[c] Leaving free peptide with the APC can lead to killing of Tc by Tc, as they can also take up peptide onto Class I and become target cells themselves.
[d] T-test ($p < 0.05$) using the coefficient of variation for the detergent release and spontaneous release for each set of APC as part of the error calculation.

5.3 Mapping of T helper epitopes with peripheral blood mononuclear cells

Peripheral blood mononuclear cells (PBMC) contain a random selection of Th and Tc, among other cell types. They have frequently been used to study responsiveness to whole antigens, but present exceptional problems for the direct mapping of Th epitopes. These problems can be summarized thus:

(a) T cells specific for a given epitope within a protein antigen will occur at low frequency in PBMC and therefore there will be a 'sampling' problem in obtaining replicate samples with a consistent number of responsive cells.

(b) Limits on the quantity of peripheral blood available, especially from human volunteers, make it difficult to test the large number of peptides needed to scan through a protein for epitopes.

(c) Whether due to pre-stimulation *in vivo* or to a response to undefined components of the culture media, 'spontaneous' proliferation of PBMC *in vitro* is often observed, making the interpretation of apparent antigen-driven proliferation less reliable.

On the positive side, PBMC readily proliferate *in vitro* in response to whole antigens or single helper epitopes, and the proliferation of one or a few specific Th cells triggers a cascade of recruitment of non-specific T cells which results in an amplified response. Thus, an aliquot of PBMC is a self-contained assay system for Th responses, needing only a suitable culture medium and supply of antigen.

The problems of low precursor frequency and large numbers of peptides have been largely solved by the following strategy:

(a) All assays are done with as many replicates of PBMC per test group as can be practically performed, within the limitation of having sufficient cell numbers in each culture to provide both the APC and the non-specific amplification function.

(b) Peptides are pooled for initial screening and only those peptide pools which are stimulatory are examined in further detail for the final identification of particular epitopes.

(c) As large a sample of peripheral blood as practical is obtained, and any surplus of PBMC is frozen for re-assay at a later date.

(d) The analysis method on the data from the proliferation assay takes account of the Poisson nature of the responses observed in replicates.

Application of these approaches to the study of several antigens (11,21,22) has given new insights into the totality of the Th epitope repertoire. *Protocol 11* describes in detail how these scans are performed.

Protocol 11. Th epitope scanning with human PBMC

Equipment and reagents

- A set of overlapping peptides designed for Th scanning (Section 5.1.1)
- A source of human PBMC: 50–100 ml whole blood equivalent per donor
- Ficoll–Paque (Pharmacia) or Nycodenz cell separation medium
- Sterile round-bottom microtitre plates (Nunc)
- Sterile 100 ml bottles containing 10–15 glass beads (3–5 mm diameter) for defibrination of whole blood
- Concanavalin A mitogen (Sigma)
- High specific activity tritiated thymidine (Amersham, 40–60 Ci/mmol)

- Culture media: 'incomplete' base medium is RPMI 1640 with 2 mM glutamine, 5 mM Hepes pH 7.4, and 20 μg/ml gentamicin (ICN)—for complete medium, a serum supplement, added to 10% (v/v), can be either autologous serum or screened, pooled, human serum, preferably from type AB donors (ICN)
- Whole protein antigen control
- Cell harvester (Skatron), beta counter (e.g. Pharmacia-Wallac Betaplate 1205)
- Computer and analysis software, e.g. the 'ALLOC' program, available free from Chiron Mimotopes

Method

1. Design a peptide pooling/decoding strategy (see Section 5.3.1) based on the number of peptides, and on the total number of replicate wells which can be prepared with the expected number of PBMC available.[a]

2. Dissolve the peptides in DMSO or in 40% (v/v) acetonitrile/water to give solutions of at least 3 mM.[b]

142

3. Pool an aliquot of each peptide into pools of a size determined in step 1, keeping the stock solutions of the single peptides for later 'decoding' of positive pools.

4. Dilute the peptide pools in saline or PBS so that each peptide is at approx. 10 μM, which is 10 × the final concentration to be used in the assay.

5. Dispense 20 μl of each pool to replicate wells in a round-bottom[c] microtitre plate. A minimum of 8 replicate wells should be used for each pool, with 16 or 24 replicates preferred.[d]

6. Draw whole venous blood from the volunteer donors. We prefer to defibrinate whole blood with glass beads. Add 50 ml whole blood to a sterile 100 ml bottle containing glass beads, and agitate gently end-over-end for 10 min or until the beads are firmly bound in the fibrin clot. Both serum and PBMC are recovered from the same blood sample.[e]

7. Dilute the whole blood 1/2 with incomplete medium and isolate the PBMC from this diluted blood by centrifugation over Ficoll–Paque, following the manufacturer's instructions. Recover the autologous serum from above the cell layer and immediately heat inactivate (56°C for 30 min).[f]

8. After washing the PBMC band twice to remove the Ficoll–Paque, resuspend the cells in a 'complete' medium containing 10% (v/v) heat inactivated autologous serum.[g]

9. Adjust the PBMC concentration to 1.1×10^6/ml with the same medium, and dispense 180 μl per well into all wells, including the negative control wells and both positive controls.

10. Incubate plates at 37°C and 5% CO_2 in a fully humidified atmosphere for four days.

11. For the last 6 h of the incubation, add to each well 20 μl of medium containing 0.25 μCi of tritiated thymidine.[h]

12. Using a multiple well cell harvester, harvest all wells onto glass fibre filter paper, washing thoroughly with water to lyse the cells and capture the tritium-labelled DNA.

13. Dry and transfer to scintillant for beta counting in the usual way for the counter you have.

14. Count all wells, preferably to a precision of better than 5% (total count of > 400) for all wells.[i]

15. To analyse the data, begin by determining a cut-off c.p.m. value to be used to separate negative (unstimulated) cultures from positive cultures.[j] A suitable basis for the calculation of the cut-off is usually the mean plus 3 SD of the c.p.m of the negative controls.

16. When each well has been scored as positive or negative, use Poisson

Protocol 11. *Continued*

statistics to calculate a precursor frequency for all experimental groups (e.g. each peptide pool), including the negative control group and each of the positive controls.[k] Estimate a confidence range for each precursor frequency value.

17. Test the significance of the difference between the precursor frequency value for each experimental group and the negative control group.[l] Experimental groups which have a highly significant precursor frequency can then be 'decoded' in another proliferation experiment (see Section 5.3.1) to identify the peptide(s) responsible for stimulation of Th precursors.

[a](Total No. of PBMC)/(No. of PBMC to be used per well). Calculate on an expected yield of 1–2 $\times 10^6$ PBMC/ml whole blood.

[b]This allows a pool of ten peptides to be diluted 300-fold to give a final concentration of 1 μM of each peptide and a final DMSO concentration of 0.3%, which is compatible with successful culture of proliferating T cells. Cells tolerate HPLC grade acetonitrile even better, proliferating readily in medium with 2% acetonitrile.

[c]Round-bottom wells keep the cells close together, enhancing antigen presentation, and uptake of cytokines released from activated cells. Flat-bottom plates give reduced responsiveness, and V-bottom plates can cause toxicity from cell overcrowding.

[d]Plates can be prepared in advance and stored frozen, then thawed just before use. Each plate should include replicates of the buffer ('cells alone' or 'no peptide') negative controls, and two types of positive control: a whole antigen control to test antigen-specific responsiveness and a mitogen control, to test non-specific responsiveness on each plate.

[e]PBMC can also be obtained from heparinized or citrated blood. As an alternative to autologous serum, pooled, screened, human serum can be used in the culture medium. Screening and choosing a suitable batch of pooled human serum is critical for success of this assay, and serum from blood type AB donors is said to be superior to other types because of the lack of anti-A or anti-B antibodies.

[f]Large numbers of leucocytes, obtained fresh from a blood bank as the 'buffy coat' sample from a unit of blood, may also be used as the starting cell suspension for PBMC isolation.

[g]The fluid recovered from above the cell layer on Ficoll–Paque centrifugation is already diluted to approximately 40% (v/v) serum by the initial dilution step. Thus, a further 1/4 dilution of this with incomplete medium brings the serum concentration to 10% (v/v).

[h]The precision of dispensing this radiolabel is vital for achieving high precision in the c.p.m. values of replicates. Use of a small dose of high specific activity thymidine is much more economical than a large dose of low specific activity thymidine. Owing to the radiotoxicity of tritiated thymidine, long labelling times are of little advantage in obtaining higher c.p.m. values.

[i]Higher precision is of little value due to the variation in replicates and the precision of the labelling step (see previous note).

[j]The 'ALLOC' software, available free from Chiron Mimotopes, performs the cut-off analysis automatically following the user's choice from a menu of four algorithms.

[k]The 'ALLOC' software also calculates the precursor frequency and confidence range for each experimental group.

[l]The difference can be tested to the $p < 0.0025$ significance level (that the two groups are the same) when the upper 5% limit of the negative control group is below the lower 5% limit of the experimental group.

5.3.1 Peptide pooling, decoding, and narrowing down to the minimal epitope

When faced with the daunting task of testing hundreds of peptides to locate T cell epitopes within a protein antigen sequence, a simplifying strategy such as a pooling/decoding process is attractive. By testing pools of ten peptides, the amount of initial screening work is reduced to one-tenth of that otherwise required. In the case of polyclonal Th cells (PBMC or Th cell lines, rather than clones), the saving in effort is maximized when the number of peptide pools which are negative (non-stimulatory) in the primary screen is maximal, and no saving in testing effort accrues at all, if all peptide pools are stimulatory. Following an initial scan with peptide pools, the peptides comprising any positive pools can be tested individually, or alternatively can be tested as a set of smaller pools prior to the final testing of individual peptides.

The idea of peptide pooling, as implemented in the scanning of PBMC for Th epitopes (Section 5.3), is equally applicable to testing of Th clones, and indeed the saving in effort can be much greater, as there are unlikely to be more than two adjacent active pools for each clone.

Acknowledgements

We thank all the wonderful people at Mimotopes who contributed over the years to the ideas and methods presented here. We also wish to express our appreciation of the leadership and inspiration of Dr Mario Geysen, inventor of the Multipin method, and to Dr Neville McCarthy for his support for research over many years.

References

1. Geysen, H. M., Meloen, R. H., and Barteling, S. J. (1984). *Proc. Natl. Acad. Sci. USA*, **81**, 3998.
2. Houghten, R. A. (1985). *Proc. Natl. Acad. Sci. USA*, **82**, 5131.
3. Jung, G. and Beck-Sickinger, A. G. (1992). *Angew. Chem.*, **31**, 367.
4. Maeji, N. J., Bray, A. M., and Geysen, H. M. (1990). *J. Immunol. Methods*, **134**, 23.
5. Bray, A. M., Maeji, N. J., Valerio, R. M., Campbell, R. A., and Geysen, H. M. (1991). *J. Org. Chem.*, **56**, 6659.
6. Valerio, R. M., Bray, A. M., Campbell, R. A., DiPasquale, A., Margellis, C., Rodda, S. J., *et al.* (1993). *Int. J. Pept. Protein Res.*, **42**, 1.
7. Weiner, A. J., Geysen, H. M., Christopherson, C., Hall, J. E., Mason, T. J, Saracco, G., *et al.* (1992). *Proc. Natl. Acad. Sci. USA*, **89**, 3468.
8. Maeji, N. J., Tribbick, G., Bray, A. M., and Geysen, H. M. (1992). *J. Immunol. Methods*, **146**, 83.
9. Geysen, H. M., Rodda, S. J., Mason, T. J., Tribbick, G., and Schoofs, P. G. (1987). *J. Immunol. Methods*, **102**, 259.
10. Worthington, J. and Morgan, K. (1994). In *Peptide antigens: a practical approach* (ed. G. B. Wisdom), p.181. IRL Press, Oxford.
11. Reece, J. C., Geysen, H. M., and Rodda, S. J. (1993). *J. Immunol.*, **151**, 1.
12. Atherton, E. and Sheppard, R. C. (ed.) (1989). *Solid phase peptide synthesis: a practical approach.* IRL Press, Oxford.

13. Bodanszky, M. and Trost, B. (ed.) (1993). *Principles of peptides synthesis.* Springer–Verlag, NY.
14. Meloen, R. H. and Barteling, S. J. (1986). *Virology*, **149**, 55.
15. Lubin, R., Schlichtholz, B., Bengoufa, D., Zalcman, G., Tredaniel, J., Hirsch, A., *et al.* (1993). *Cancer Res.*, **53**, 5872.
16. Geerligs, H. J., Weijer, W. J., Bloemhoff, W., Welling, G. W., and Welling-Wester, S. (1988). *J. Immunol. Methods*, **106**, 239.
17. Tribbick, G., Triantafyllou, B., Lauricella, R., Rodda, S. J., Mason, T. J., and Geysen, H. M. (1991). *J. Immunol. Methods*, **139**, 155.
18. Busch, R. and Rothbard, J. B. (1990). *J. Immunol. Methods*, **134**, 1.
19. De Magistris, M. T., Alexander, J., Coggeshall, M., Altman, A., Gaeta, F. C. A., Grey, H. M., *et al.* (1992). *Cell*, **68**, 625.
20. Schumacher, T. N. M., De Bruin, M. L. H., Vernie, L. N., Kast, W. M., Melief, C. J. M., Neefjes, J. J., *et al.* (1991). *Nature*, **350**, 703.
21. Bungy Poor Fard, G. A., Latchman, Y., Rodda, S., Geysen, M., Roitt, I., and Brostoff, J. (1993). *Clin. Exp. Immunol.*, **94**, 111.
22. Rodda, S. J., Benstead, M., and Geysen, H. M. (1993). In *Options for the control of influenza II* (ed. C. Hannoun, A. P. Kendal, H. D. Klenk, F. L. Ruben), p. 237. Excerpta Medica, Amsterdam, London, New York, Tokyo.

Enzyme-linked immunoassays

D. M. KEMENY

1. Introduction

Enzyme-linked immunoassays are in widespread use in all areas of biological science today. Their popularity stems from their ease of use and safety. In this chapter I will review the current status of ELISA including assay format, reagent choice, assay optimization, and quantification. Other variations on the method such as the ELISA plaque assay, signal amplification systems, and affinity measurement will be addressed by other authors in these volumes. My comments will be restricted to assays set-up and developed in the user's own laboratory and to plastic microtitre plate-based assays (which apply equally to plastic tubes or pins) as these are the most widely used. However readers should be aware of their limitations and alternatives which have been reviewed extensively elsewhere (1,2).

2. Assay format

There are many different ways of configuring ELISAs and the range of assay formats that are available can be bewildering. The choice depends on the nature of the sample, availability of reagents, and the precision and sensitivity required. Non-competitive assays in which all the constituents, other than the sample, are in excess will always be more precise than their competitive counterparts. In the interests of space only the main categories of assay and their relative merits will be considered.

2.1 Non-competitive ELISA
2.1.1 Antigen-coated plate ELISA
These are probably the simplest of the ELISA procedures. They are variably described as indirect ELISA or dirty plate assays. Microtitre plates are coated with antigen, sample containing antibody is added, and the antibody binds to the antigen on the plate. Finally the bound antibody is detected by addition of an enzyme-labelled antibody specific for the bound antibody (*Figure 1a*), and the presence of bound enzyme detected with a substrate that

(a)   (b)

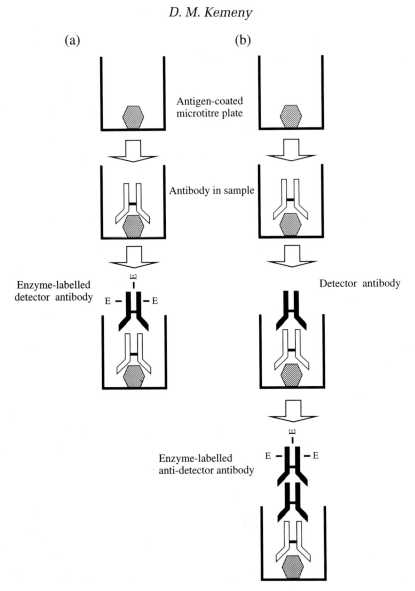

Figure 1. The antigen-coated plate ELISA. Antigen-coated microtitre plates are incubated with sample containing antibody which binds to the antigen on the plate and is subsequently detected by addition of an enzyme-labelled antibody specific for the bound antibody (a), or may be detected following addition of an unlabelled antibody specific for the bound antibody followed by a second enzyme-labelled anti-globulin antibody (b).

generates colour or light. This enzyme-labelled antibody is referred to as the detector antibody which may be labelled directly (*Figure 1a*) or may be detected following addition of an unlabelled antibody specific for the bound antibody followed by a second enzyme-labelled anti-globulin antibody

(*Figure 1b*). Alternatively the detector antibody may be biotinylated and subsequently detected with avidin-labelled enzyme.

2.1.2 Antibody-coated plate (two-site or sandwich) ELISA

Sometimes referred to as the two-site or sandwich assay, microtitre plates are coated with capture antibody (so-called because it captures the antigen). The sample containing antigen is added and this antigen binds to the antibody on the plate. Finally the bound antigen is detected by addition of an enzyme-labelled antibody specific for the bound antigen (*Figure 2a*) and this is detected by addition of substrate as before. As above, the detector antibody may be labelled directly (*Figure 2a*), or may be detected following addition of an unlabelled antibody specific for the bound antibody followed by a second enzyme-labelled anti-globulin antibody (*Figure 2b*). Alternatively the detector antibody may be biotinylated and subsequently detected with avidin-labelled enzyme. The coating antibody and the detector antibody may be the same or different, and they may also be polyclonal or monoclonal, although in the latter case it is important that both monoclonals are able to bind to the antigen at the same time. This is usually accomplished by using two monoclonal antibodies that recognize different epitopes although there are examples of monoclonal antibodies that can be used for both capture and detection in the same assay (3).

2.1.3 Class capture ELISA

There are times when it is necessary to separate one class of antibody from another before detection of specific antibody activity. For example, IgM antibodies are diagnostic of recent infection with a number of pathogens such as German measles (Rubella) but if the patient has been previously exposed to Rubella there may be many times more IgG anti-rubella antibodies which are of little diagnostic value. The problem is compounded by the fact that the IgG antibodies present will usually be of much higher affinity and so may disproportionately compete with IgM antibodies for binding to the Rubella antigen. In IgM class capture assays the microtitre plate is coated with polyclonal or monoclonal antibody specific for IgM. This anti-IgM-coated plate is used to capture IgM in the sample and the anti-Rubella antibody activity of the bound IgM is determined by addition of labelled antigen. It is important that most, or at least a substantial proportion, of the IgM in the samples is captured and this may be difficult due to the limited capacity of the plastic microtitre plate. In such cases the capture antibodies must be carefully selected (4).

2.2 Competitive ELISA

Competitive assays make it possible to obtain an estimate of the amount of a particular antibody or antigen, even when these cannot be isolated from the

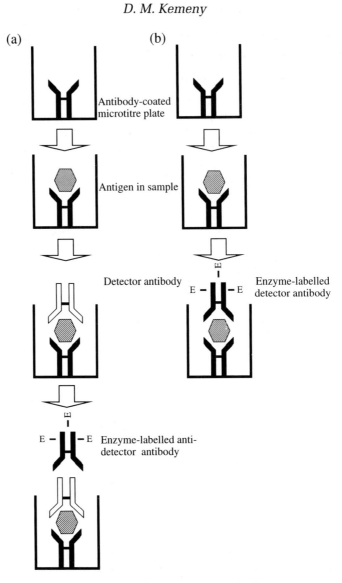

Figure 2. The two-site or sandwich ELISA. Antibody-coated microtitre plates are incubated with sample containing antigen which binds to the antibody on the plate and is subsequently detected by addition of an enzyme-labelled antibody specific for the bound antigen (a), or may be detected following addition of an unlabelled antibody specific for the bound antigen followed by a second enzyme-labelled anti-globulin antibody (b).

medium in which they are found. They are less precise than non-competitive assays as more than one component is present at a limiting concentration.

2.2.1 Competitive ELISA for antibody

Antibody can be measured by competition with a fixed amount of labelled antibody for solid phase antigen (*Figure 3a*) or in competition with solid phase antibody for a fixed amount of labelled antigen (*Figure 3b*). The amount of fixed and labelled reagents must be limited so that small differences in the amount of sample antibody can be seen. Furthermore, any dissociated antigen or antibody will also act as an inhibitor.

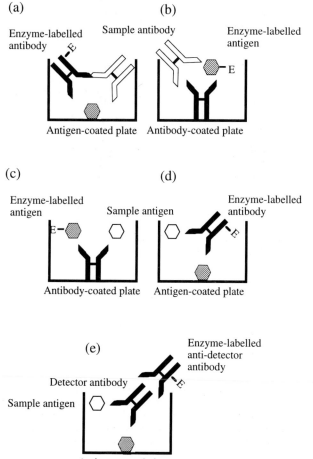

Figure 3. The competitive ELISA for detection of antibody (a, b) or antigen (c, d, e). In this assay the competition of a fixed amount of labelled antigen or antibody for a fixed amount of antigen or antibody bound to the microtitre plates is determined. The amount of sample is inversely proportional to the amount of enzyme bound.

151

2.2.2 Competitive ELISA for antigen

Antigen can be measured in a competitive ELISA in two ways. In the first, sample antigen competes with a fixed amount of labelled antigen for binding to solid phase antibody (*Figure 3c*), in the second, sample antigen and solid phase antigen compete for binding to a fixed amount of labelled antibody (*Figure 3d*). Indeed, the antibody in this system may be unlabelled and its presence detected with a second labelled antibody (*Figure 3e*). The drawback to the antigen-coated plate method is that a precise, limiting quantity of antigen must be bound to the microtitre plate. Competitive assays work best when competition is carried out in solution where the association and dissociation of antibody and antigen approximates to the law of mass action. The use of coated pins to remove one or other of the components of the assay facilitates this (*Figure 4*).

3. Enzymes and substrates

A fundamental and sometimes difficult choice is which enzyme and substrate to use. The label used should be safe, inexpensive, and generate an adequate amount of colour at an acceptable rate.

3.1 Alkaline phosphatase (EC 3.1.3.1)

In many ways this might be regarded as the enzyme of choice for ELISA. Conjugated to protein it is stable over long periods of time and is resistant to

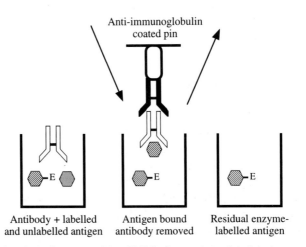

Figure 4. The microtitre pin competitive ELISA. Separation of solid phase antibody or antigen in competitive assays such as those shown in *Figure 3* can be effected by removal of the coated pins which obviates the need for washing.

bacteriostatic agents (5,6). Alkaline phosphatase is normally present at a low level in serum but is elevated in patients with liver disease and this can give rise to false positive results (7). Alkaline phosphatase conjugates are normally stored at 4 °C in buffer containing: 1–2% (w/v) protein, 50% glycerol, and azide. The substrate used for spectrophotometric measurements is *p*-nitrophenyl phosphate (PNP) which is stable, safe, and available commercially in convenient tablet form (8,9). The enzyme is, however, much more expensive than horse-radish peroxidase (see Section 10.2). The most common sources are calf intestine (10,11) and *Escherichia coli*. Enzymes prepared from these two sources have different properties, the bovine enzyme having a pH optimum of 10.3 while that obtained from *E. coli* has a pH optimum of 8.0. The substrate, *p*-nitrophenyl phosphate (PNP), is easy to use and produces linear colour development with time. The most widely applicable and simplest procedure for coupling enzymes to proteins is the one-step glutaraldehyde method (12). It is highly efficient (60–70%) and is simple to perform. The conjugates prepared in this way tend to be of the high molecular weight and are heterogeneous with different numbers of enzyme molecules coupled to the antigen or antibody. Furthermore, since horse-radish peroxidase cannot be readily conjugated using this method a two-step glutaraldehyde procedure has been described in which the protein is first reacted with glutaraldehyde and then, after dialysis, with the enzyme (12). The coupling efficiency of this procedure is lower (5–10%) but the majority of the conjugates formed have an enzyme:protein ratio of 1:1. These conjugates give lower background binding (13)—an advantage when developing very sensitive assays. The conjugation of alkaline phosphatase has also been described using periodate (14).

A number of substrates which yield insoluble products are available for alkaline phosphatase. One such substrate is 5-bromo-4-chloro-3-indolyl phosphate (BCIP). We have used this in the ELISA plaque assay and in the dot blot assay where it works well. The rate of colour development with alkaline phosphatase can be increased by using an amplified substrate system (15) and will not be discussed here as it is the subject of another chapter in this book.

3.2 Horse-radish peroxidase (EC 1.11.1.7)

Horse-radish peroxidase is the most widely used enzyme in ELISA. In the field of histochemistry, horse-radish peroxidase is unrivalled because of the insoluble substrates such as diamino benzidine (DAB) that can be used (16–18). It is cheap, readily available, and high quality conjugates can be prepared easily. It can be coupled via carbohydrate determinants with little loss of immunological or antigenic activity (19–21) and using the periodate method a 1:1 molar ratio of enzyme to antigen or antibody can be accomplished (6). The efficiency of coupling is, however, much lower than for the glutaraldehyde procedure and as little as 10% of the antigen or antibody added may be bound to the enzyme. It has a high turnover rate and many of the substrates that are used, such as *o*-phenylene diamine (OPD), have end-points that can be determined visually. The rate at which colour is generated is, however, not always linear. pH is critical and this can change when the chromogen salt (OPD) is dissolved in the substrate buffer although it can readily be corrected with Na_2HPO_4. Hydrogen peroxide concentration too is critical. Too much and the enzyme is inactivated, too little and sensitivity is lost. Optimal substrate conditions are given in Section 10.2 but each laboratory should test individual batches to confirm optimum working conditions in terms of sensitivity and linearity of colour development. Hydrogen peroxide can be stored in aliquots at $4°C$; alternatively, urea peroxide tablets can be used.

Another commonly used chromogen is 2,2'-azino-bis (3-ethylbenzthiazoline-6-sulfonic acid) (ABTS) which gives a dark, visually readable, end-point. Other substrates that have been used include 5-amino salicylic acid (5-AS), *o*-diansidine (3,3'-diemethoxylbenzidine) and 3,3',5,5'-tetramethyl benzidine (TMB). TMB is much safer than OPD and a suitable and reliable substrate is now available in kit form (Pierce Chemical Co). Some substrates for peroxidase, such as OPD, have been reported to be mutagenic (22–24) but if handled with care they need not represent a serious health hazard. Fluorogenic substrates have been reported for peroxidase but these are less stable compared with those for other enzymes such as alkaline phosphatase and β-galactosidase (25). Peroxidase is sensitive to micro-organisms, and to antimicrobial agents such as sodium azide (26) and methanol (27), and can be inactivated by plastic—although the latter can be prevented by the addition of Tween 20 to the diluent (28).

3.3 β-D-Galactosidase (EC 3.2.1.23)

The cleavage of *o*-nitrophenyl-β-D-galactoside (*o*-NPG) to *o*-nitrophenol (*o*-NP) can be used to measure the activity of β-D-galactosidase. The concentration of *o*-NP is measured at 405 nm. Although it has a somewhat slower turnover rate compared with horse-radish peroxidase and alkaline phosphatase, β-D-galactosidase has the advantage that it is not normally found in plasma or other body fluids, although it can be found in some micro-organisms. Coloured substrates such as *p*-nitrophenyl-β-D-galactoside, as well as stable fluorogenic substrates such as 4-methylumbelliferyl-β-D-galactoside (MUG), are available. With MUG, as little as one attomole (10^{-18}) of enzyme can be detected per hour (29). β-D-Galactosidase can be coupled to proteins using the one-step glutaraldehyde (30) and maleimide procedures (20,31,32).

4. Purification of conjugate

Most labelled conjugates will perform better if conjugated and unbound enzyme and detector are removed. Indeed, this is essential for the periodate and two-step glutaraldehyde methods where a minority of enzyme molecules have been conjugated. A wide range of different chromatography media are suitable. When dealing with small amounts of conjugates such techniques as high-performance liquid chromatography (HPLC) and fast protein liquid chromatography (FPLC) are ideal. If such facilities are not available, gel filtration and ion exchange chromatography can be used. It is sensible to keep the size of the column to a minimum in order to reduce loss of sample. Gel filtration columns should be equilibrated with protein-containing buffers.

5. Assay optimization

The reliability, sensitivity, and specificity of an ELISA is dictated by the quality of the reagents used and the extent to which the assay has been optimized. Such aspects as sample volume (typically 100 μl—sometimes reduced to 50 μl for economy), temperature (typically 21 °C), incubation time (1–2 h), and the type of plate used will be constant. However other aspects should be tested and checked periodically.

6. Coating of the solid phase

This is of fundamental importance since failure at this stage renders all other steps useless. Coating can be evaluated in two ways. To test whether a mouse monoclonal antibody has bound to the plate enzyme-labelled anti-mouse antibody can be used; to test whether the bound antibody is functional the appropriate labelled antigen can be added. If the coating material can be labelled it is possible to determine both association (binding) and dissociation (elution).

The most obvious aspect of the coating to be optimized is the concentration of antibody or antigen used to coat the plate. The capacity of plastic microtitre plates to bind protein is limited. At high concentrations there is a tendency for protein molecules to bind to each other. Such protein–protein interactions are generally weaker than those between protein and plastic and can result in dissociation of supposedly bound protein during the assay. The range of protein concentrations at which there is no interference with binding to the plastic is called the zone of independent binding (33). In practice this is typically 1 μg/ml although as much as 10 μg/ml can be used without difficulty. A typical result is shown in *Figure 5*. The best concentration to use is the one that gives a reasonably steep dilution curve and where a small change in coating concentration, either up or down, has little effect.

By far the easiest method of coating the microtitre plate is by passive adsorption to the surface of the plastic. However, proteins may be damaged by binding to the plate (34) and may be specifically orientated, for example through the binding of hydrophobic regions to the plastic, so that some anti-

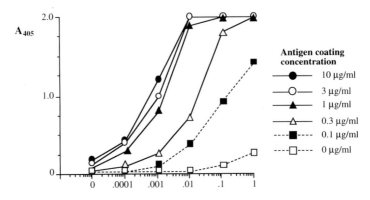

Figure 5. The optimal concentration of coating antigen is determined by coating the microtitre plate with increasing concentrations of antigen, here bee venom, according to *Protocol 5*. The optimal concentration is 3 μg/ml as a small increase or decrease has little effect on the amount of signal generated.

genic determinants fail to be expressed. Antigen can be desorbed or leached off the plastic (35–37). Such difficulties may be overcome by polymerization of the coating material with cross-linking agents such as glutaraldehyde or carbodiimide (38–41), or with specific antibody, which also has advantages in terms of reproducibility, sensitivity, and specificity (35,42,43) (*Figure 6*) (*Protocol 5*). Most antibodies will retain their ability to bind antigen when coated to plastic, although there is little doubt that a substantial amount of antibody activity is lost. Furthermore, some monoclonal antibodies perform poorly once coated. Under such circumstances the antibody itself can be bound to the plate via a linking agent. Any of the methods, described above for antigen, will work.

Alternatively, antibodies can be biotinylated and immobilized with avidin-coated plates. Plates can be coated with antibodies to haptens (low molecular weight chemicals that bind to antibodies) such as trinitrophenyl phosphate (TNP) (44) are able to bind antigens labelled with these small chemical determinants. The mechanism by which proteins stick to plastic is poorly understood but charge is believed to be important. The charge expressed by a protein depends on the pH of the buffer in which it is dissolved. By using buffers of different pH it is possible to create optimum conditions for a specific coating material. To reduce non-specific binding of the sample to the plate some authors have added a blocking step. It has been reported that smaller molecules such as casein make better blocking proteins (45) but we and others have observed that blocking can either increase or decrease background activity (46). Part of the problem may be that proteins do not evenly coat the plastic surface but exist as small islands (47) and additional protein simply binds to these foci.

(a)

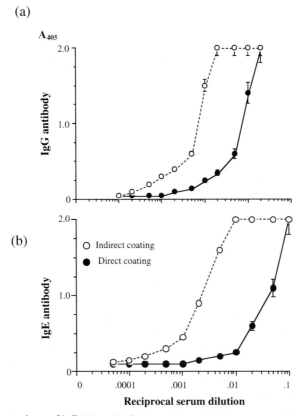

(b)

Figure 6. Comparison of IgE (a) and IgG (b) antibody binding to microtitre plates coated directly (open symbols) and indirectly (closed symbols) with antigen (bee venom phospholipase A_2) according to *Protocol 6*. The amount of antibody in the reference serum bound by plates coated with antigen indirectly is much greater than with plates coated with antigen directly.

7. Sample

Just as there is a limited range of protein concentrations that bind effectively to plastic, there is a limit to the amount of sample that can be measured. In practice this is rather similar between assays and typically covers a concentration range of 1–1000 ng/ml. Very low concentrations require special modifications and higher concentrations need dilution. In practice it is rarely possible to assay serum samples at a dilution of less than 1/50 because the high protein content of serum often interferes with the assay. The composition of the sample buffer too is important. The addition of 1% animal serum and 0.1% Tween 20 helps to reduce non-antibody–antigen interactions. To ensure low assay backgrounds it is a good idea to use serum from the same species as the enzyme-labelled antibody. To make it easier to check which wells have been filled I find it particularly helpful to add coloured dyes to the assay diluent (buffer 9) (1).

8. Labelled detector

In immunoassay, a 'detector' is any molecule that has specificity for the sample and which is used to detect it. It can be labelled directly with enzyme or in-directly, as described below. Methods for preparing enzyme conjugates have

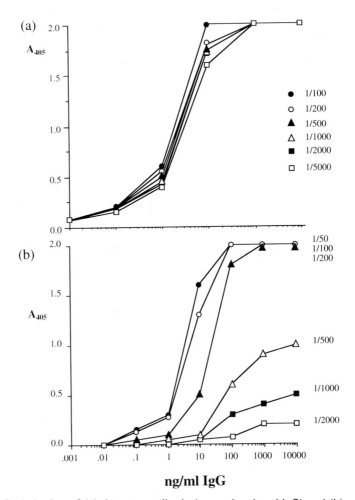

Figure 7. Optimization of (a) detector antibody (monoclonal anti-IgG) and (b) enzyme-labelled anti-mouse antibody (b) according to *Protocol 5*. Different concentrations of anti-human IgG monoclonal antibody are used in the assay. All give similar results and 1/1000 is chosen. Once this has been optimized, the second enzyme-labelled antibody that detects the bound monoclonal antibody optimized in (a) is added at different concentrations. Here there is a much bigger effect of concentration and 1/100 is optimal.

been extensively reviewed elsewhere (11,12,19,48,49). In non-competitive assays the detector antibody or antigen should be present in excess. Where more than one detector is used, each should be optimized in turn (*Figure 7*) (*Protocol 5*). The low non-specific binding of enzyme-labelled reagents to plastic means that they can be used at a relatively high concentration in ELISA. The diluent should normally be the same as for the sample. The detector can be labelled directly, or indirectly using an enzyme-labelled anti-detector antibody (58,59) or other ligands such as biotin/avidin (50) or FITC/anti-FITC (51). Removal of the Fc portion of the antibody yields conjugates with increased tissue penetration (for histology) and lower background binding (3,52) and greater sensitivity when amplified substrate systems are used (*Figure 8*) (*Protocol 6*). The rate of colour development with alkaline phosphatase can be increased using an amplified substrate system (3,15,53,54) and this is described in detail elsewhere (Chapter 9).

9. Equipment

Of all laboratory immunological techniques the apparatus needed for ELISA is probably the simplest. However, there is now a considerable amount of dedicated equipment available. The most important item is an automated plate reader with computerized data processing and there are a number of excellent machines on the market produced by such companies as Molecular

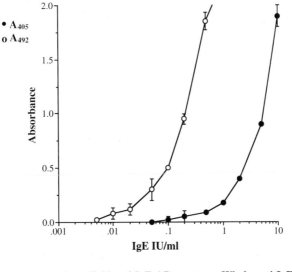

Figure 8. Detection of IgE by a two-site ELISA using whole anti-IgE-AP (•) and Fab′ anti-IgE-AP (o) conjugates according to *Protocol 6* and as shown in *Figure 2b*. The data is shown as the mean ± standard deviation.

Devices, Crawley, UK; Life Sciences International, Basingstoke, UK; and Denley, Billinghurst, UK.

Washing microtitre plates can be carried out with a wash bottle and a bucket and for some assays requiring different wash solutions, this is perfectly adequate. There are, however, a number of automated washing machines on the market today which work well. Nevertheless, whenever we have reviewed such products in the past we have come to the conclusion that the simple set-up shown in *Figure 9* is adequate. This consists of a wash reservoir placed on a shelf one to two metres above the bench which is connected to a 12 channel dispenser and aspirator that can be purchased from a number of suppliers.

10. Buffers and substrates

10.1 Buffers

(a) Buffer 1. Carbonate/bicarbonate (coating buffer) pH 9.6, 0.1 M. Take 4.24 g Na_2CO_3 and 5.04 g $NaHCO_3$, make up to 1 litre with distilled H_2O, and check pH. Store at $4\,^{\circ}C$ for no more than a few weeks.

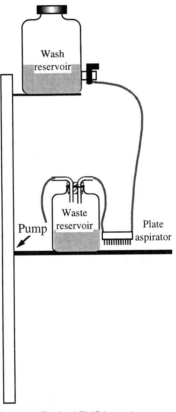

Figure 9. Typical ELISA wash set-up.

(b) Buffer 2. Citrate/phosphate buffer pH 5.0, 0.2 M. Make up stock solutions of 21.01 g citric acid ($C_6H_8O_7.H_2O$, 0.1 M) and 28.4 g of Na_2HPO_4 (0.2 M) each in 1 litre of H_2O. Add 48.5 ml of 0.1 M citric acid to 51.5 ml of 0.2 M Na_2HPO_4 producing a pH of approximately 5.0. Check and adjust as necessary.

(c) Buffer 3. Diethanolamine buffer pH 9.8, 0.05 M. Dissolve 101 mg of $MgCl_2$ in 800 ml of distilled water. When dissolved, add 97 ml of diethanolamine, mix thoroughly, and adjust the pH to 9.8 with concentrated HCl. Make up to 1 litre with H_2O, add 200 mg of NaN_3, and store in the dark at 4 °C.

(d) Buffer 4. Phosphate-buffered saline (PBS) pH 7.4, 0.05 M. Dissolve 16.7 g Na_2HPO_4, 5.7 g NaH_2PO_4, 85 g NaCl, and 100 mg of NaN_3 in distilled H_2O. Make up to 10 litres with H_2O and check pH. Store at room temperature.

(e) Buffer 5. Tris–HCl pH 8.0, 0.5 M. Dissolve the Tris in 800 ml and adjust the pH to 8.0 using 1 M HCl. Make up to 1 litre and store at 4 °C.

(f) Buffer 6. Tris–HCl pH 7.4, 0.1 M. Dissolve the Tris in 800 ml and adjust the pH to 7.4 using 1 M HCl. Make up to 1 litre and store at 4 °C.

(g) Buffer 7. Amplified ELISA substrate buffer. Prepare 50 mM diethanolamine buffer at pH 9.5. Add 1 mM $MgCl_2$, 4% (v/v) ethanol, 0.1% NaN_3, and check pH.

(h) Buffer 8. Amplified ELISA amplifier buffer. Prepare 20 mM sodium phosphate pH 7.2. Add 1 mM INT violet, 0.1% NaN_3, and check pH.

(i) Buffer 9. Assay diluent. PBS (buffer 4), or Tris–HCl (buffer 6)/1% animal serum/0.1% Tween 20. To 98.5 ml of PBS or Tris–HCl add 1 ml animal serum, 0.5 ml Tween 20, and 0.5 ml of the red and blue solutions in a 100 ml bottle (see below). The particular animal serum used will depend on the reagents and antibodies used in the assay. Horse serum or rabbit serum are commonly used but cross-reactivity between serum and reagent antibodies may make it necessary to use another serum such as goat. Phenol red (40 mg/100 ml) and Amido black (60 mg/100 ml) made up in H_2O can be used as coloured additives. Add 0.5 ml of each to 100 ml of the buffer being used. Store for no more than two to three days. At neutral pH it should be a green colour.

(j) Buffer 10. Washing buffer. Add 5 ml of Tween 20 to 10 litres of PBS (buffer 4) or Tris–HCl (buffer 6) located appropriately to provide gravity feed.

10.2 Enzyme substrates

10.2.1 Peroxidase

(a) Substrate 1. Dissolve 0.4 mg *o*-phenylenediamine (OPD) hydrochloride in 10 ml of citrate/phosphate (buffer 2) and adjust the pH to 5.0 with

0.2 M Na_2HPO_4 or citric acid as necessary. The substrate should be made up no earlier than 10 min before use and hydrogen peroxide (4 μl of 3% (v/v) H_2O_2) added just prior to use. It is well worth optimizing the substrate conditions for each batch of reagents. Incubation time with the substrate is about 15 min after which the reaction is stopped by the addition of 50 μl of 2 M H_2SO_4. Absorbance must be recorded (at 492 nm) as soon as possible as the substrate product is unstable.

(b) Substrate 2 (49). Dissolve 5 mg of TMB in 2.5 ml of absolute ethanol (warming to 40°C may be necessary). Prepare 0.2 M acetate buffer pH 3.3, and add 5 ml to 92.5 ml of water containing 100 mg of nitroferricyanide. Mix the two solutions just prior to use. Substrate kits for peroxidase using such substrates as TMB are available commercially (Pierce Chemical Co.) and work well.

10.2.2 Alkaline phosphatase

(a) Substrate 3: *p*-nitrophenyl phosphate (PNP). Add one PNP tablet to 5 ml of diethanolamine buffer and allow it to dissolve, mixing for 10 min. The substrate is stable at 4°C for at least 1 h. Substrate incubation time is up to 2 h at 37°C or 24 h at 4°C. The reaction is stopped using 50 μl of 3 M NaOH. The coloured substrate product is very stable at 4°C and plates can be stored for several days in the dark. Absorbance should be read at 405 nm.

(b) Substrate 4. Amplified substrate for alkaline phosphatase. The substrate NADP is purified with a suitable ion exchange resin, such as DEAE or QAE Sephadex to remove contaminating NAD, and made up as a 100 μM solution in amplified ELISA substrate diluent. Alcohol dehydrogenase 70 mg plus 1% BSA or similar protein stabilizer is dissolved in 7 ml amplifier diluent and dialysed extensively against amplifier diluent at 4°C. 10 mg of diaphorase (NADH dye oxidoreductase EC 1.6.4.3) are dissolved in 5 ml of 50 mM Tris–HCl pH 8.0 and 1% BSA or similar protein stabilizer added. This is then dialysed against amplifier diluent. The two amplifier enzymes are diluted tenfold and mixed together immediately prior to use. The optimum ratio can be determined by mixing different ratios of the enzymes from 10:1 to 1:10 although 1:1 is generally satisfactory.

11. Enzyme labelling of proteins

Protocols 1–4 give details of four commonly used procedures for labelling antibodies and antigens. The one-step glutaraldehyde method is easiest but can give high backgrounds. The periodate method gives particularly low backgrounds.

Protocol 1. One-step glutaraldehyde method (11)

Equipment and reagents
• See Sections 9 and 10

Method

1. Dialyse a mixture of 2 mg of antibody (2 mg/ml) and 5 mg of alkaline phosphatase (activity > 1000 U/ml) against PBS (buffer 4), overnight at 4°C with several changes of PBS.
2. Remove mixture from the dialysis tubing and measure volume. Calculate amount of glutaraldehyde required to make a final concentration of 2% (v/v) glutaraldehyde, add to the mixture and incubate, continuously mixing, for 2 h at 4°C.
3. Transfer mixture to dialysis tubing and dialyse against PBS (buffer 4) overnight at 4°C with two changes of PBS.
4. Transfer dialysis bag to 0.5 M Tris–HCl pH 8 (buffer 5) and dialyse overnight at 4°C with two changes of Tris–HCl buffer.
5. Remove the mixture from the dialysis bag and dilute the final conjugate to 4 ml with Tris–HCl (buffer 5) containing 1% BSA and 0.2% NaN_3.
6. Store in the dark at 4^°C.

Protocol 2. Two-step glutaraldehyde method (12)

Equipment and reagents
• See Sections 9 and 10

Method

1. Dissolve 10 mg of horse-radish peroxidase in 200 μl of phosphate buffer (0.1 M pH 6.8) containing 1.25% (v/v) glutaraldehyde. The solution is then left to mix overnight at room temperature.
2. The unbound glutaraldehyde is then removed either by dialysis overnight against normal saline, changing the saline twice, or by passing the mixture down a Sephadex G-25 or an ACA 34 column, monitoring the absorbance of fractions at 280 nm, and pooling the 'activated' HRP peak.
3. The affinity purified antibody is made up to a 5 mg/ml solution with normal saline.
4. 1 ml of activated HRP is mixed with 1 ml of antibody solution. 0.1 ml of 1 M carbonate/bicarbonate buffer pH 9.6 is then added and the mixture incubated at 4°C overnight.

Protocol 2. *Continued*

5. Add 0.1 ml of 0.2 M lysine and mix at room temperature for 2 h.

6. Dialyse the mixture against PBS pH 7.4 (buffer 4) overnight at 4°C.

7. The HRP-conjugated antibody is then precipitated by adding an equal volume of saturated ammonium sulfate solution. The precipitate is spun down and supernatant removed. The precipitate is then washed twice using half-saturated ammonium sulfate. The final precipitate is then resuspended in 1 ml of PBS (buffer 4).

8. Dialyse the conjugate extensively against PBS (buffer 4) overnight at 4°C.

9. The conjugate is then spun at 10000 *g* for 30 min to remove any sediment. The supernatant is removed and BSA or HSA is added to a final concentration of 1%.

10. Filter the conjugated antibody through a 0.22 µm membrane.

11. Store the conjugate either at −20°C or, if made up to a 50% glycerol concentration using equal volumes of glycerol and conjugate, it can be stored at 4°C.

Protocol 3. Periodate method (19)

Equipment and reagents
• See Sections 9 and 10

Method

1. Dialyse 5 ml of alkaline phosphatase against 0.3 M Na_2CO_3 pH 8.0 to remove all $(NH_4)_2SO_4$ and stabilizers. Add 100 µl of a 1% solution of 1-fluoro-2,4-dinitrobenzene, dissolved in absolute ethanol, to the dialysed suspension, and mix gently for 1 h at room temperature.

2. Then add 1 ml of 0.08 M sodium periodate (freshly made up) and gently mix for 30 min at room temperature.

3. Add 1 ml of 0.16 M ethylene glycol and mix gently for a further hour at room temperature.

4. Dialyse the enzyme–aldehyde solution against 0.01 M Na_2CO_3 pH 9.5 for 24 h at 4°C with at least three changes of buffer.

5. The anti-IgG immunoglobulin is prepared by dialysing 5 mg against 0.01 M Na_2CO_3 pH 9.5, overnight at 4°C.

6. Add the dialysed immunoglobulin to the 3 ml solution of enzyme–aldehyde and mix for 2–3 h at room temperature.

7. 5 mg of NaBH$_4$ is then added, dissolved, and then left to stand at 4°C for 3 h.

8. The conjugate is then dialysed against PBS pH 7.4 (buffer 4), at 4°C for 24 h. Any precipitate that forms should be spun down and removed.

9. The remaining conjugate should then be applied to the 85 × 1.5 cm Sephadex G-200 column and eluted using PBS pH 7.4 (buffer 4) with a flow rate of 5 ml/h. Collect 2 ml fractions and monitor the absorbance at 403 nm, the first peak should contain the conjugate.

10. Store conjugate at –20°C in 1% BSA. Do not thaw and re-freeze.

Protocol 4. Biotinylation of antibodies

Equipment and reagents
- See Sections 9 and 10

Method

1. The antibody solution (approx. 3 mg/ml) is washed twice though a Centricon™ membrane with bicarbonate coating buffer pH 9.6 (buffer 1).

2. Biotin (30 mg/ml) in the same pH 9.6 bicarbonate coating buffer (buffer 1) is added to the antibody solution in a ratio of approx. 0.2 mg biotin to 1 mg antibody, and mixed for 1 h 30 min.

3. Remove free biotin by washing though the Centricon membrane with PBS (buffer 4) twice. Add 1–2% (w/v) protein for storage.

---

12. Assays

Protocols 5–8 give four examples of ELISAs commonly used in my laboratory. *Protocol 5* measures IgG antibodies to ovalbumin in rats that we immunize (*Figure 10*). It can readily be adapted to measure human antibodies of different classes and subclasses. As long as the antigen binds well to the plate this is a highly reliable procedure. *Protocol 6* and *7* detail typical two-site ELISAs for IgE (*Figure 8*) and interferon-γ (*Figure 11*). These are the most reliable ELISA formats with interassay coefficients of variation as low as 5%. *Protocol 8* is a competitive ELISA which we have used to assay ovalbumin (*Figure 12*). It can also be used to determine whether the antigen is question is present as a contaminant in another protein preparation.

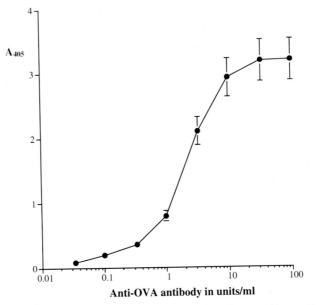

Figure 10. Measurement of rat IgG antibodies to ovalbumin according to *Protocol 5* and as detailed in *Figure 1a*. The data is shown as the mean ± standard deviation.

Protocol 5. Antigen-coated plate ELISA for detection of rat IgG antibody to ovalbumin (35,43,56,57)

Equipment and reagents
- See Sections 9 and 10

Method
1. Coat the plate with antigen (ovalbumin) overnight at 4°C or for 1 h at 37°C at a concentration of between 10 and 3 μg/ml in bicarbonate coating buffer (buffer 1). Alternatively plates can be coated with anti-phospholipase A_2 at 1 μg/ml overnight in bicarbonate coating buffer (buffer 1), washed as below, 10 ng/ml phospholipase A_2 in assay diluent (buffer 9) added next day, incubated for 1 h, and the assay continued as below.
2. Wash three times with 300 μl of wash buffer (buffer 10) leaving 30 sec to 1 min between each wash.
3. Add dilutions of reference sample pool (e.g. 1/50–1/50 000), quality controls (e.g. 1/100, 1/1000, 1/10 000), and sample at 1/500 in assay diluent (buffer 9). Any samples outside the range of the standard curve are repeated at up to 1/50 dilution or as low as necessary and incubate for 2 h at room temperature.

4. Wash three times with 300 μl of wash buffer (buffer 10) leaving 30 sec to 1 min between each wash.

5. Add AP-labelled rabbit anti-rat IgG at 1/300 in assay diluent (buffer 9) and incubate for 1 h at 4°C.

6. Wash three times with 300 μl of wash buffer (buffer 10) leaving 30 sec to 1 min between each wash.

7. Make up PNP (substrate 3), in diethanolamine buffer (buffer 3) as described above. Add 100 μl/well and incubate at 37°C for 1–2 h.

8. Stop enzyme reaction by adding 50 μl 3 M NaOH/well for PNP or 50 μl 2 M H_2SO_4/well for OPD or TMB.

9. Measure absorbance at 405 nm for PNP, 492 nm for OPD, and 450 nm for TMB.

10. Read results from the reference curve described above.

Protocol 6. Two-site ELISA for the detection of IgE (3,55)

Equipment and reagents
• See Sections 9 and 10

A. *Standard assay*

1. Coat the plate with monoclonal anti-IgE at 3 μg/ml overnight at 4°C or for 1 h at 37°C in bicarbonate coating buffer (buffer 1).

2. Wash three times with 300 μl of wash buffer (buffer 10) leaving 30 sec to 1 min between each wash.

3. Add IgE standards, quality controls, and patient sera at 1/10 and 1/100 in assay diluent (buffer 9). Any samples outside the range of the standard curve are re-assayed using dilutions as necessary. Incubate for 3 h or overnight.

4. Wash three times with 300 μl of wash buffer (buffer 10) leaving 30 sec to 1 min between each wash.

5. Add AP or HRP–anti-IgE or Fab' anti-human IgE at 1/300 or 1/3000 in assay diluent respectively, and incubate for 3 h or overnight.

6. Wash three times with 300 μl of wash buffer (buffer 10) leaving 30 sec to 1 min between each wash.

7. For alkaline phosphatase make up substrate, PNP (substrate 3), in diethanolamine buffer (buffer 4) as described above. For peroxidase make up substrate, OPD (substrate 1) in citrate/phosphate buffer (buffer 2) or in acetate buffer for TMB (substrate 2) as described above. Add 100 μl/well and incubate at 37°C for 1–2 h for PNP (substrate 3) or 15 min for OPD (substrate 1) and TMB (substrate 2).

Protocol 6. *Continued*

8. Stop enzyme reaction by adding 50 μl 3 M NaOH/well for PNP (substrate 3) or 50 μl 2 M H_2SO_4/well for OPD (substrate 1) or TMB (substrate 2).

9. Measure absorbance at 405 nm for PNP, 492 nm for OPD, or 450 nm for TMB.

10. Read results from the standard curve and express as IU/ml (1 IU of IgE is approximately equal to 2.4 ng).

B. *For ultra-sensitive assay*

1. All buffers must be phosphate free.

2. Substrate 4: 100 μl of NADP are added to the wells and incubated for 10 min (this can be increased or decreased depending on the sensitivity required).

3. Equal volumes of the amplifier enzymes are mixed together as detailed above and 200 μl added.

4. The colour produced can be monitored and the results read at 492 nm kinetically or after stopping with 50 μl of 0.5 M H_2SO_4.

5. Read results from the standard curve and express as IU/ml (1 IU of IgE is approximately equal to 2.4 ng).

Protocol 7. Two-site ELISA for the detection of human IFN-γ

Equipment and reagents
• See Sections 9 and 10

Method

1. Coat the plate with anti-IFN-γ monoclonal antibody at 1 μg/ml overnight at 4°C or for 1 h at 37°C in coating buffer (buffer 1).

2. Wash three times with 300 μl of wash buffer (buffer 10) leaving 30 sec to 1 min between each wash.

3. Add IFN-γ standards, quality controls, and samples in assay diluent PNP (buffer 9). Any samples outside the range of the standard curve are re-assayed at an appropriate dilution. Incubate for 1 h at room temperature.

4. Wash three times with 300 μl of wash buffer (buffer 10) leaving 30 sec to 1 min between each wash.

5. Add biotin-labelled anti-IFN-γ at 1 μg/ml in assay diluent (buffer 9) for 1 h at room temperature.

6. Wash three times with 300 μl of wash buffer (buffer 10) leaving 30 sec to 1 min between each wash.

7. Add streptavidin-labelled alkaline phosphatase at 1/500 in assay diluent (buffer 9) for 1 h at room temperature.

8. Make up substrate PNP (substrate 3) in diethanolamine buffer pH 9.8 as described above. Add 100 μl to each well and incubate at 37°C for 1–2 h.

9. Stop enzyme reaction by adding 50 μl 3 M NaOH/well.

10. Measure absorbance at 405 nm.

11. Record the results as units, ng, μg, or mg/ml from the standard curve.

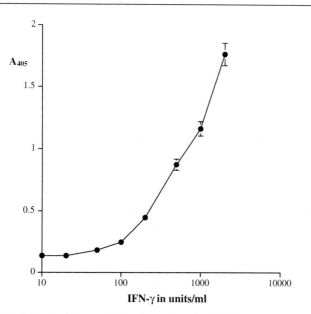

Figure 11. Measurement of human IFN-γ by a two-site ELISA according to *Protocol 7* and as shown in *Figure 2b*. The data is shown as the mean ± standard deviation.

Protocol 8. Competitive ELISA for antigen

Equipment and reagents
• See Sections 9 and 10

Method

1. Determine the concentration of antigen that gives 50–75% maximum OD. Coat the plate with this concentration overnight at 4°C or for 1 h at 37°C in the appropriate coating buffer (buffer 1).

Protocol 8. *Continued*

2. Wash three times with 300 μl of wash buffer (buffer 10) leaving 30 sec to 1 min between each wash.

3. Add dilutions (typically 200–0.02 μg/ml) of reference antigen (the same as the coating antigen) or unknown samples in assay diluent (buffer 9).

4. Add antibody to the coating antigen diluted (buffer 9) at 2 × the concentration that gave 50–75% of maximal binding and incubate for 1 h.

5. Wash three times with 300 μl of wash buffer (buffer 10) leaving 30 sec to 1 min between each wash.

6. Add AP-labelled species-specific antibody against the immunoglobulin used in step 4 at an optimum dilution in assay diluent (buffer 9) and incubate for 1 h.

7. Wash three times with 300 μl of wash buffer (buffer 10) leaving 30 sec to 1 min between each wash.

8. Make up substrate PNP (substrate 3) in diethanolamine buffer pH 9.8 as described above. Add 100 μl to each well and incubate at 37 °C for 1–2 h.

9. Stop enzyme reaction by adding 50 μl 3 M NaOH/well.

10. Measure absorbance at 405 nm.

11. Plot the results as per cent inhibition. The per cent inhibition obtained with the homologous antigen (the one bound to the plate) is the reference. The per cent inhibition obtained with the standard indicates the amount of antigen present whereas the maximum inhibition achieved represents the degree of immunological homology between the test antigen and the standard.

13. Quantification

The ways in which we express the results of ELISA procedures should be fairly straightforward. During optimization experiments and methodological development it is acceptable to express the data as units of optical density (colour). But for most ELISAs there is not a simple numerical relationship between signal (colour) and analyte concentration and there can be variation in enzyme activity from day to day. To take account of this, the amount of colour generated must be calibrated against a standard—a sample whose content is known, or has been assigned arbitrary units. Because we know that the relationship between the amount of colour generated and the concentration of a standard or sample is not linear, different amounts of standard must

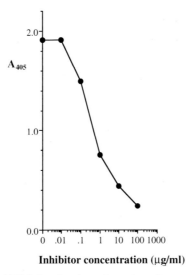

Figure 12. A competitive ELISA for the detection of ovalbumin according to *Protocol 8* and as shown in *Figure 3d*. The data is shown as the mean.

be used. Once plotted, the data forms a reference or standard curve. The reference material can be a pool of positive sera, cell culture supernatant, or a recombinant protein. It is usually stored frozen in aliquots to reduce damage due to freeze–thawing.

Many ELISA plate readers come with suitable software for plotting curves and calculating results. Where such facilities are not available the results can be plotted by hand on semi-logarithmic paper as most reference curves cover a relatively wide (2–$3 \times \log_{10}$) concentration range. The data are plotted as optical density against the reciprocal of the reference material (*Figure 13*) such that a 1/10 dilution is plotted at 0.1, 1/20 at 0.05, and 1/50 at 0.02, etc. Results are then read from the reference curve. They will then need to be multiplied by the appropriate dilution factor. The lower limit of the assay must be defined. In our laboratory we use a lower limit of 1.5 times the background value. If the background is high then the working range of the assay will be inadequate and the assay will clearly require further development in order to reduce this.

Another way in which assay results can be expressed is as a titre. This is the lowest dilution of the sample that can be detected at a pre-determined cut-off. Originally used to score red cell agglutination assays, titres can provide a considerable amount of information about the sample. Unlike using a reference curve, where values will be recorded over a range of antibody/antigen ratios, the end-point titre will have a constant antibody/antigen ratio. End-points that are too close to the background are unreliable because very small differences in concentration give equally small changes in colour. An end-

171

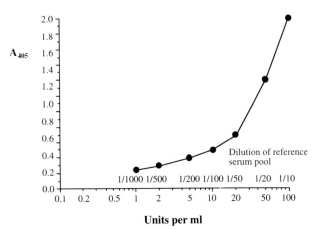

Units per ml

Figure 13. Calibration of the reference curve. For convenience the results are plotted on semi-log graph paper. Assuming the reference has an arbitrary value of 1000 U/ml then the value 1 corresponds to a 1/1000 dilution and a value of 20 to a 1/50 dilution. The results can be calculated by comparing the amount of colour in the sample wells against the curve and reading the units on the horizontal axis.

point 0.2 absorbance units above background (wells with diluent alone) is usually adequate.

13.1 Quality controls

In order to ensure that comparable results are obtained in different assays it is necessary to run quality controls. These are simply the same samples which are run in each assay. They should be run at a minimum of three dilutions to cover the upper, middle, and bottom parts of the standard curve (*Figure 14*). If the results obtained with these standards fall outside acceptable limits then

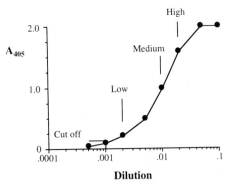

Dilution

Figure 14. Quality controls are independent samples which correspond to different parts of the standard curve which are run in each assay and make it possible to assess the reliability of the assay. The cut-off is the least value that can be reproducibly measured, typically 1.5–2 times the background.

172

the assay will need to be repeated. Acceptable limits will differ from assay to assay but are typically within 10%.

13.2 Coefficient of variation

By running quality controls in each assay it is possible to determine the coefficient of variation (CV). This is a measure of the variation between assays and the formula for this is:

$$CV = \frac{\text{Standard deviation}}{\text{Mean}} \times 100$$

Intra- (within) assay and inter- (between) assay variation should be monitored. At least four or five assays need to be run before the coefficient of variation can be properly determined. With new batches of reagent one sometimes sees a change in the CV so if comparisons of the data are to be made it is a good idea to try and use the same batch of reagents throughout. With many automated ELISA plate readers the CV for each of the duplicate or triplicate samples is given but this does not give information regarding the inter- or intra-assay variation.

References

1. Kemeny, D. M. (1991). *A practical guide to ELISA*. Pergamon Press.
2. Kemeny, D. M. (1994). *Anwendung des enzyme linked immunosorbent assay im biologisch/medizinischen labor to ELISA*. Gustav Fisher Verlag.
3. Kemeny, D. M., Richards, D., Johannsson, A., and Durnin, S. (1989). *J. Immunol. Methods*, **120**, 251.
4. Chantler, S. and Diment, J. A. (1981). In *Immunoassays for the 80s* (ed. Voller, Bartlett, and Bidwell). MTP Press.
5. Belanger, L. (1978). *Scand. J. Immunol.*, **8**, 33.
6. Avrameas, S., Ternynck, T., and Guesdon, J.-L. (1978). *Scand. J. Immunol.*, **8**, 7.
7. Underdown, B. J., James, O., and Knight, A. (1983). In *Recent developments in RAST and other solid-phase immunoassay systems* (ed. Kemeny, D. M. and Lessof, M. H.). Excerptia Medica, Elsevier.
8. Fenerley, H. N. and Walker, P. G. (1965). *Biochem. J.*, **97**, 95.
9. Garen, A. and Lenvinthal, C. (1960). *Biochim. Biophys. Acta*, **38**, 470.
10. Engvall, E. and Perlmann, P. (1971). *Immunochemistry*, **8**, 871.
11. Avrameas, S. (1969). *Immunochemistry*, **5**, 43.
12. Avrameas, S. and Ternynck, T. (1971). *Immunochemistry*, **8**, 1175.
13. Boorsma, D. M. and Kalsbeeck, G. L. (1975). *J. Histochem. Cytochem.*, **23**, 200.
14. Hazlett, D. T. G. and Garner, P. (1981). *J. Clin. Exp. Immunol.*, **2**, 325.
15. Self, C. H. (1985). *J. Immunol. Methods*, **83**, 89.
16. Nakane, P. and Pierce, G. B. (1966). *J. Histochem. Cytochem.*, **14**, 929.
17. Avrameas, S. (1970). *Int. Rev. Cytol.*, **27**, 349.
18. Avrameas, S. (1972). *Histochem. J.*, **4**, 321.
19. Nakane, P. K. and Kawaoi, A. (1974). *J. Histochem. Cytochem.*, **22**, 1084.

20. Kato, K., Fukin, H., Hamaguchi, Y., and Ishikawa, E. (1976). *J. Immunol.*, **116**, 1554.
21. Porstmann, B., Porstmann, T., Nugel, E., and Evers, U. (1985). *J. Immunol. Methods*, **79**, 27.
22. Sharpe, S. L., Cooreman, W. M., Bloome, W. J., and Lackman, G. M. (1976). *Clin. Chem.*, **22**, 733.
23. Holland, U. R., Saunders, B. C., Rose, F. L., and Walpole, A. L. (1974). *Tetrahedron*, **30**, 3299.
24. Venitt, S. and Searle, C. E. (1976). *Inserm Symposium Series* (ed. C. Rosenfeld and W. Davis), Vol. 52. IARC Scientific Publication no. 13, Inserm, Paris; 263.
25. Barman, T. E. (1969). *Enzyme handbook.* Springer–Verlag.
26. Schonbaum, G. R. (1973). *J. Biol. Chem.*, **248**, 502.
27. Straus, W. (1971). *J. Histochem. Cytochem.*, **19**, 682.
28. Berkowitz, D. M. and Webert, D. W. (1981). *J. Immunol. Methods*, **47**, 121.
29. Ishakawa, E. and Kato, K. (1978). *Scand. J. Immunol.*, **8**, 43.
30. Cameron, D. J. and Erlanger, B. F. (1976). *J. Immunol.*, **116**, 1313.
31. Kitagawa, T. (1981). *Enzyme immunoassay* (ed. E. Ishikawa, T. Kawai, and K. Miyai). Igaku-SHoin.
32. Yoshitake, S., Imagawa, M., and Ishikawa, E. (1982). *Anal. Lett.*, **15**, 147.
33. Cantarero, L. A., Butler, J. E., and Osborne, J. W. (1980). *Anal. Biochem.*, **105**, 375.
34. Pesce, A. J., Ford, D. J., Gaizutis, M., and Pollak, V. E. (1977). *Biohem. Biophys. Acta*, **49**, 399.
35. Kemeny, D. M., Urbanek, R., Samuel, D., and Richards, D. (1985). *J. Immunol. Methods*, **78**, 217.
36. Rubin, R. L., Hardtke, M. A., and Carr, R. I. (1980). *J. Immunol. Methods*, **33**, 277.
37. Salonen, E. M. and Vaheri, A. (1979). *J. Immunol. Methods*, **30**, 209.
38. Herremann, J. E. and Collins, M. F. (1976). *J. Immunol. Methods*, **10**, 363.
39. Howell, E. E., Nasser, J., and Schay, K. J. (1981). *J. Immunoassay*, **2**, 205.
40. Lehtonen, O. P. and Viljaken, M. K. (1980). *J. Immunol. Methods*, **34**, 61.
41. Rotmans, J. P. and Scheven, B. A. A. (1984). *J. Immunol. Methods*, **70**, 53.
42. Kemeny, D. M., Urbanek, R., Samuel, D., and Richards, D. (1985). *Int. Arch. All. Appl. Immunol.*, **77**, 198.
43. Urbanek, R., Kemeny, D. M., and Samuel, D. (1985). *J. Immunol. Methods*, **79**, 123.
44. Aalberse, R. C., Van Zoonen, M., Clemens, J. G. J., and Winkel, I. (1986). *J. Immunol. Methods*, **87**, 51.
45. Mauracher, C. A., Mitchell, L. A., and Tingle, A. J. (1991). *J. Immunol. Methods*, **145**, 251.
46. Mohammad, K. and Esen, A. (1989). *J. Immunol. Methods*, **117**, 141.
47. Butler, J. E., Ni, L., Nessler, R., Joshi, K. S., Suter, M., Rosenberg, B., *et al.* (1992). *J. Immunol. Methods*, **150**, 77.
48. Nakane, P. and Pierce, G. B. (1966). *J. Histochem. Cytochem.*, **14**, 92.
49. Tijssen, P. (1985). In *Laboratory techniques in biochemistry and molecular biology* (ed. Burdon and Knippenberg), Vol. 15. Elsevier.
50. Guesdon, J.-L., Ternynck, T., and Avrameas, S. (1979). *J. Histochem. Cytochem.*, **27**, 1131.

51. Harmer, I. J. and Samuel, D. (1989). *J. Immunol. Methods*, **122**, 115.
52. Ishikawa, E., Imagawa, M., Yoshitake, S., Niitsu, Y., Urushizaki, I., Inada, N., *et al.* (1982). *Ann. Clin. Biochem.*, **19**, 379.
53. Johannsson, A., Ellis, D. H., Bates, D. L., Plumb, A. M., and Stanley, C. J. (1986). *J. Immunol. Methods*, **87**, 7.
54. Macy, E., Kemeny, D. M., and Saxon, A. (1988). *FASEB J.*, **2**, 3003.
55. Kemeny, D. M. and Richards, D. (1987). In *Immunological techniques in microbiology* (ed. Grange, Fox, and Morgan), Vol. 24, p. 47.
56. Diaz-Sanchez, D. and Kemeny, D. M. (1990). *Immunology*, **69**, 307.
57. Diaz-Sanchez, D. and Kemeny, D. M. (1991). *Immunology*, **72**, 297.
58. Kemeny, D. M., Urbanek, R., Amlot, P. L., Ciclitira, P. J., Richards, D., and Lessof, M. H. (1986). *Clin. Allergy*, **16**, 571.
59. Kemeny, D. M., Urbanek, R., Richards, D., and Greenall, C. (1986). *J. Immunol. Methods*, **96**, 47.

9

Enzyme amplification systems in ELISA

COLIN H. SELF, DAVID L. BATES and DAVID B. COOK

1. Introduction

Enzyme amplification was devised to allow the development of highly sensitive and rapid immunoassays (1).

Enzyme labels were amongst the earliest alternative labels to radioisotopes introduced to immunoassay. With the limited specific activity of radioisotopes, it had been realized that, theoretically, enzymes had a much more sensitive detection capability because of their capacity for continuous catalysis of transformation of substrate to deliver some form of signal; for example colour, fluorescence, or electric current. The original concept of enzyme amplification was developed in response to the fact that, in routinely used tests, enzyme labels were not able to generate sufficient signal to compete with the then much more widely employed radioactive labels. Increasing substrate incubation times to days rather than minutes or hours simply served to increase background signals. Enzyme labels capable of much greater powers of signal generation were required to enable the development of not only more sensitive, but faster immunoassays. The major advantage of employing a sensitive enzyme measurement even in a competitive method with its inherently limited sensitivity (2), was seen to be the ability to deliver a very rapid assay system (3).

The aim of enzyme amplification was to increase the signal generating capacity of the labelling enzyme by making it part of a system in which it gave rise to a product, which was not itself detected, but rather which acted as a catalytic activator for a secondary system to produce a much greater detectable signal. Thus the signal was the product of two catalysts, the labelling enzyme and the activator. A number of approaches were originally described (1). The earliest involved the choice of the labelling enzyme to be one that gave rise to an activator for a secondary system based on an allosteric enzyme. Thus phosphofructokinase was used as a labelling enzyme to produce fructose 1,6, biphosphate which activated the *E. coli* type 1 isoenzyme of pyruvate kinase when assayed at a limiting concentration of

Colin H. Self, David L. Bates, and David B. Cook

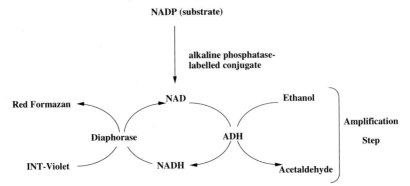

Figure 1. Principle of enzyme amplification for the detection of alkaline phosphatase label by the production of formazan.

phosphoenolpyruvate substrate. A critical feature of the enzyme amplification system is that the activator is not consumed as a result of the activity of the secondary system. However, a key factor in the development of the system was the realization that the activator could be consumed as long as it was re-formed. This led to the development of cyclic amplifiers. In these the labelling enzyme produces a substance, such as a cofactor, which is recycled by a cycling system, the operation of which produces the substance to be detected (1).

The most widely used example of this are the alkaline phosphatase-based systems employing NAD cofactor microcycles (4). As shown in *Figure 1*, in this system the alkaline phosphatase may be quantitated by its activity in dephosphorylating NADP to produce NAD. The NAD formed completes a cycle in a second step catalysed by alcohol dehydrogenase and diaphorase, in the presence of ethanol and INT-violet, giving rise to an intensely coloured purple formazan. Thus the NAD produced by the action of the immuno-chemical labelling enzyme, alkaline phosphatase, is converted to NADH by the action of the alcohol dehydrogenase, and when re-formed by the action of diaphorase INT-violet is reduced to form the highly coloured formazan. One molecule of formazan is produced in each 'turn' of the cycle and the speed at which the cycling reactions operate depends on the concentration of the two enzymes and of the NAD. This has been likened to an amplifier and pre-amplifier in sound reproduction, where the alkaline phosphatase whose activity delivers a missing cofactor, represents the preamplifier, and the subsequent cycling system represents the amplifier.

The application of enzyme amplification to alkaline phosphatase has resulted in systems which can detect as little as one-thousandth of an atto-mole of the enzyme either fluorimetrically (5) or colorimetrically (6). Another application of the general approach, employing FAD formed from FADP as the recycled component (7), has been developed subsequently and has shown the expected benefits of enzyme amplification.

As far as the high performance NAD systems are concerned, alkaline phosphatase is particularly applicable to cycling systems since it is capable of employing NADP or NADPH as substrates yielding NAD or NADH (8).

As alternatives to the production of a coloured end-product the redox cycle can be designed so that it generates a fluorescent (5) or electrochemical signal (9, 10). The substrate cycle is a redox process catalysing the transfer of electrons from ethanol to INT. It was apparent early in the development of the technology that the final biochemical step could be bypassed and the process monitored amperometrically with an appropriate mediator and a suitable electrode. A system using hexacyanoferrate as electron acceptor, reoxidized at an electrode surface was described by Stanley *et al.* (9). More recently Athey and McNeill (10) described an adaptation to the substrate cycle whereby the diaphorase was replaced with NADH oxidase which generated H_2O_2 that was detected at an electrode set at 600 mV.

Amperometric systems hold the potential for new applications for enzyme amplification with important additional advantages. Instruments could be simplified and new formats and matrices incompatible with optical measurements could be introduced.

The enzyme amplification cascade and cycle function equally well if NADPH is used as substrate, provided that pig heart diaphorase is used. Diaphorase from microbiological sources (typically *Clostridium kluyveri*) has a much greater reaction with NADPH. If the latter source of enzyme is used (which is essential in the fluorescence cycle employing resazurin as diaphorase substrate) NADP and not NADPH must be employed.

The system has been extensively used in research applications and, as far as clinical applications are concerned, enzyme amplification has been widely employed commercially especially in the infectious disease area, for example, with the production of high performance kits for the determination of HIV I and II, hepatitis, and chlamydia. The technology has been applied to genetic testing. For example, Minter (11) has applied the technology to DNA determination whereby a biotinylated oligonucleotide probe is hybridized to immobilized DNA and reacted with streptavidin–alkaline phosphatase conjugate. The enzyme label is then quantified by enzyme amplification. In this present description we shall refer to the use of the technology in ELISA, but identical principle and techniques apply in the detection phase of all the various applications.

2. Enzyme-amplified assay with colorimetric determination

An ultrasensitive detection system for alkaline phosphatase has applications in many assay configurations, but it can be exploited to the greatest advantage in a two-site ELISA employing high concentrations of ligands under conditions of reagent excess. One such application, an ultrasensitive assay for TSH, is shown below (*Protocol 3*). Although this example uses reagents supplied in kit form, exactly the same principles apply to any ELISA, provided that adequate care is taken in the design of the assay. The following practical considerations apply to many types of assay, but especially in the design of enzyme-amplified ELISAs. Although we place emphasis on the need for care in the manipulative steps of the assay, it should be appreciated that the level required is that called for by good laboratory practice.

(a) The quality of the conjugate is very important to obtaining optimal performance of the assay and to the elimination of non-specific binding. Heterobifunctional coupling chemistries, such as those described by Ishikawa *et al.* (12) using SMCC, SPDP, etc. (Pierce and Warriner, Chester, UK) are greatly preferred to one-step methods using glutaraldehyde.

(b) Conjugate performance is improved by careful fractionation and selection of individual fractions. A convenient fraction method is high pressure gel permeation chromatography on TSK-3000 or TSK-4000 columns (Anachem).

(c) The use of antibody fragments (Fab, Fab', or $F(ab')_2$) may be preferable to whole antibodies.

(d) Conjugate buffers require careful optimization; factors to be optimized include pH, ionic strength, buffer type, detergent concentration, and protein concentration. The following buffer is suitable for the storage and use of conjugates in amplified alkaline phosphatase assays: 100 mM triethanolamine–HCl pH 7.5, 6% (w/v) BSA, 1 mM $MgCl_2$, 150 mM NaCl, 0.05% (w/v) Triton X-100, and 15 mM sodium azide as preservative.

(e) Efficient washing is essential for the elimination of background and non-specific binding in obtaining an optimized amplified assay. The pH, ionic strength, detergent concentration, and buffer type may all require separate optimization. The use of phosphate buffers in wash solutions immediately prior to alkaline phosphatase incubation must be avoided as this will cause product inhibition of the enzyme. Conjugate buffer should preferably not contain phosphate; the use of Tris buffers is recommended.

(f) Great care should be taken to avoid contamination of reagents. Use a separate pipette and reagent trough for conjugate and ensure that all pipettes and containers that contact the reagents are sterile if storage of reagents for further use is contemplated.

(g) The signal antibody may be labelled directly with alkaline phosphatase, or labelled with biotin which is subsequently detected with avidin (or preferably streptavidin)-conjugated alkaline phosphatase.

Note, diluted antiserum may be suitable for use without purification, but it will be necessary to determine an appropriate dilution for each individual antiserum used. The use of ammonium sulphate precipitated fractions is preferable. Affinity purified polyclonal antibodies may usually be used at a dilution of 1 μg/ml as for monoclonal antibodies.

Add 0.1–0.25 ml of coating antibody in carbonate buffer to the appropriate number of wells and incubate overnight at ambient temperature, or at 4 °C. Before assay, the plates may be incubated with a solution of bovine serum albumin, 10 g/litre in deionized water, to block non-specific binding of protein to the plastic surfaces. The plates or strips are finally washed three times with a 0.1% (v/v) solution of Tween-20 in deionized water.

We have found that some batches of albumin contain considerable alkaline phosphatase activity making them unsuitable for use in ELISA assays with alkaline phosphatase conjugate labels. All batches of albumin consequently require to be checked for such activity prior to use.

Protocol 1. ELISA

Equipment and Reagents

- Microtitre plates or strips (Nunc Maxisorp, or Dynatech Immulon 2 or Immulon 4)
  (We have experience of the above plates and strips. Other manufacturers will also be suitable. The manufacturer's recommendations regarding antibody binding should be consulted.)
- Eight or twelve-channel multipipettes, adjustable volume. Single channel digital pipettes (e.g. Gilson, Finnpipette)
- 0.05 M Tris–HCl buffer, pH 7.4–8.0 (diluent for alkaline phosphatase conjugates, containing 0.1% (v/v) Tween-20 (Sigma) and 0.1% (w/v) human or bovine albumin and 0.015% (w/v) sodium azide
- 0.1 M Phosphate buffer, pH 7.4 (suitable for other dilutions)

- 0.1 M carbonate buffer, pH 8.65 for coating plastic solid phase with capture antibody
- Bovine serum albumin, free of alkaline phosphatase activity (Sigma)
- Tween-20 (Sigma)
- Capture antibody. Monoclonal antibody at 1 μg/ml in carbonate buffer.
  Polyclonal antibody (ammonium sulphate fraction of IgG) diluted in carbonate buffer. (For many commercial and other polyclonal antiserum preparations we have found a concentration of 10 μg/ml[1] to be suitable but this will need to be checked with individual antisera, and will depend on the antibody concentration in the serum.)

Method

1. Perform ELISA using microtitre plates in the usual manner, using highly pure reagents. Avoid phosphate buffer when incubating the assays with the alkaline phosphatase conjugates; traces of phosphate may inhibit the enzyme. Use Tris buffer at pH 8, though phosphate buffer may be used when incubating sample with the capture antibody alone.

2. Perform all immunochemical and enzyme detection incubation steps with the microtitre plate enclosed in a suitable container to minimize temperature variation across the plate (13) to which highly sensitive assays are prone. (Some workers prefer to avoid the use of wells at the edge of the plate to nullify the potential for such 'edge effects'. A small plastic box is convenient.)

Protocol 1. *Continued*

3. At the various wash stages employ at least six wash changes for best results as regards sensitivity and precision. (Some workers have preferred eight, to minimize background.)

Protocol 2. Use of enzyme amplification reagents

Equipment and Reagents

- NADPH, NADP (Boehringer-Mannheim)
- INT-violet (Sigma)
- 50 mM diethanolamine buffer, pH 9.5 containing 1 mM $MgCl_2$
- Ethanol HPLC grade (Rathburn Chemicals)
- 20 mM Sodium phosphate buffer, pH 7.2
- Triton X-100 Scintran grade (BDH-Merck)
- 50 mM Tris-HCl buffer, pH 8.0

- Yeast alcohol dehydrogenase (Boehringer Mannheim or Sigma)
- Pig heart diaphorase (Boehringer Mannheim or Sigma)
- Bovine Serum Albumin (Sigma)
- Dialysis tubing, or sacks (cellulose tubing) (Sigma)

Method

1. Prepare substrate diluent: 50 mM diethanolamine pH 9.5, 1 mM $MgCl_2$, 0.7 M ethanol.

2. Dissolve NADPH in substrate diluent at a concentration of 0.1 mM.

   (a) The purest available grade of NADPH should be used, but even 'ultrapure' reagent may be contaminated with low concentrations (typically < 0.1%) of NAD(H) leading to a significant background. For many purposes NADP described as NAD-free (Boehringer Mannheim) is satisfactory. However if necessary NADPH can be repurified by conventional ion exchange chromatography on a suitable matrix such as Q–Sepharose.

   (b) Prepared reagent may be stored for one week at 4°C or for longer periods at −20°C or below. Sodium azide (1 g/litre) may be added as preservative. Great care should be taken to avoid contact with phosphatase-containing solutions, including bacterially contaminated solutions.

3. Prepare amplifier diluent: 20 mM sodium phosphate pH 7.2, 1 mM INT-violet, 0.05% (w/v) Triton X-100. Store at 4°C.

4. Dissolve alcohol dehydrogenase from yeast (EC 1.1.1.1), 70 mg in 7 ml of 20 mM sodium phosphate buffer pH 7.2. Extensively dialyse against four or five changes of the same buffer at 4°C.

5. Resuspend pig heart diaphorase (NADH : dye oxidoreductase) (EC 1.6.4.3), 10 mg in 5 ml of 50 mM Tris–HCl buffer pH 8.0. Extensively dialyse against four or five changes of 20 mM sodium phosphate buffer pH 7.2.

(a) Before dialysis an inert carrier protein such as BSA may be added (to 10 g/litre) to improve stability. After dialysis the reagents should be centrifuged or filtered to produce clear solutions.

(b) Prepared enzyme reagents should be stored at $-20\,°C$ or below. Sodium azide (1 g/litre) may be added as preservative.

6. Prepare working amplifier solution by adding both alcohol dehydrogenase and diaphorase to amplifier diluent. A suitable strength reagent is produced by adding 50 μl of each enzyme to 900 μl of diluent. The activity of the amplifier can be adjusted by varying these volumes.

Note: The same reagents in a stable freeze-dried kit form are available commercially (Dako Diagnostics Ltd., Ely, UK). A liquid-stable version will become available shortly. An example of the use of these reagents as commercially supplied follows.

Protocol 3. Enzyme-amplified ELISA for TSH in microtitre plates (and comparison with unamplified detection)

Equipment and reagents

- TSH ELISA (Dako Ltd)
- 5 mM pNPP in 1 M diethanolamine buffer pH 9.8
- Eight channel multipipette, sterile pipette tips, and reagent troughs
- Absorbent paper
- Plate reader (405 nm and 492 nm wavelength filters)
- Stopping solution: 0.46 mol/litre sulfuric acid

A. Capture of TSH

1. Add 75 μl of anti-TSH/alkaline phosphatase into each well.

2. Add 25 μl of TSH to each well. Use different concentrations (calibrators) in the range 0.1–500 mU/litre.

3. Incubate at 25°C (ambient temperature) for 2 h on a plate shaker.

4. Discard the contents of the wells and drain the plate on absorbent paper.

5. Wash the plate three times with diluted wash buffer.

6. Drain the plate on absorbent paper, slapping hard to remove the last traces of solution.

B. Amplified enzyme detection

1. Add 100 μl of prepared substrate to each well.

2. Incubate for 15 min at 25°C.[a]

3. Add 100 μl of prepared amplifier to each well.

Protocol 3. *Continued*

4. Incubate for 15 min at 25 °C.

5. Add 100 µl of stopping solution to each well.

6. Ensure the contents of the wells are mixed and free of bubbles.

7. Read the plate at 495 nm.

C. Detection with pNPP[b]

1. Add 200 µl of pNPP to each well.

2. Incubate for 60 min at 25 °C.

3. Add 100 µl of 1 M NaOH to each well.

4. Ensure the contents of the wells are mixed and free of bubbles.

5. Read the plate at 405 nm.

[a] In the example shown in *Figure 2*, incubations were for 15 or 30 min.
[b] An alternative, unamplified detection procedure.
[c] In the example shown in *Figure 2*, incubations were for 60 to 120 min.

The TSH assay shown in *Figure 2* illustrates the key features of an enzyme-amplified assay. First, for equal detection periods of 60 min, the signal amplification factor is about 250-fold. Secondly, compared to the pNPP assay where the signal approximately doubles with doubling incubation time, the amplified assay increased approximately fourfold, as predicted by the kinetics of the two-step process (14).

3. Application of fluorescence to enzyme amplification

As a final detection step, fluorimetry has been shown to be more sensitive than equivalent spectrophotometric methods, often by orders of magnitude (15). Such impressive improvements of sensitivity are not however directly achievable with the ELISA format, largely on account of the background problems that arise from non-specific binding in the system (16). None the less fluorescence has been usefully applied in conventional ELISA by the use of such substrates as umbelliferone phosphate which can be hydrolysed by alkaline phosphatase to the fluorescent product umbelliferone and this technique has been applied to a commercial immunoassay analyser system ('Vidas', bioMérieux sa, Marcy-l'Étoile, France).

Similarly with enzyme amplification, a small but significant improvement of sensitivity can be achieved by the application of fluorescence in circumstances where even further sensitivity may be required. A further theoretical advantage of using fluorescence is that the operational range is very large.

Amplifier vs pNPP Substrates

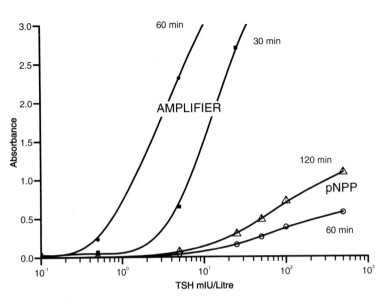

Figure 2. A comparison of enzyme amplified assay and pNPP in an immunoassay for TSH. For equal detection periods of 60 min, the amplification factor is 250-fold.

While the ideal fluorogenic substrate has not been reported, resazurin can be usefully exploited.

The dye resazurin has long been in use in the dairy industry where it has been used to detect contamination from dehydrogenase-producing bacteria, and it was applied to a general method for the detection of dehydrogenases by Guibault and Kramer in the 1960s (17). Upon reduction it produces the highly fluorescent pink product, resorufin. Consequently resazurin can be substituted for INT-violet in the cycling system described above.

It is, however, important to note that, in this case, the diaphorase used must be the microbiological type (*Clostridium kluyveri*) because the pig heart enzyme does not operate well with resazurin as substrate. Since the *Clostridium* enzyme has a greater residual reaction with NADPH than the pig heart (18) the cycle must, therefore, be initiated using NADP and not NADPH (with the pig heart diaphorase/INT-violet system either NADP or NADPH may be used).

The major drawback however is that the resazurin, being blue quenches the pink fluorescence produced so that a balance has to be chosen between the sensitivity desired and the dynamic range of the assay. Sensitivity requires that the starting concentration of the resazurin be limited to minimize quenching for the achievement of high sensitivity. As a result the dynamic range is limited as at higher concentrations of analyte the resazurin

185

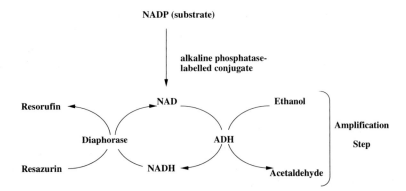

Figure 3. Application of fluorescence to the enzyme amplification principle. Resorufin is quantified fluorimetrically at an activation wavelength of 560 nm and an emission wavelength of 580 nm.

is quickly exhausted by the cycling reaction. Conversely, increasing the concentration of resazurin results in the production of a greater amount of fluorescence, especially at higher analyte concentrations, but the resulting quenching limits the sensitivity achieved which defeats the object of employing enzyme amplification for high sensitivity. Consequently the sensitivities obtained are comparable to those obtaining by the colorimetric approach (one-thousandth of an attomole). None the less, the large dynamic range over which the fluorimetric system can operate may be of value in specific instances. The authors have described an assay for proinsulin over a range of 0.017 pmol/litre to 8.5 pmol/litre, a range of about 500-fold (5).

3.1 Water sources and buffers

The optimization of enzyme amplification requires the use of ultrapure water and especially with the additional demands imposed by the use of fluorimetry. We have found that good results can only be obtained when buffers are freshly prepared from water freshly drawn from a suitable deionizer, such as the Millipore Milli-Q system, which includes a final pass through a 0.22 μm filter to remove bacterial and other particulate matter. Use of inferior water, or water or buffers which have been left standing in the laboratory for only two or three days, is liable to produce higher background values presumably due to bacterial enzyme contamination of one or more of the components of the highly sensitive cycling system. The problem can invariably be attributed to the primary incubation with alkaline phosphatase implying contaminating phosphatases as the source of the problem. It is of course impossible to construct highly sensitive analyses if reagents are compromised in this way though such water sources may be adequate for less sensitive work (this applies also to the use of INT-violet). However, our experience is that it is not only enzyme-amplified assays which require careful attention to water purity. We have also observed that inferior water, even double distilled from the laboratory supply is capable of interfering with the immunochemical reaction in the case of certain analytes such as proinsulin and PTH-related peptide and for this reason we prefer to use fresh high quality (Milli-Q) water at all times.

3.2 Preparation of enzymes

With the fluorimetric technique it is necessary to remove traces of NAD or NADH from commercial enzyme preparations by further purification, and we have found that adequate results are achieved by dialysis as described above.

Suitable enzymes are obtainable from a number of sources. We have used *Clostridium* diaphorase from Sigma and Biogenesis (dialysed at 50 U/ml) and yeast alcohol dehydrogenase from Sigma and Boehringer Mannheim (dialysed at 4000 U/ml. After dialysis, enzymes are stored at $-20\,°C$ or below in aliquots of 100–200 μl.

3.3 Alkaline phosphatase substrate

Johannsson and Bates have purified NADPH to a quality adequate for this work by ion exchange chromatography (14). We have found that NADP described as 'NAD-free', crystallized monopotassium salt (Boehringer Mannheim reagent no. 1179 969), to be satisfactory for fluorescence enzyme amplification without further purification.

3.4 Balance of cycling enzyme concentrations

As with the INT-violet method the optimal amounts of alcohol dehydrogenase and diaphorase in the reaction mixture must be determined as previously described in detail (14). We have found that the optimal ratios of units of alcohol dehydrogenase to *Clostridium* diaphorase to be 40:1 but it is advisable that this be assessed with individual batches.

Protocol 4. Practical steps in enzyme amplification with fluorescence

Equipment and reagents

The type of microtitre plate employed depends on the microplate fluorimeter available. We have experience of three manufacturers.

- Transparent microtitre plates (Nunc Maxisorp or Dynatech Immulon 2 or Immulon 4)
- White or black microtitre plates (Nunc FluoroNunc or Labsystems (Life Sciences International))

- Millipore Cytofluor: transparent plates must be used—excitation is from below the wells
- Life Sciences International Fluoroskan: excitation is from above—transparent plates may be used but better results are obtained with opaque black plates
- 50 mM diethanolamine buffer. pH 9.5, containing 1 mM $MgCl_2$ and 1 μM ZnCl,*

187

Protocol 4. *Continued*

- Dynatech Fluorolite: excitation from above—more sensitive detection of fluorescence in other applications has been observed with opaque black, and even better with opaque white plates intended for luminometry
- Freshly prepared solution of resazurin (Sigma) 6.3 mg in 5 ml ethanol (Rathburn)
- 0.1 M sodium phosphate buffer, pH 7.4
- *Clostridium kluyveri* diaphorase (Sigma or Biogenesis)
- Yeast alcohol dehydrogenase (Sigma or Boehringer Mannheim)

- NADP (NAD-free grade) (Boehringer-Mannheim). (NADPH *may not* be used).
- * Alkaline phosphatase has a requirement for zinc for optimal activity. Many preparations contain sufficient zinc when it becomes unnecessary to add further zinc, because at higher zinc concentrations the enzyme is inhibited. A requirement for zinc in dilution buffers therefore needs to be tested with individual preparations of conjugates.

Method

1. Prepare diethanolamine buffer; add 4.8 ml diethanolamine to 1 litre pure water. Bring to pH 9.5 with pure HCl.

2. Prepare substrate solution: 100 μM NADP in 50 mM diethanolamine buffer pH 9.5, containing 1 mM $MgCl_2$ and 1 μM $ZnSO_4$ It is convenient to keep stock solutions of 1 M $MgCl_2$ and 1 mM $ZnSO_4$ for this purpose when 10 μl of each can be added to 10 ml of the solution of NADP in buffer.

3. Prepare a fresh solution of 6.3 mg resazurin (Sigma) in 5 ml AR ethanol.

4. Prepare solution of amplifying mixture in 0.1 M phosphate buffer pH 7.4 as follows:

   - diaphorase 7.5 U[a]
   - alcohol dehydrogenase 302 U[a]
   - alcoholic resazurin solution 580 μl (final concentration of resazurin ~ 0.29 mM)
   - 0.1 M phosphate buffer pH 7.4 to 10 ml (at this stage inhibition of alkaline phosphatase is immaterial, and possibly desirable)

   Make up the amount of solution required for the number of wells employed. For a 96-well plate about 11–12 ml will be required. Unlike INT-violet there is no need to include Triton X-100 detergent in this solution in order to keep the product soluble.

5. Perform ELISA as above. Finally wash wells six times with wash buffer.

6. Drain the plate on absorbent paper, slapping hard to remove the last traces of solution.

7. Add the same volume of substrate reagent as used for coating the plastic with capture antibody so that all the surface captured enzyme is detected. This will usually be 100–200 μl.

8. Incubate for 30 min at ambient temperature (within a box). This incubation time may be extended to 60 min or more for greater sensitivity.

9. Add 100 μl amplification solution to each well. Incubate at room temperature for development of fluorescence (this will be visible).

10. Make fluorescence readings from about 10 min onwards if necessary for 1–2 h as fluorescence increases at the lower concentrations. The fluorescence is stable for this period of time and we have found that even during extended periods of incubation blank readings do not increase unduly providing scrupulously pure reagents have been employed.

11. Read the fluorescence on a suitable microplate reader. The excitation and fluorescence spectra of resorufin overlap considerably (5) and this places demands on choice of filters for isolation of excitation and fluorescent wavelengths. None the less the instruments of which we have experience contain filter sets which are suitable for the quantitation of resorufin fluorescence. These are:

   (a) Fluoroskan: Filter set 3—excitation spectrum bandwidth 524–564 nm, fluorescence spectrum bandwidth 572–608 nm.

   (b) Cytofluor: Filter set C—excitation spectrum bandwidth 505–555 nm, fluorescence spectrum bandwidth 555–625 nm. Sensitivity setting 1 or 2 is usually satisfactory.

   (c) We have not investigated the Dynatech Fluorolite or other microplate fluorimeters as to their suitability for measuring resorufin.

[a] The exact amounts will need to be checked for each batch of enzyme.

The proinsulin assay shown in *Figure 4* demonstrates the signals achieved with 0.29 mM resazurin and with 0.875 mM resazurin. Greater fluorescence is observed at low proinsulin concentrations with the lesser concentration of dye which is consequently less quenched. The greater range is illustrated at the higher resazurin concentration.

4. Concluding remarks

While several other high performance non-isotopic methods of quantitating immunoassays have now been described, enzyme amplification has the particular advantage that the degree of laboratory expertise required to make the reagents for fast highly sensitive assays already exists in competent laboratories and the equipment needed is no more than the colorimeters already possessed by most laboratories involved in this type of work. For those wishing to bypass actual reagent preparation, highly effective kits are available

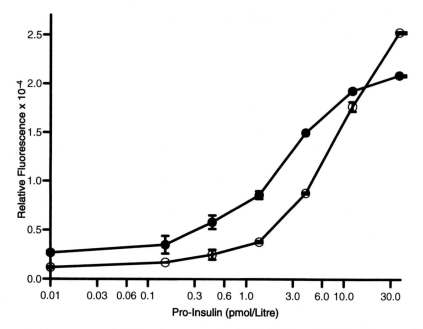

Figure 4. Fluorescence detection in enzyme-amplified ELISA with 0.29 mM resazurin (•——•) and 0.875 mM resazurin (○——○). Sensitivity is greater with 0.29 mM dye which has been exhausted at 4 pmol/litre proinsulin. The greater dye concentration results in quenching at lower analyte concentration but produces more fluorescence at higher analyte levels as the quenching is relieved as the resazurin is consumed. The wider working range is truncated at 3.0×10^4 RFU which is the limit the instrument could record.

for increasing the sensitivity and speed of alkaline phosphatase-based assays (AmPak, Dako Diagnostics Ltd., Ely, UK). For the fluorimetric approach a microplate fluorimeter is, however, more expensive and unusual, though being increasingly found for a variety of applications in biochemical and other laboratories.

Finally, an important aspect of enzyme amplification methods described above is that the signal is detectable visually. For some applications even this may suffice. For example, in screening programmes, such as the measurement of TSH in screening for congenital hypothyroidism, where no further quantitification of the signal may be necessary. Thus the signal from the one positive sample in thousands can be easily observed by eye in the microtitre plate whether colour or resorufin fluorescence is used as the signal.

References

1. Self, C. H. (198). European Patent Publication number 0027036.
2. Jackson, T. M. and Ekins, R. P. (1986). *J. Immunol. Methods*, **87**, 13.
3. Stanley, C. J., Johannsson, A., and Self, C. H. (1985). *J. Immunol. Methods*, **83**, 89.

4. Self, C. H. (1985). *J. Immunol. Methods*, **76**, 389.
5. Cook, D. B. and Self, C. H. (1993). *Clin. Chem.*, **39**, 965.
6. Bates, D. L. (1995). *Int. Labmate*, **20**, 11.
7. Obzhansky, D. M., Rabin, B. R., Simons, D. M., Tseng, S. Y., Severino, D. M., Eggelte, H., *et al.* (1991). *Clin. Chem*,m **37**, 1513.
8. Morton, R. K. (1955). *Biochem. J.*, **61**, 240.
9. Stanley, C. J., Cox, R. B., Cardiso, M. F., and Turner, A. P. F. (1988). *J. Immunol. Methods*, **112**, 153.
10. Athey, D. and McNeill, C. J. (1994). *J. Immunol. Methods*, **176**, 153.
11. Minter, S. (1995). *Novel Amplification Technologies Conference*. Washington DC, USA.
12. Ishikawa, E., Imagawa, M., Hashida, S., Yoshitake, S., Hamaguchi, Y., and Ueno, T. (1983). *J. Immunoassay*, **4**, 209.
13. Kemeny, D. M. (1991). *A practical guide to ELISA*, p. 79. Pergamon Press, Oxford, UK.
14. Johannsson, A. and Bates, D. L. (1988). In *ELISA and other solid phase immunoassays* (ed. D. M. Kemeny and S. J. Challacombe), pp. 85–106. John Wiley, Colchester, UK.
15. Guilbault, G. C. (1990). *Practical fluorescence*, 2nd edn. Marcel Dekker, New York.
16. Porstmann, T. and Kiessig, S. T. (1992). *J. Immunol. Methods*, **150**, 5.
17. Guilbault, G. C. and Kramer, D. M. (1965). *Anal. Chem.*, **37**, 120.
18. Bergmeyer, H. U. (ed.) (1983). In *Methods of enzymatic analysis*, 3rd edn, Vol II, pp. 179–80. Verlag Chemie, Weinheim, Germany.

Photoluminescence immunoassays

I. A. HEMMILÄ

1. Introduction

In biospecific assay technologies the detection of a label is most often based on measurement of emitted light. The emission can originate, for example, from radiation (radioimmunoassays, radioluminescence, RIA, IRMA), from a fluorescent compound (fluoroimmunoassays, FIA, IFMA), or the fluorescent end-product of an enzymatic reaction (fluorometric EIAs, ELFIA), from a luminogenic label (luminoimmunoassay, LIA), or luminogenic substrate after enzymatic triggering (enzyme-linked chemiluminescence immunoassay, ECLIA), from a phosphorescent compound (phosphoro-immunoassay), or from photoluminescent lanthanide chelate (time-resolved fluoroimmuno-assay). In the case of lanthanides the emission emanates directly from the long life-time excited ionic state of the metal and is today classified with the generic term luminescence (the term fluorescence being defined as emission emanating from excited singlet level of π-electrons and hence is restricted to aromatic hydrocarbons). In immunoassays the lanthanide emission is measured with a fluorometer using time resolution. These assays are generally named fluoroimmunoassays, or more specifically time-resolved fluoroimmunoassays, and to avoid confusion, never luminoimmunoassays.

Fluorometry provides a powerful tool in a variety of assay technologies, because of its potential sensitivity and the multitude of information it can provide. The signal can give information, not only on the existence of a compound in a sample, but also on its exact localization, its rotational (polarization) or lateral movement (membrane mobility by bleaching recovery measurement), its environment (pH, polarity, viscosity, etc.), or critical distances from energy accepting or donating groups (energy transfer). Fluorometric methods and the wide variety of fluorescent compounds available (1) are gaining increasing applications in molecular biology, cytology and histology, in micro- and macroscale imaging, in stationary and flow cytometry, microscopy, sensor technology, automated immunoassay systems, and so on.

In the field of bioanalytical analysis, where extreme sensitivities are

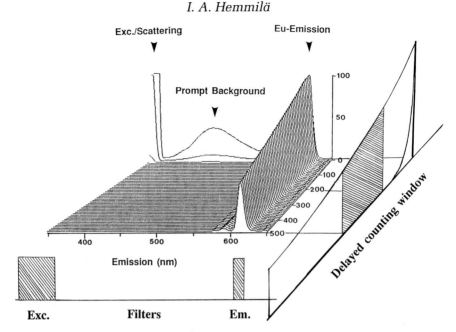

Figure 1. The emission profile of Eu^{3+} in enhancement solution after a pulsed excitation. Positions of filters are given on the wavelength scale and the counting window on the time-scale.

required, conventional fluorometry does not, however, perform satisfactorily due to its serious background limitations. In addition to the spectral parameters, the exploitation of temporal parameters, such as emission rise and decay times, can be used to give additional information and create higher detection specificity, robustness, and sensitivity when using a time-resolved fluorometer. In particular, with long decay time luminescent lanthanide chelates as labels, the long-lasting specific signal can be distinguished from any short-lived background interference and quantitated at a sensitivity level of single photons (*Figure 1*).

The development of methods to label analytes with lanthanide chelates and to measure labelled reagents using time-resolved fluorometry (2) has made it possible to employ lanthanides, as their fluorescent or fluorogenic chelates, as labels in different types of assays. In particular, the DELFIA system, which is based on a bioaffinity recognition reaction and a microtitration plate format, has been widely exploited in diagnostic immunoassays (3) as well as in screening, in *in vitro* hybridizations, and other solid phase-based bioaffinity assays. Furthermore the advent of new types of lanthanide chelates optimized to various applications has opened new perspectives to future assay technologies with features such as more sensitive homogeneous assays (4), miniaturized multiparametric assays (5), sensitive fibre sensors, macro- and microimaging (6), and so on.

2. Lanthanides as probes

Lanthanides are 4f-transition metals, which in their tervalent ionic state have an electronic configuration, where the optically active electrons are well shielded from their surroundings. That results in very typical line-type emissions characterized by the ion. Because of their electronic configuration lanthanides have found a number of biomedical applications (7). Gadolinium, for example, is widely used as a contrast enhancing agent in MRI (magnetic resonance imaging), Tb^{3+} can be used as an electrogenerated luminogenic label (8), and the photoluminescent lanthanides, Eu^{3+}, Sm^{3+}, Tb^{3+}, and Dy^{3+} as fluorescent labels (2,3). The energy transfer-based photoluminescence makes fluorescent lanthanide chelates particularly suitable for routine use in bioanalytical assays. In this process energy collection is accomplished by the organic chelator which subsequently transfers the energy with high efficiency to the central ion, which in turn produces the ion-characterized emission (*Figure 2*) having an exceptionally long excited state lifetime (from about 1–10 μsec, with Dy^{3+} to up to 1.5 msec with some Tb^{3+} and Eu^{3+} chelates). In *Figure 3* the emission profile, i.e. the emission spectrum of a mixture of Eu^{3+} and Sm^{3+} in a β-naphthoyltrifluoroacetone-based solution is given as a function of increasing delay time.

In addition to the unique fluorescent properties the specific advantages of lanthanide chelate labels include their small size, high water solubility, their negative net charge (polycarboxylate-based complexes), chemical and biological inertness, and low toxicity. When chelated with appropriate complexones, preferably with seven to nine dentate polycarboxylates, those chelates are stable, allowing, among other things, their use in very different assay matrices and almost indefinite storage of labelled reagents under suitable storage conditions (for antibodies, this is generally as a stock solution at 4 °C), thus avoiding frequent labellings and batch-to-batch variations.

3. DELFIA as a label system for bioanalytical assays

In the fluorescence enhancement-based technology, the label used is composed of a luminogenic metal ion. In the labelling, the ion is coupled to the analyte-specific reagent with a bifunctional chelating agent composed of a derivative of polyaminopolycarboxylic acid. These chelates are stable, hydrophilic, and biocompatible, but as such do not involve energy collecting or transferring functions, are non-fluorescent and thus require an additional step to develop fluorescence. On the other hand, incorporating a bulky aromatic structure into a labelling reagent would have an adverse effect on the affinity of labelled antibodies and their tendency to stick non-specifically to assay cuvette surfaces by hydrophobic interactions and giving rise to increased backgrounds.

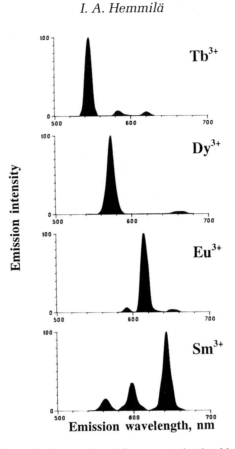

Figure 2. The major emission peaks of fluorescent lanthanides as their β-diketone chelates.

In DELFIA the enhancement is performed after the specific binding reaction is completed and the non-bound fraction of the label has been efficiently washed away. Fluorescence development is made possible by adjusting to a low pH, 3.2, at which the conditional stability of polycarboxylate-based chelates are strongly decreased in comparison to β-diketones present in the solution in large excess (9) (*Figure 4*). Under these conditions the ligand exchange reaction is completed within less than one minute. Immediately after dissociation a new chelate is formed with UV light absorbing β-diketone and the remaining empty coordination sites are occupied with tri-*n*-octylphosphineoxide (TOPO), which prevents aqueous quenching. The hydrophobic chelate formed is solubilized in a micellar detergent solution, in which it can be measured with high reproducibility and precision using a time-resolved fluorometer dedicated for this purpose.

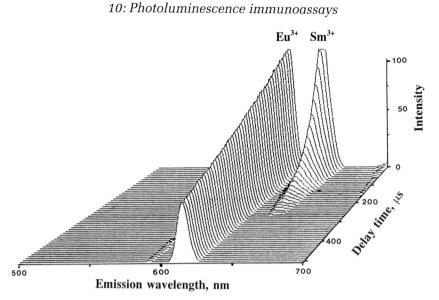

Figure 3. The emission profile of a mixture of lanthanides (only Eu^{3+} and Sm^{3+} are emitting) in DELFIA enhancement solution.

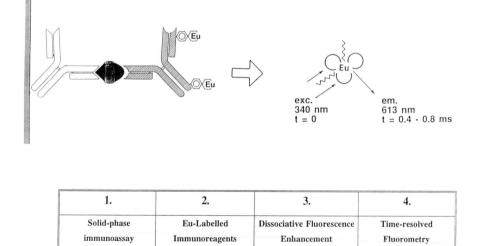

1.	2.	3.	4.
Solid-phase immunoassay	Eu-Labelled Immunoreagents	Dissociative Fluorescence Enhancement	Time-resolved Fluorometry

Figure 4. The dissociative fluorescence enhancement principle exemplified in a two-site sandwich assay.

3.1 Microtitration plates as solid phase matrix

Today, microtitration plates are the most commonly used solid phases in immunoassays, and are also used in increasing numbers in other assay formats, such as *in vitro* hybridization assays, receptor–ligand assays, a number of enzymatic assays based on photometry, fluorometry, or luminometry—and even β-particle counting assays utilizing solidified scintillant or solid phase scintillant in the form of particles or plastic. This development has resulted in a number of automated plate dispensing devices, washing devices, and incubators, in addition to plate counting devices, including plate photometers, plate fluorometers, plate luminometers, and even plate β-counting instruments.

DELFIA is generally based on microtitration plates (usually in the form of strips) as a conveniently applied solid phase manifold. For fluorometric use, the raw material (generally polystyrene) and the injection moulding process has to result in a high binding capacity (generally improved by irradiation) and a low luminescence background. The background can be further decreased by including black or scattering (white) particles into the plastic, but it can also be controlled by using pure monomeric components in the production of the polystyrene or by increasing UV-protecting compounds at concentrations that prevents excitation light penetration into the plastic material. In DELFIA, the plates used have to be transparent to the emission measured. In the case of plate fluorometers, such as the 1232 and 1234 DELFIA fluorometers, this meant wavelengths above 545 nm and for the older Wallac instrument, the 1230 Arcus, plates transparent to UV 340 nm were required for excitation through the sides of wells. The modern plate counting instruments (such as the Wallac 1420 Multilabel reader or the SLT SpectraFluor) (Austria) use an epi-illuminating optical arrangement and plates do not have to be transparent at all, thereby permitting the use of even white or black plates.

Coating of assay plates with the capture reagent is often a critical step in the assay development. Coating has to create a maximum binding capacity, a stable surface sustaining washings, even with detergent solution, and a consistent quality. A number of simple coating protocols can been used, e.g. Barnard (10). In *Protocol 1*, some further hints are given in order to help optimize the coating procedure for a particular use.

Protocol 1.   Coating of microtitration plates with antibodies

Equipment and reagents

- Microtitration plates or strips of high binding capacity and low fluorescence background (e.g. from Wallac)—low background plates are particularly required for Tb^{3+} and Dy^{3+} measurements; black or white plates can be used in epifluorometric instruments (1420 Multilabel reader)
- HCl for pre-treatment of antibodies

- Coating buffer: 0.2 M NaH_2PO_4 or 0.1 M Tris–HCl pH 7.5
- Saturation buffer: coating buffer containing 0.1–0.5% BSA; addition of suitable sugars, e.g. 6% sorbitol to the saturation buffer permits plates to be dried after coating—in assays where BSA is not suitable, gelatin or non-fat milk powder can be used instead

Method

1. Pre-treat the required amount of the monoclonal antibody at a pH of approx. 2.5. For example, incubate the antibody for 5–10 min in

---

1/1000 diluted HCl prior to dilution into coating buffer. Note: acid treatment has to be optimized for each particular antibody.

2. Dilute the acid treated antibody solution into the coating buffer to a final concentration of 4 μg/ml, and dispense 200 μl per well.

3. Incubate plates in humid chambers at 37°C overnight.

4. Aspirate wells and dispense 250–300 μl of saturation buffer, incubate for 2–8 h at room temperature.

5. If stored wet, aspirate wells and store plates sealed in bags to maintain a wet surface. With sugar-containing saturation buffer plates can be air dried (e.g. 2 h 35°C) and stored dry. When stored dry, insert a bag of desiccant to keep the plate dry during storage.

3.2 Labelling of immunoreagents with lanthanide ions

The exploitation of the numerous useful features of the metallic ions requires the development of bifunctional chelating agents suitable for labelling bio-analytical reagents with the ions of interest. Lanthanides are known to form kinetically labile chelates; consequently, to ensure adequate stability, the binding chelate has to contain at least seven to nine chelating groups composed of nitrogen or carboxylic acid. The stability requirement depends on the intended application. When the conjugate is to be used *in vivo*, e.g. when Gd^{3+} is used as a magnetic resonance imaging (MRI) contrast agent, its chelate must be very stable—both to ensure signal production and to avoid toxic side-effects. *In vivo* chelates generally consist of DTPA, either as such (a soluble tracer), as cyclic anhydride (for coupling), or preferably as isothio-cyanatobenzyl-derivatized penta-acetate with five preserved carboxylic acids. Similarly, in cellular assays based on lanthanide-labelled target cells, the label used has to be both very stable to avoid decomplexing and binding into cellular components and highly hydrophilic to decrease spontaneous release through the intact cell membrane. For that purpose a simple underivatized DTPA is recommended (see Section 3.7).

For *in vitro* immunoassays, where the sample is diluted and the assay time restricted, the chelate can be optimized to give rapid and complete dissociation in the detection phase. Therefore the optimal choice for immunoassays is an asymmetric DTTA derivative (11), most conveniently derivatized with an aromatic isothiocyanato group. This chelate is available from Wallac both in the form of a labelling kit and as a labelling reagent. In hybridization assays the labelling reagent is a seven-dentate structure containing one pyridine ring within the complexing group (3). Despite its greatly improved kinetic stability, this chelate gives rapid and complete fluorescence enhancement within ten minutes of enhancement.

In stable fluorescent chelates, the energy transfer needed to excite the central ion takes place inside the labelling chelate and accordingly no dissociation is needed for fluorescence enhancement. Thus the fluorescent chelates can be made as stable as possible. For most demanding conditions a fluorescent nine-dentate chelate, composed of, e.g. terpyridine–tetraacetate (12), provides a useful reagent.

3.2.1 Labelling protein with a lanthanide

In labelling of antibodies, antigens, or haptens (haptens coupled to carrier) an isothiocyanate coupling reaction has been found to be most suitable, particularly when labelling sensitive monoclonal antibodies, whose high affinity, specificity, and low non-specific binding are essential for high sensitivity assays. High temperature, high pH, and high concentrations enhance the coupling reaction. However, when optimizing assays using sensitive proteins, such as monoclonal antibodies, both temperature and pH often have to be lowered—and reagent concentration increased and/or reaction times prolonged. Customarily the labelling conditions are calculated according to the required molar excess of reagent to protein—even though in reagent excess conditions, a reproducible labelling primarily requires the use of a reproducible reagent concentration.

When labelling a single fraction of an antibody (or other protein), the labelling kit (either with Eu^{3+} or Sm^{3+}) provides a simplified procedure (instructions described in the kit insert). If high labelling yield is desired or if the labelling has to be performed at low pH with high molar excess of reagent, or a larger amount of protein has to be labelled, the Eu-labelling reagent is recommended (another possibility is to request Wallac customer labelling service to do the labelling). One vial of labelling reagent contains 1 mg of activated chelate (about 1500 nmols) giving, after reconstitution, 3 mM reagent concentration. For example, if several different antibodies have to be screened to find a working pair for two-site assay, labelling of up to 20 antibodies (0.5 mg of each with 25-fold excess) can be performed with one vial.

Protocol 2 gives some general guidelines for the labelling of proteins with a bifunctional chelating agent.

Protocol 2. Protein labelling using a bifunctional chelating agent

Equipment and reagents

- Activated chelate: isothiocyanate derivative of DTTA chelated with Eu^{3+} or Sm^{3+} (from Wallac Oy)
- Activated chelate: isothiocyanate activated free DTPA (see ref. 13) and free ions, e.g. Gd^{3+} atomic absorption standard, which for quantitation purposes can be spiked with 0.1–0.01% of Eu^{3+} to facilitate calculations
- Labelling buffer: 0.1 M carbonate buffer pH 9.3, containing 0.9% NaCl
- DELFIA system, including fluorometer and enhancement solution
- Chromatographic system: e.g. Sepharose 6B or Sephacryl S300 columns (Pharmacia) including fraction collector, UV monitoring system, peristaltic pump, and recorder

---

Method

1. The protein to be labelled should be in a non-amine containing buffer, such as carbonate, borate, or phosphate. Bacteriostatic agents, such as NaN_3 interfere with most coupling reactions. If pre-purification is needed, it can be simply carried out with a disposable PD-10, NAP-5/10, or Nick column (Pharmacia) equilibrated with the labelling buffer.

2. Weigh the required (exact or approximate) amount of labelling chelate and dissolve in a small volume (200–500 μl) of purified water (Wallac labelling reagent can be directly reconstituted into water to give 3 mM reagent concentration).

3. (a) Calculate the reagent concentration by measuring the complexed lanthanides (if not using Wallac labelling kit or reagent) in the DELFIA system. After dilution 1 in 10^{6-7} into enhancement solution measure against Eu^{3+} or Sm^{3+} standard (1/100 diluted Eu^{3+} or Sm^{3+} standard, available from Wallac).

   (b) Tb^{3+} and Dy^{3+} can be measured using enhancement solution for the dissociation and Tb–Enhancer for enhancing Tb^{3+} or Dy^{3+} fluorescence.

   (c) When using a very stable fluorescent chelate for labelling, a photometric quantitation is recommended using molar absorptivity if it is known.

   (d) With free chelator, ligand quantitation is based on weighing, where the ligand is saturated by adding a stoichiometric amount of free ion.

4. Mix the protein with the calculated amount of chelate. For example, for 1 mg of IgG in 500 μl of labelling buffer, a 50-fold molar excess (i.e. 0.6 mM concentration of labelling reagent) can be used initially. Labelling conditions (time, temperature, pH, and reagent concentrations) may have to be optimized for each assay, particularly if very high sensitivity is needed.

5. Incubate overnight at room temperature.

6. Separate the labelled conjugate from uncoupled chelate; this can be performed simply by gel filtration. For antibodies, a gel filtration column is recommended to separate aggregated IgG (may be formed during the labelling process) from the monomeric IgG fraction. Separation through a Sepharose 6B column, about 1 × 30 cm, eluted with Tris–HCl pH 7.7 containing 0.9% NaCl and 0.05% NaN_3, is recommended. For simple tracer screening (e.g. when choosing a combination of antibodies) a brief passage through a disposable PD-10 column is satisfactory; any aggregates can be removed by filtration through a 0.2 μm membrane.

Protocol 2. *Continued*

7. The labelled product can be collected according to the UV profile or according to Eu^{3+} content. Avoid collecting aggregate or free chelates with labelled antibodies. Calculate the protein concentration by absorbance at 280 nm (an aromatic thiourea bond increases absorbance by 0.008/per μM of chelate), by calculating the percentage pooled, or using a suitable protein measuring system (e.g. dye binding assay, chelating compounds may interfere with the Lowry assay).

8. Stabilize the conjugate by adding a suitable carrier protein. For most antibodies pure (heavy metal-free) bovine serum albumin is suitable (purified BSA is available from Wallac). Labelled antibodies can be stored at 4–8°C at concentrations greater than 20 μg/ml.

3.3 Dissociative fluorescence enhancement of lanthanides and their chelates

The bifunctional chelating agents best suited to the labelling of antibodies are composed of non-aromatic polyaminopolycarboxylates and absorb negligible light. Thus, fluorescence enhancement is required to detect the label. A dissociative enhancement provides a convenient, rapid, and efficient method by which the labelling chelate is initially dissociated by a low pH solution and the free ion instantly forms highly fluorescent chelates with fluorogenic ligands present in large excess (9). β-Diketones are the most suitable light absorbing and energy mediating ligands—aromatic diketones for Eu^{3+} and Sm^{3+} and aliphatic β-diketones for Tb^{3+} and Dy^{3+} (14). However, due to the short excitation wavelength and unexpectedly short decay times found with Tb β-diketonates, other absorbing ligands have been found to be optimal for Tb^{3+} and Dy^{3+}—e.g. those composed of dicarboxylic acid derivatives of pyridine, bipyridine, or terpyridine (15). Accordingly, when measuring more than two lanthanides in a single well, a sequential enhancement system is recommended (see Section 3.6).

3.4 Time-resolved fluorometry of lanthanides

Depending on the chelate employed (metal ion and the ligand) and on the plastic material used, the fluorometric protocol should be optimized to achieve the maximal sensitivity for each ion used. With the commercial DELFIA system, the instrument default protocols give satisfactory measuring parameters without any need for further optimization.

It is important to follow good pipetting practices in order to keep contamination levels as low as possible. It is advisable to use plate shaking devices in order to ensure good mixing and efficient lanthanide release from the surfaces with rapid fluorescence stabilization.

Table 1. Fluorometric properties of lanthanides Eu^{3+}, Sm^{3+}, Tb^{3+}, and Dy^{3+} in the DELFIA system

Ion	Enhancement	Step	Exc. (nm)	Em. (nm)	Decay (μsec)
Eu^{3+}	DELFIA ES	I	340	613	730
Sm^{3+}	DELFIA ES	I	340	643	50
Tb^{3+}	Tb–Enhancer	II	324	544	1154
Dy^{3+}	Tb–Enhancer	II	324	572	16

The fluorometric measurement of lanthanide ions in the DELFIA system, or their stable fluorescent chelates in fluorometric analysis, can be optimized according to their photoluminescence properties (excitation and emission maxima and decay time) and to the background (primarily the plastic photoluminescence of the microtitration plate used). *Table 1* gives the properties of Eu^{3+}, Sm^{3+}, Tb^{3+}, and Dy^{3+} in the recommended commercial enhancement systems. *Protocol 3* gives general guidelines for optimizing the measurement of any lanthanide chelate within a given assay plate.

Protocol 3. Measurement of lanthanide chelates

Equipment and reagents

- Plate fluorometer with time-resolving measuring mode
- Microtitration plates with low background, suitable for intended application (see *Protocol 1*)

- Lanthanide chelate solutions: when using an enhancement system, commercial atomic absorbance standard preparations can be used as a standard source diluted in enhancement solution to about 1–10 nM with Eu^{3+} and Tb^{3+} and about 10–50 nM with Sm^{3+} and Dy^{3+}—with other fluorescent chelates measured in solution, the respective concentrations depend on their fluorescence intensities

Method

1. Take the microtitration plate, fill one line or strip with water and a second line/strip with the label solution.

2. From the measuring protocol (protocols 95–100 in Wallac 1232, communication protocol when used with MultiCalc in 1234, measuring parameters in 1420), choose a narrow counting window, i.e. a time approximately 1/4–1/5 of the emission decay time (e.g. for Eu^{3+} a counting window of 100–200 μsec).

3. Measure strips with repeated protocols using various delay times starting from the shortest (1/10 of decay) to about 2 × decay (i.e. for Eu^{3+}, from 50 μsec to 2 msec).

4. Plot signal to background ratios as a function of delay time.

Protocol 3. *Continued*

5. Choose a value for delay at which the signal-to-noise ratio approaches its maximum value. For a counting window choose a time frame giving optimum signal–noise ratios (generally a window extending to one to two decays is enough; signal collecting after the emission has decreased below 10% is unnecessary).

6. Store the optimized parameter for future use.

3.5 DELFIA multilabel measurement

DELFIA can incorporate simultaneously up to four labels. The 1234 DELFIA plate reader contains three pre-set filters (Eu, Tb, Sm); the 1420 Multilabel reader contains filters for all the four fluorescent lanthanides. When the detection is split into two separate measuring cycles separated by Tb–Enhancer dispensing, intersignal spill-over is minimized. Eu^{3+} and Sm^{3+} measurement can be performed with pre-set protocols, which upon request in cases of high Eu^{3+} excess subtract the trace spill-over originating from the minor emission line of Eu^{3+} (emission line at 650 nm) from the Sm^{3+} channel. This correction is based on a measurement performed with the Sm^{3+} filter and using the Eu^{3+} time window, during which the Sm^{3+} emission has totally decayed. During the first phase, neither Tb^{3+} nor Dy^{3+} produce any emission (see *Figure 3*), and in the second phase the remaining Eu^{3+} emission takes place only above 585 nm and does not interfere with Tb^{3+} or Dy^{3+} counting. Using a short counting window for Dy^{3+} restricts the spill-over of any remaining Sm^{3+} emission.

For any other 'in-house' system, e.g. those based on one single enhancer, all corrections between the emissions has to be controlled in assays requiring a wide dynamic range. Such an experiment is shown in *Table 2*, performed with a β-diketone-based common enhancer in a quadruple label assay in neonatal screening. This exemplifies the use of aliphatic β-diketone-based co-fluorescence enhancement (16).

Table 2. An example of spectral overlaps between lanthanides in pivaloyl trifluoro-acetone-based co-fluorescence enhancement[a]

Interfering ion	Measured ion			
	Eu^{3+}	Tb^{3+}	Sm^{3+}	Dy^{3+}
Eu^{3+}	–	0.0017	0.0032	0.0033
Tb^{3+}	0.0336	–	0.0049	0.0169
Sm^{3+}	0.0167	0.0000	–	0.0593
Dy^{3+}	0.0000	0.0000	0.0031	–

[a] Values given as ratios of the signal obtained with the same concentration of interfering ion to that of measured signal.

Protocol 4 describes the measurement of three or four lanthanides from a single assay well using the DELFIA system.

Protocol 4. Measurement of three or four lanthanides

Equipment and reagents

- Lanthanide standard solutions: use only very pure lanthanide preparations, Sm^{3+} and Dy^{3+}, in particlular, should not contain any traces of Eu^{3+} (purity of at least 99.99% is recommended)—atomic absorbance standards (containing lanthanides at concentration of 1 mg/ml) in are a convenient source of lanthanides even though purity is not normally guaranteed, and the cross-talk figures may be misleadingly high

- Time-resolved fluorometer (e.g. Wallac 1234 Plate fluorometer for three labels, Wallac 1420 Multilabel reader for four labels)
- DELFIA enhancement solution and Tb–Enhancer (Wallac)

Method

1. Dilute each standard with 2.5–3 ml of DELFIA enhancement solution—Eu^{3+} and Tb^{3+} to 1 nM; Sm^{3+} and Dy^{3+} to 10 nM.

2. Dispense 200 µl of standards into separate strips (separate strip for each lanthanide).

3. For Tb^{3+} and Dy^{3+} strips, dispense 50 µl of Tb–Enhancer, shake strips for 10 min.

4. Count all strips with pre-set counting programs for Eu^{3+}, Tb^{3+}, and Sm^{3+} (and Dy^{3+} in 1420).

5. Calculate the signal cross-talk figures. The cross-talk between the Tb^{3+} to Eu^{3+} channels, and the mutual interference between Sm^{3+} and Dy^{3+}, does not apply in assays because the ions are measured as different chelates in different steps.

3.6 DELFIA FIA and IFMA: assay optimization

For non-competitive IFMA assay optimization is a relatively straightforward process. After coating the plates, incubations can be performed in two steps (separate incubation for analyte and labelled antibody), sequentially (addition of labelled antibody without prior washing), or simultaneously in one step (addition of all components simultaneously). Practical hints can be found in the earlier volume of this series (10). When optimized assay speed antibody affinity is of prime importance. Other factors to be considered relate to buffer, possible additives (chelating agents, detergents, proteins), pH, temperature, shaking, and incubation times (17).

Some guidelines for the optimization of a typical sandwich IFMA with Eu^{3+}-labelled antibodies are given in *Protocol 5*. Additional features which have to be taken into consideration are the avoidance of interference from either rheumatoid factor or heterophilic antibodies. The addition of mouse IgG is recommended, particularly to assays based solely on monoclonal antibodies. The only DELFIA-specific requirements relate to the use of suitable chelating capacity in the buffer (20 μM DTPA used in DELFIA assay buffer) and the avoidance of phosphate, which tends to precipitate lanthanides.

Protocol 5.   Optimization of a typical two-site sandwich assay

Equipment and reagents

- DELFIA system
- Coated plates (*Protocol 1*)
- Antigen with two or multiple epitopes
- Eu-labelled antibody (*Protocol 2*), known to form a two-site sandwich with the coating antibody and antigen to be analysed

Method

1. To evaluate the practical working range, dilute standard with assay buffer from about 100 ng/ml to about 1 pg/ml.

2. Dilute labelled antibody (Eu–MAb) with assay buffer to a concentration of 500 ng/ml.

3. (a) One-step protocol. Add 50–100 μl of standard and zero calibrator (assay buffer), with duplicates, to coated strips, dilute with 100 μl of assay buffer.

   (b) Two-step protocol. Add 50–100 μl of standards and assay buffer to strips, add 100 μl of diluted labelled antibody.

4. Shake plate at room temperature for 1 h.

5. (a) One-step protocol. Wash plate three times with plate washer.

   (b) Two-step protocol. Wash plate six times with plate washer.

6. (a) One-step protocol. Add 100 μl of tracer and 100 μl of assay buffer to washed strips and incubate further, with shaking, for 1 h.

   (b) Two-step protocol. Add 200 μl of DELFIA enhancement solution to strips, shake for 10 min, and count with the fluorometer.

7. One-step protocol. Wash strips six times and continue as in step 6(b).

8. Evaluate results. A good monoclonal antibody should give a zero-dose value around 200–1500 counts per second; polyclonal antibodies tend to give higher backgrounds. The maximum signal (at about 10 ng of antigen) should exceed one million c.p.s. The one-step assay often gives somewhat higher signals, but at high doses, it suffers from a high-dose hook effect. Depending on the hook and the clinical range of antigen in the sample, a one- or two-step assay can be chosen.

Sandwich assays can also be constructed using an indirect approach with generic reagents labelled with the lanthanides (some of which, like

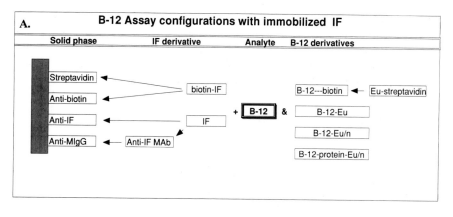

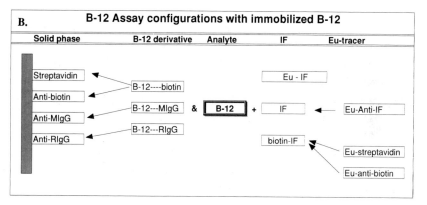

Figure 5. Assay configurations in developing a competitive assay based on either (A) immobilized binding protein, or (B) immobilized antigen derivative. Abbreviations: IF, intrinsic factor; anti-IF, monoclonal anti-intrinsic factor antibody; Anti-MIgG, anti-mouse IgG antibody; B-12, vitamin B_{12}; B-12—biotin, biotinylated B-12 with long spacer; B-12-Eu, Eu-labelled B-12; B-12-Eu/n, B-12 polyglutamate derivative labelled with several Eu chelates; B-12-protein-Eu/n, a carrier protein labelled with B-12 and with several Eu chelates; B-12—M/RIgG, rabbit or mouse IgG conjugated with B-12 using long spacer arm.

Eu-labelled streptavidin, protein G, anti-mouse, -human, and -rabbit IgG, are made available by Wallac). *Figure 6* shows schematically some established procedures for indirect competitive and non-competitive IFMA/FIAs with generic labelled reagents.

The construction of a competitive assay is much more complicated, requiring, among other things, titration of all the reagent concentrations to find the clinically relevant assay range. In addition, the sequence of reagent additions can be used to alter the sensitivity of the assay (simultaneous, sequential addition, or back-titration). Depending on the availability of reagents, access to synthetic chemistry, the sensitivity required, and signal level desired,

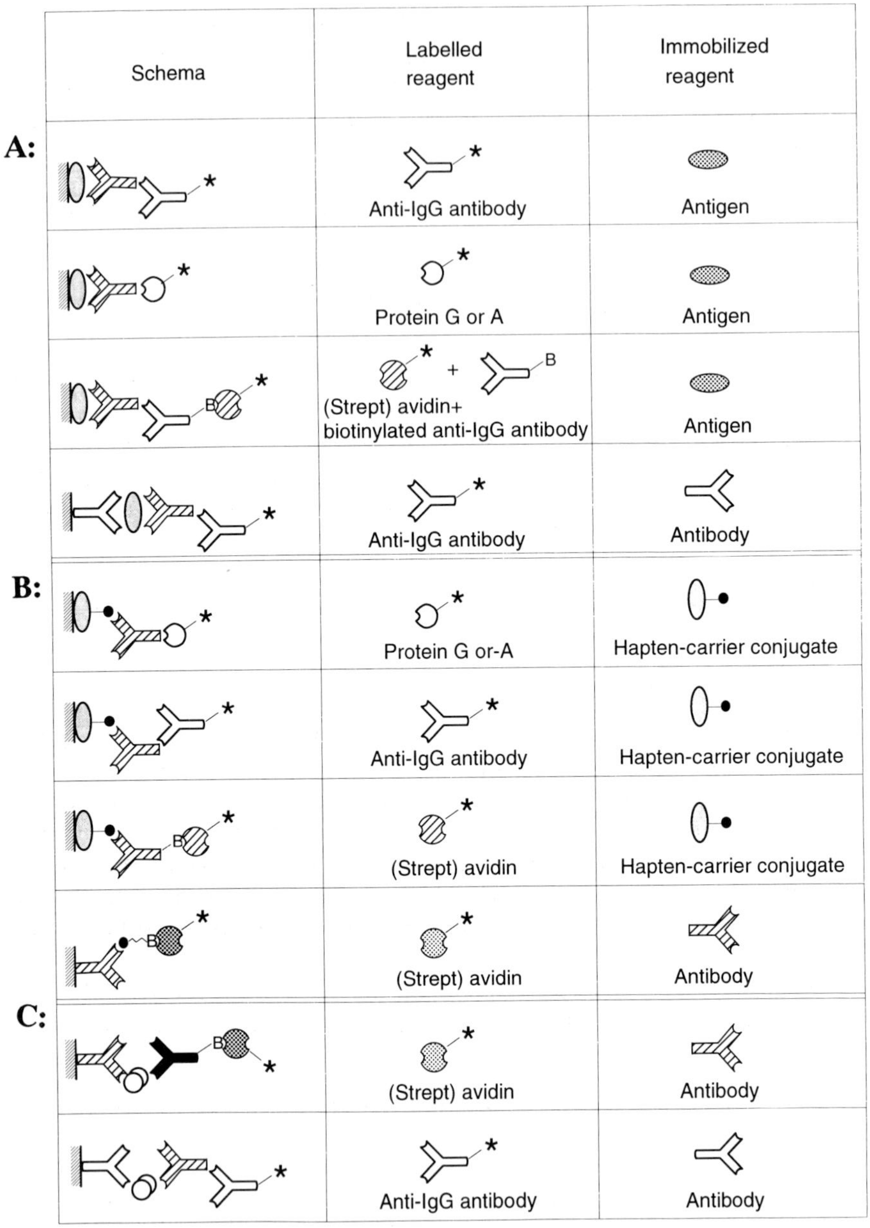

Figure 6. Use of labelled generic reagents in the construction of an assay for (A) circulating antibodies, (B) a competitive assay of haptenic antigens, and (C) a non-competitive two-site assay.

competitive assays can be approached in a number of ways. A general rule is to avoid using either of the critical components (primary antibody or hapten) on the solid phase, but try to provide excess binding capacity on the surface with secondary antibodies or other binding reagents. Some examples of the experimental set-up used in competitive assays are described by Barnard (10). *Figure 5* shows some approaches which have been tested when optimizing a competitive specific binding assay for vitamin B_{12}, which is technically demanding because of its low serum concentration.

3.7 Receptor binding assay with DELFIA technology

Receptor binding assays are traditionally performed with impure tissue homogenates or membrane fragments. Different separation systems, such as precipitation, gradient centrifugation, or harvesting on a suitable filter matrix can be used. Any receptor assay which can be performed on the surface of a microtitration well (e.g. through immobilization of soluble receptor protein with coated antibodies, using recombinant receptors having a suitable attachment site or membrane receptors on adherent cells) can be directly transferred into DELFIA technology. To obtain labelled ligands (peptides or small molecular weight ligands) it is advisable to contact Wallac labelling service.

3.8 DELFIA cytotoxic release assays

In cytotoxicity assays, target cell killing either by natural killer cells (NK), antibody-mediated cytotoxic cells (ADCC), or lymphokine activated (LAK) killers is evaluated by measuring the release of cytoplasmic markers. Traditionally the label used is radioactive chromium but because of its instability and toxicity, there is a desire to replace it with non-toxic labels. Cytotoxicity assays can also be monitored by DELFIA technology using target cells labelled with a stable and water soluble chelate, which does not penetrate the intact cell membrane of vital cells (18).

Protocol 6 describes the labelling of a target cell line with Eu–DTPA and *Protocol 7* the use of Eu^{3+}-labelled target cells in a cytotoxicity assay.

Protocol 6. Labelling of cells with Eu–DTPA

Equipment and reagents

- Hepes buffer: 50 mM Hepes, 93 mM NaCl, 5 mM KCl, and 2 mM $MgCl_2$ pH 7.4
- Repairing buffer: take 294 mg of $CaCl_2$ and 1.8 g of glucose and dissolve in one litre of Hepes buffer (2 mM Ca, 10 mM glucose)
- 0.5 M DTPA stock solution: dissolve DTPA (3.93 g) in Hepes buffer (dissolving the penta-acid can be speeded up by adding two equivalents of Na^+, e.g. as $NaHCO_3$)
- Dextran sulfate stock solution: dissolve 50 mg dextran sulfate in 10 ml of Hepes buffer
- 1 mM Eu^{3+} stock solution: mix 1.52 ml of Eu standard solution (1 mg/ml) into 8.0 ml of Hepes buffer and add 0.5 ml of DTPA solution
- Labelling buffer (optimized for K562 cell line): mix together 20 μl of Eu^{3+} stock solution and 100 μl of dextran sulfate with 800 μl of Hepes buffer

Protocol 6. *Continued*

Method

1. Wash cells once with Hepes buffer.

2. Suspend 5–10 $\times$ 10^6 cells in 1 ml of labelling buffer.

3. Incubate cells at appropriate temperature (4°C–37°C) for 15–20 min, keeping the cells in suspension by occasional gentle shaking.

4. Suspend cells with 1–2 ml of repairing buffer, fill the tube with repairing buffer, and spin down (500 *g*, 5 min).

5. Wash cells once with repairing buffer and four times with medium (e.g. RPMI 1640 containing 10% fetal calf serum).

6. Suspend cells in a small volume of medium. Transfer the pellet into a clean tube and dilute to appropriate concentration.

7. If not used immediately, cells may be stored frozen or lyophilized.

Protocol 7. Eu release test

Equipment and reagents

- DELFIA system (Wallac)
- Centrifuge
- Sterile round-bottomed microtitration plate

Method

1. Take 0.1 ml of labelled target cells, containing about 5 $\times$ 10^3 cells.

2. Add 0.1 ml of NK cells (effector cells) at various concentrations to achieve effector-to-target (E:T) ratios from 1:1 to 1:100.

3. Incubate in round-bottom microtitration plates for 2 h in a humidified 5% CO_2 atmosphere at 37°C.

4. Centrifuge cells for 5 min at 500 *g*.

5. Take 20 μl aliquot of supernatant and dilute with 200 μl of DELFIA enhancement solution.

6. Incubate for 5 min at room temperature on a shaker to allow fluorescence to stabilize.

7. Measure fluorescence in time-resolved fluorometer and calculate specific release by comparing it to spontaneous release.

New chelates (stable fluorescent chelates) permit alternative assay approaches. The label technology based on dissociative enhancement is still the most sensitive, since it permits cells to be labelled with a high amount of Eu^{3+} thereby allowing a single cell to be measured. Furthermore three labels can be incorporated simultaneously (19). The high hydrophilicity of Eu–DTPA (a feature which results in low spontaneous leakage and rapid release) requires controlled opening of the cell membrane and this is usually performed either by dextran sulfate treatment or by electroporation (20). A similar release assay can also be demonstrated with stable fluorescent Eu^{3+} and Tb^{3+} chelates (21), which do not need a separate enhancement step, but cell membrane permeabilization is still needed. A fluorogenic chelating ligand composed of esterified dicarboxylic acid derivative will permit a further simplified labelling procedure. The esterified ligand will penetrate cell membranes and stay in vital cells after esterase hydrolysis, the fluorescence being developed extracelluarly by incubating with free Eu^{3+}.

4. Time-resolved fluorometric assays with stable fluorescent chelates

An increasing number of fluorescent chelates suitable for labels have been developed and described in the literature, including fluorogenic ligands (CyberFluor system), cryptates applied in homogeneous energy transfer-based systems (4), and stable chelates used in immunoassays (22), sensors, cellular assays (23), imaging (24), and in homogeneous assays (25). Even though compromises have to be made in respect of fluorescence quantum yield and hydrophilicity properties when trying to combine both high molar absorptivity, efficient protection, and energy transfer to a single chelating compound, considerable progress has occurred, and the new generation of chelate labels clearly opens up new applications for time-resolved fluorometry.

One of the opportunities provided by stable fluorescent chelate labels is the application of time-resolved fluorometry in homogeneous assays. Because the sample constituents are present during the fluorometric measurement, the background rejection by temporal resolution is particularly important. One approach is to develop an environmentally sensitive chelate (25) and use it in direct homogeneous assays (26). A homogeneous assay for estrone-3-glucuronide has been described (10). Another approach developed by Oris Industries (4) is based on two different labels—Eu–cryptate is used as the energy donating group and an allophycocyanin used as acceptor. The measurement of energy transfer by calculating the ratio of emission is used to avoid possible excitation attenuation under UV caused by the sample.

Another important future application is in the field of microimaging of cells or tissues stained with chelate-labelled antibodies (QIF) or with DNA (FISH). The use of chelates and time resolution can totally avoid the interference of tissue autofluorescence, make data more easily accessible by image processing programs, give quantitative responses with a wide dynamic range, and increase the number of measurable parameters. *Protocol 8* gives an example of the use of chelate-labelled streptavidin in the determination of cell surface antigens with a time-resolved fluorescence microscope applied to a cancer cell line and its surface antigen, CA 242.

Protocol 8. Quantitation of cell surface antigen with lanthanide chelate-labelled antibodies

Equipment and reagents

- Cell culture: in this example, a colon cancer cell line (COLO 205) was used because of its high expression of surface antigen (CA242)—alternatively mononuclear cells from whole blood can be analysed for leucocyte surface markers, e.g. CD4, CD8, etc.
- Streptavidin labelled with a fluorescent Eu chelate (W8044), obtained from Wallac customer labelling service
- Centrifuge
- Mounting medium, e.g. Merckoglas, from Merck
- Fluorescence microscope or microfluorometer, with time-resolving capability—detailed descriptions have been provided by Seveus *et al.* (24) and by Tanke *et al.* (27) using Leica microscopes or by Herman *et al.* using laser excitation (28)

Method

1. Suspend cells in warm PBS and transfer to 1 ml Eppendorf tubes.
2. Add specific antibody (CA 242 monoclonal antibody), 1 μg/ml, 500 μl in PBS, and incubate with occasional shaking at room temperature for 1 h.
3. Wash cells three times with PBS.
4. Incubate cell suspension with 500 μl (10 μg/ml) of biotinylated rabbit-anti-mouse IgG (Sigma) in PBS for 30 min.
5. Wash cells three times with PBS.
6. Incubate with 500 μl of a solution containing 5 μg/ml of a fluorescent Eu–chelate-labelled streptavidin.
7. Wash cells three times.
8. Spin cells onto glass slide with a cytocentrifuge.
9. Rinse with distilled water and mount with aqueous mounting medium (e.g. PBS–glycerol–gelatin). Higher signal level and better stability will be obtained when mounted into non-aqueous media; dehydrate slide with ethanol series (50, 75, and 99%) and xylene and mount with Merckoglas.
10. Cover slides with a cover glass and analyse cells for fluorescence using fluorescence microscope or a microfluorometer.

---

5. Conclusion

When suitable lanthanide chelates (e.g. an Eu–chelate activated with a protein reactive isothiocyanato group) are available, in-house time-resolved fluorometric research applications become accessible. The DELFIA system, which so far is the most widely used, can be applied in a number of specific binding reactions, not only in clinical immunoassays in the form of diagnostic kits, but also in various forms of research immunoassays, high throughput DNA hybridization assays with chelate-labelled DNA probes, cytotoxicity assays with labelled target cells, enzyme activities with labelled substrates, receptor binding assays with labelled ligands, and so on. The introduction of labelling reagents of different stabilities chelated with different ions makes this wide range of uses possible.

In addition to DELFIA, other chelate technologies are also being developed. Once commercialized for research applications, these will open up new opportunities—such as homogeneous approaches for immunoassays, hybridization assays, receptor–ligand interactions, and so on. The advent of a variety of highly fluorescent stable chelates would help in increasing the number of parameters in *in situ* technologies with two to three colours in the millisecond time domain (545 with Tb, 613 with Eu, 665 with Pt or Pd coproporphyrins) in addition to colours available in the nanosecond and microsecond time domains. In microimaging, the time-resolved principle made possible by long lifetime probes, would permit the construction of more specific, more sensitive, and quantitative analysis.

References

1. Haugland, R. P. (1996). *Handbook of fluorescent probes and research chemicals.* Molecular Probes, OR.
2. Hemmilä, I. A. (1990). *Applications of fluorescence in immunoassays.* Wiley Interscience, NY.
3. Hemmilä, I., Ståhlberg, T., and Mottram, P. (ed.) (1994). *Bioanalytical applications of labelling technologies.* Wallac Oy, Finland.
4. Mathis, G. (1993). *Clin. Chem.,* **39,** 1953.
5. Iitiä, A., Kuusisto, I., Pettersson, K., and Lövgren, T. (1995). *Clin. Chem.,* **41,** S81.
6. Seveus, L., Väisälä, M., Hemmilä, I., Kojola, H., Roomans, G. M., and Soini, E. (1994). *Microscopy Res. Tech.,* **28,** 149.
7. Hemmilä, I. (1995). *J. Alloys Compounds,* **225,** 480.
8. Kankare, J., Haapakka, K., Kulmala, S., Näntö, V., Eskola, J., and Takalo, H. (1992). *Anal. Chim. Acta,* **266,** 205.
9. Hemmilä, I., Dakubu, S., Mukkala, V.-M., Siitari, H., and Lövgren, T. (1984). *Anal. Biochem.,* **137,** 335.
10. Barnard, G. (1996) In *Immunoassays in clinical biochemistry: a practical approach* (ed. I. Gow). IRL Press, Oxford University Press.
11. Mukkala, V.-M., Mikola, H., and Hemmilä, I. (1989). *Anal. Biochem.,* **176,** 319.
12. Mukkala, V.-M., Takalo, H., Liitti, P., and Hemmilä, I. (1995). *J. Alloys Compounds,* **225,** 507.
13. Brechbiel, M. W., Gansow, O. A., Atcher, R. W., Schlom, J., Esteban, J., Simpson, D. E., *et al.* (1986). *Inorg. Chem.,* **25,** 2772.
14. Hemmilä, I. (1985). *Anal. Chem.,* **57,** 1676.

15. Hemmilä, I., Mukkala, V.-M., Latva, M., and Kiilholma, P. (1993). *J. Biochem. Biophys. Methods*, **26**, 283.
16. Xu, Y.-Y., Pettersson, K., Blomberg, K., Hemmilä, I., Mikola, H., and Lövgren, T. (1992). *Clin. Chem.*, **38**, 2038.
17. Lövgren, T., Leivo, P., Siitari, H., and Pettersson, K. (1991). In *Rapid methods and automation in microbiology and immunology* (ed. A. Vaheri, R. C. Tilton, and A. Balows), p. 84. Springer–Verlag, Berlin, Germany.
18. Blomberg, K., Granberg, C., Hemmilä, I., and Lövgren, T. (1986). *J. Immunol. Methods*, **86**, 225.
19. Blomberg, K. (1993). *J. Immunol. Methods*, **168**, 267.
20. Bohlen, M., Manzke, O., Engert, A., Hertel, M., Hippler-Altenburg, R., Diehl, V., *et al.* (1994). *J. Immunol. Methods*, **173**, 55.
21. Lövgren, J. and Blomberg, K. (1994). *J. Immunol. Methods*, **173**, 119.
22. Wortberg, M., Middendorf, C., Katerkamp, A., Rump, T., Krause, J., and Cammann, K. (1994). *Anal. Chim. Acta*, **289**, 177.
23. Gu, Y. J., van Oeveren, W., Boonstra, W. P., de Haan, J., and Wildevuur, C. R. H. (1992). *Ann. Thorac. Surg.*, **53**, 839.
24. Seveus, L., Väisälä, M., Syrjänen, S., Sandberg, M., Kuusisto, A., Harju, R., *et al.* (1992). *Cytometry*, **13**, 329.
25. Mikola, H., Takalo, H., and Hemmilä, I. (1995). *Bioconjugate Chem.*, **6**, 235.
26. Hemmilä, I., Malminen, O., Mikola, H., and Lövgren, T. (1988). *Clin. Chem.*, **34**, 2320.
27. Verwoerd, N. P., Hennink, E. J., Bonnet, J., Van der Geest, C. R. G., and Tanke, H. (1994). *Cytometry*, **16**, 113.
28. Periasamy, A., Siadat-Pajouh, M., Wodnicki, P., Wang, F. W., and Herman, B. (1995). *Microsc. Anal.*, **March**, 33.

11

Enzyme immunoassays for detection of cell surface molecules, single cell products, and proliferation

N. W. PEARCE and J. D. SEDGWICK

1. Introduction

For more than 25 years, immunoenzymatic techniques have been applied extensively to the localization of antigens and antibodies in tissue sections as well as to the quantitation of immunoglobulins (Ig) and specific antibodies present in body fluids or in cell culture supernatants. The enzyme-linked immunosorbent assay (ELISA) is a sensitive, rapid, and versatile assay system for the quantitation of antigens and antibodies and is now used extensively for the measurement of soluble molecules. However aside from an early report by Avrameas and Guilbert in 1971 (1) on the application of immuno-enzymatic technology to the analysis of cell surface molecules on isolated lymphoid cells, it is only in the past 15 years that the principles have routinely been applied to the analysis of the varied immunological and biochemical properties of cells in suspension. It is these latter procedures that will be the subject of this chapter.

Relevant *in vitro* cellular immunoenzymatic techniques can be divided into three broad groups. The first, as mentioned above, is concerned with the analysis of cell surface molecules and is often referred to as cell–ELISA or CELISA. There can be little doubt that the major impetus behind the more recent development of these procedures stemmed from the need for a rapid technique to analyse monoclonal antibody (mAb) specificities. In many laboratories, cell–ELISA is used as a rapid screening test prior to further analysis by more discriminatory methods such as flow cytometry. The second group of techniques is concerned with analysis of secreted products of cells at the single cell level. The method termed enzyme-linked immunospot (ELISPOT) is a modification of the classical ELISA that allows localized antigen–antibody reactions to be visualized in the immediate vicinity of individual

ISC. The ELISPOT has also been termed ELISA–plaque assay, ELISA–spot, or spot–ELISA. The ELISPOT method is designed for the enumeration of individual cells secreting a given metabolite. This methodology was originally developed as an alternative to Jerne's haemolytic plaque-forming cell (PFC) assay (2) for enumeration of antibody-secreting cells. The detection and enumeration of cells secreting immunoglobulin or cytokines has now been reported. There are a variety of these assays in current use, all of which are based on similar principles. A related technique is called cell-blotting. The third and most recently described cellular immunoenzymatic technique is used for the analysis of cellular proliferation *in vitro*, and is therefore a potential replacement for standard tritiated [³H]thymidine incorporation assays.

2. Cell–ELISA

2.1 Background to the technique

Traditionally, the detection and quantitation of cell surface molecules has been achieved employing radiolabelled materials (such as antibody or lectin) which bind to the molecule of interest. Such techniques are generally quite sensitive and quantitative but there are numerous problems associated with these techniques. They include: the cost and handling of radioactive materials, the short shelf-life of the reagents, and the difficulty in performing a large number of analyses. The advent of mAb technology produced a need for alternative techniques which could screen the hundreds of clones produced by a single fusion procedure. Enzyme immunoassay technology is ideally suited to this task, both for the detection of mAb against soluble antigens adsorbed to the surface of plastic 96-well plates by conventional ELISA and for the screening of mAb specific for cell antigens by cell–ELISA.

2.2 Protocol for cell–ELISA

The main features of the cell–ELISA assay detailed in *Protocol 2* are illustrated in *Figure 1*. Viable cells expressing surface antigens are added to the microwells (*Figure 1A*). Test antibody, usually monoclonal, is added (*Figure 1B*). Antibody binds to specific cell surface antigens. Unbound antibody is removed. Enzyme-linked secondary antibody is added and binds to cell-bound test antibody (*Figure 1C*). Following addition of substrate, soluble colour formation reveals the presence of the specific cell-bound antibody (*Figure 1D*).

Two important considerations when setting-up a cell–ELISA are:

- using fixed or viable cells
- the type of enzyme-linked antibody employed

The cells used in the cell–ELISA can be fresh, frozen in liquid nitrogen and thawed for the assay, fixed in 1% paraformaldehyde, glutaraldehyde,

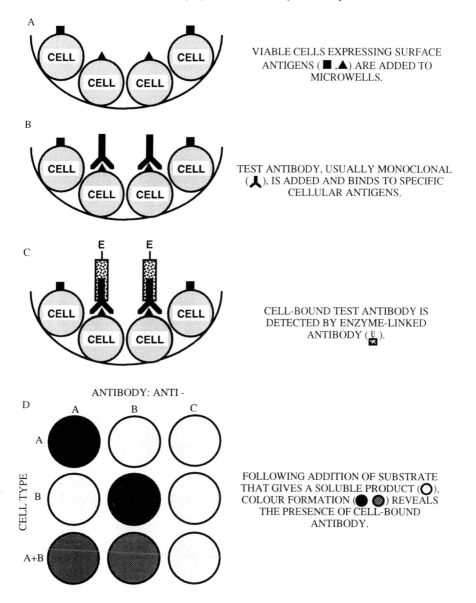

Figure 1. Detection and quantitation of cell surface molecules by cell–ELISA. From ref. 3 with permission.

formalin, or ethanol, and attached to plates by desiccation or poly-L-lysine. Cells fixed in ethanol, glutaraldehyde, formalin, or by desiccation can be stored for future use whereas cells fixed in 1% paraformaldehyde and stored at 4°C for one week do not yield results above background. An important limitation when using glutaraldehyde, ethanol, and formalin fixed cells is that some cell surface antigens are destroyed by fixatives and non-specific antibody binding can increase, leading to false positives (4,5). The sensitivity of the cell–ELISA assay is greater when fresh rather than fixed cells are used (4). A higher number of clones secreting mAbs to the desired cell surface antigen may be detected when viable cells are used to screen hybridoma clones (6). To avoid any potential problems due to antigen alteration it is best to use either fresh or frozen cells when the antigen of interest is cell surface expressed. When the antigen is expressed in the cytoplasm, it is usually necessary to fix and permeabilize the cells.

Early cell–ELISA assays used horse-radish peroxidase (HRP) detection systems. A problem encountered when using either HRP or alkaline phosphatase (AP) is the high colour formation due to endogenous cellular enzyme activity. Attempts to inhibit the endogenous enzyme activity have had limited success (7). An alternative approach is to use enzymes not found in eukaryotic cells, such as bacterial β-galactosidase (8,9) or fungal glucose oxidase (10). The use of these enzymes eliminates the potential high background due to endogenous cellular enzyme activity.

The cell–ELISA is commonly used for the initial rapid screening of new mAb clones to cell surface antigens. It is also particularly valuable for quality control of mAbs, including titration of mouse mAb against rat lymphocyte cell surface antigen using fresh rat spleen cells. In *Protocol 1* preparation of fresh rat splenocytes is described, while in *Protocol 2* these cells are used to detect mouse anti-rat lymphocyte cell surface antigen mAbs. Sheep anti-mouse Ig conjugated to bacterial β-galactosidase is used to detect the bound mouse mAb. Note that for the detection of rat mAb against mouse cell surface antigens, such as those present on lymphoid cells, the cells used in the assay would be mouse lymphocytes and sheep anti-rat Ig β-galactosidase would be used.

Protocol 1. Production of a single cell suspension from rat spleen

Equipment and reagents
- PBS/BSA/Az:[a] PBS, 1% (w/v) BSA, 10 mM NaN$_3$
- Fine aluminium sieves (Sigma CD 1)
- 1.8% NaCl

Method

1. Sacrifice a rat by CO$_2$ asphyxiation.

2. Remove the spleen.

3. Dissociate the cells by forcing the tissue through the wire mesh of the sieve into a Petri dish containing 10 ml of PBS/BSA/Az using the rubber end of a 5 or 10 ml syringe plunger.

4. Wash the wire mesh with a small volume of PBS/BSA/Az.

5. Transfer the cell suspension into a 50 ml centrifuge tube.

6. Centrifuge the tube at 400 g for 5 min.

7. Aspirate the PBS/BSA/Az.

8. Add 1 ml of distilled H_2O to lyse erythrocytes.

9. Vortex the cell pellet for approx. 6 sec.

10. While vortexing add 1 ml of 1.8% NaCl.

11. Add 40 ml of PBS/BSA/Az.

12. Filter the cells through loose cotton wool plug.

13. Centrifuge the tube at 400 g for 5 min.

14. Aspirate the PBS/BSA/Az.

15. Add 1 ml of distilled H_2O.

16. Vortex the cell pellet for approx. 3 sec.

17. While vortexing add 1 ml of 1.8% NaCl.

18. Add 40 ml of PBS/BSA/Az.

19. Filter the cells through loose cotton wool plug.

20. Centrifuge the tube at 400 g for 5 min.

21. Aspirate the PBS/BSA/Az.

22. Resuspend the cells in 10 ml of PBS/BSA/Az.

23. Count the cells and adjust the volume to give the desired concentration (usually 9×10^6 cells/ml).

[a] Can be stored for at least six months at 4°C.

Protocol 2. Cell–ELISA for titrating mouse mAb against rat lymphocyte cell surface antigens

Equipment and reagents

- Flexible acrylic 96-well U-bottom plates (ICN, 77–176–05), chosen for lowest level of non-specific antibody binding
- Biotinylated rabbit F(ab')$_2$ anti-mouse IgG (rat absorbed) (Serotec STAR44)
- Streptavidin–β-galactosidase (Boehringer Mannheim 1 112 481)
- Ortho nitrophenyl-β-D-galactopyranoside (ONPG) substrate (Boehringer Mannheim 810 088)[a]
- PBS/BSA/Az[b] (see *Protocol 1*)

- Wash buffer:[b] PBS, 0.2% BSA (w/v), 10 mM NaN$_3$
- Rabbit anti-mouse IgG diluent:[b] 45 ml PBS/BSA/Az, 3.5 µl 2-mercaptoethanol, 2.5 ml normal rat serum, adjust volume to 50 ml with PBS/BSA/Az
- ONPG substrate diluent:[b] 45 ml PBS, 0.1 g MgCl$_2$.6H$_2$O, 0.5 ml 1 M NaN$_3$, 3.5 µl 2-mercaptoethanol pH 7.5, adjust volume to 50 ml with PBS
- Mouse monoclonal antibodies (mAb)

219

Protocol 2. *Continued*

- 96-well plate reader (ICN Flow, MCC/340 MKII)
- Rat spleen cells (see *Protocol 1*)
- 12 channel multichannel pipette (Sigma, P 3549)

- Microplate carriers (Heraeus 8082) for Heraeus bench-top Megafuge 2.0R centrifuge or microplate carriers (Beckman 362394) for Beckman bench-top GPR centrifuge

Method

1. Block the plate by adding 100 μl PBS/BSA/Az to the inner 70 wells.[c]

2. Incubate at 4°C for 1–2 h.

3. Aspirate the PBS/BSA/Az from the plate.[d]

4. Add 9×10^5 cells in 100 μl of to each well (cells at 9×10^6/ml—see *Protocol 1*).

5. Spin the plate for 4 min at 200 *g*.

6. Using a 10–200 μl plastic disposable pipette tip attached to a suction line to give maximal control of suction aspirate the media from the plate.

7. Add 50 μl of serially diluted mouse mAb[e] specific for cell surface antigens to each well.

8. Resuspend cells carefully but thoroughly using a multichannel pipette.

9. Incubate the plate at 4°C for 1 h by placing on crushed ice.[f]

10. Add 200 μl of cold wash buffer.

11. Centrifuge the plate at 4°C for 5 min at 200 *g*.

12. Aspirate the media from the plate.

13. Repeat steps 10–12 twice.

14. Resuspend cells in 50 μl of biotinylated rabbit F(ab')₂ anti-mouse IgG (diluted 20 μg/ml in rabbit anti-mouse IgG diluent).

15. Incubate the plate on ice for 1 h.[f]

16. Repeat steps 10–13.

17. Add 50 μl of streptavidin–β-galactosidase diluted between 1/1000 and 1/5000 in PBS/BSA/Az.

18. Incubate the plate at 4°C for 1 h by placing on crushed ice.[f]

19. Repeat steps 10–13.

20. Add 50 μl of ONPG substrate.[g]

21. Incubate the plate at room temperature for up to 4 h.[h]

22. Add 20 µl of 1 M Na_2CO_3. (This stops the reaction and enhances the colour.)

23. Measure absorbance at 405 nm on 96-well plate reader.

[a] ONPG substrate comes as a white lyophilized powder. Make stock solution at 30 mM in PBS (0.1 g in 11 ml PBS). Dissolve the powder by warming the solution to 40–50 °C. Store 1 ml aliquots at –20 °C.

[b] Can be stored for at least six months at 4 °C.

[c] Do not use the outer 36 wells of the plate to avoid edge effects.

[d] Can be reused two or three times.

[e] mAbs are pre-diluted in PBS/BSA/Az in a second 96-well plate (any type of plate can be used) in an identical format to that for the cell–ELISA, allowing sufficient volume to enable transfer of 50 µl to cell-containing wells. Tissue culture supernatants are initially diluted 1/2 whereas mAbs in ascites are initially diluted 1/100. Each different mAb sample is run in duplicate and is serially diluted five times 1/5 down the plate—that is through rows B (1/2 or 1/100) to G (1/6.25 $\times$ 10^3 or 1/3.125 $\times$ 10^5). It is usually a good idea to also have a negative control mAb (such as mouse mAb against a human antigen) to establish background absorbance levels. Additionally, titration of a known concentration of mAb would enable the construction of a standard curve for the quantitation of the concentration of mAb in tissue culture supernatant or ascites.

[f] Avoid crushing the soft U-bottom well of the plate when pushing the plate into the ice.

[g] Prepare just before use by thawing a 1 ml aliquot of 30 mM ONPG substrate and adding 9 ml ONPG diluent.

[h] Monitor every hour, stop the reaction when sufficient development has occurred (for example maximal absorbance reading approx. 1–1.5).

2.3 Advantages and disadvantages

Detection and quantitation of cellular antigens with radiolabelled reagents is not as versatile as enzyme-based techniques and fails to offer significant advantages in terms of sensitivity. When screening for monoclonal antibodies the cell–ELISA performed in 96-well plates permits the analysis of several hundred samples in about six hours. In contrast, flow cytometric analysis of the same number of samples is very time-consuming and does not provide the same quantitative data in terms of mAb titration. Flow cytometry is more suited to qualitative and quantitative analysis of a limited number of samples. Sensitivity of the cell–ELISA is similar to immunofluorescence staining and subsequent flow cytometric analysis (4), but flow cytometry is superior when the percentage of cells in the test population is small or the antibody has a low affinity. Flow cytometry is the most discriminating method for the detection of cellular antigens in which the parameters such as cell size and heterogeneity in addition to antigen expression can be easily assessed. The cell–ELISA assay requires no specialized equipment to screen hybridoma clones apart from an ELISA plate reader. Thus, laboratories with no specialized flow cytometry equipment or equipment for measuring radioactivity can produce and screen mAbs readily.

2.4 Applications of the technique

The cell–ELISA is predominantly used to screen hybridoma clones where the mAb of interest binds to a cell surface antigen. The assay is sensitive

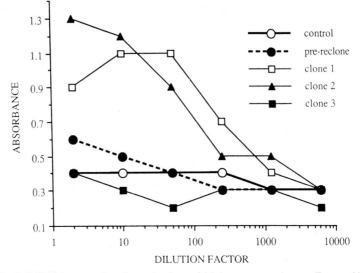

Figure 2. Cell–ELISA screening for selection of high secretor clones. Test mAb was a mouse anti-rat CD4 mAb (MRC OX35). Control mAb was a mouse anti-human mAb that did not bind to rat cells.

enough to screen for very rare clones (11). The assay is also useful for testing mAb concentration (*Protocol 2*), screening positive clones for high mAb secretors (*Figure 2*), and determining high affinity mAbs (12). The results in *Figure 2* show that test clones 1 and 2 are secreting high levels of mAb whereas clone 3 is either a non-secretor or is only secreting non-specific Ig. We have also used the technique to investigate the effect of radiolabelling a mAb on its capacity to bind antigen.

The method has been adapted to screen for mAbs against intracellular antigens (11). In this study cells were fixed with 2% paraformaldehyde and permeabilized using digitonin to enable binding of mAb. The mild fixation retained excellent morphology of the cells.

The species origin and cell types which have been used in the cell–ELISA is extensive as are the potential applications and these have been reviewed elsewhere (3).

3. ELISPOT

3.1 Background to the technique

For 30 years the haemolytic PFC assay (13) has been used widely to study immunoglobulin secretion and to analyse the regulatory mechanisms that control the expansion of immunoglobulin-secreting cells (ISC). The technique has been a fundamental tool in the clonal analysis of antibody formation. The fact that the assay is still used today attests to its versatility. However, the indicator system used, the localized complement-mediated lysis of erythrocytes, is not always easy to work with and often it is difficult to couple the required antigens to erythrocytes (14). The reliability of the PFC assay has also been questioned. Furthermore, the method does not permit quantification of antibody molecules secreted. In 1983 two groups described an assay for the detection and enumeration of antigen-specific ISC based on solid phase immunoassay technology (15,16).

The ELISPOT technique is in principle applicable to the detection of cells secreting antibodies against any antigen, soluble or particulate, that can be adsorbed either directly or indirectly onto a solid surface. A modification of the ELISPOT technique (17,18), the reverse ELISPOT, using surfaces coated with antibodies can be used to detect virtually any cell-secreted product provided that an antibody is available and the ligand is secreted in sufficient quantities to be detected.

The original description of the assay by Sedgwick and Holt and Czerkinsky used plastic plates with agarose in the wells to aid the localization of a single reaction product. The standard and reverse assay have changed little from their inception in 1983. Two modifications of note are the use of nitrocellulose as the solid phase support (19) and the ability to simultaneously detect two cell metabolites such as distinct isotypes of antibodies (20) or cytokines (21).

The ELISPOT contrasts to the cell–ELISA and cell proliferation (see below) assays in two important ways:

- the cells are removed from the wells
- the substrate forms an insoluble product

3.2 Protocol for ELISPOT

The main features of the standard technique for the detection of ISC and the reverse ELISPOT for the detection of metabolite-secreting cells are illustrated in *Figure 3*. The cells of interest are added to 24- or 96-well plates which have been pre-coated with either antigen or antibody. The cells secrete Ig, cytokine, or some other cell metabolite which binds locally to the antigen or is captured by the antibody (*Figure 3A*). The cells are removed from the plate and an enzyme-conjugated antibody specific for the secreted product is added (*Figure 3B*). The site of the secreted product is revealed by the addition of enzyme substrate which produces an insoluble coloured product (*Figure 3C*). The spots produced by the assay represent the secreted product of a single cell which is counted either by eye or more often under low magnification (*Figure 3D*).

Both plastic and nitrocellulose solid phase carrier systems have been used. Plastic support systems although cheaper than nitrocellulose, require higher concentrations of antigen coating solution. The lower antigen concentration required when using nitrocellulose results from its higher protein binding

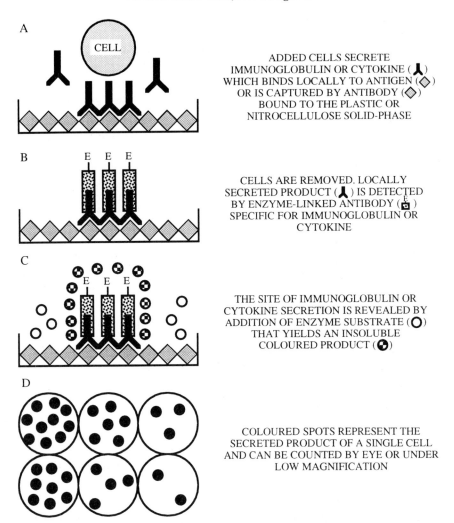

Figure 3. Detection and enumeration of secreting cells by ELISPOT. From ref. 3 with permission.

capacity compared to plastic (approx. 80 $\mu g/cm^2$ versus 10 $\mu g/cm^2$) (19). The concentration for most coating antigens will range between 1–20 $\mu g/ml$ when using nitrocellulose plates, whereas the concentration required for effective coating of plastic plates is much higher. For example 10 $\mu g/ml$ keyhole limpet haemocyanin (KLH) is sufficient to coat a nitrocellulose plate whereas a concentration of at least 200 $\mu g/ml$ is required to coat plastic plates. When using plastic surfaces it is advisable to incorporate the substrate in agarose to aid in product location as definition of spots is poor in the absence of agarose. The use of agarose is not necessary when using a nitrocellulose solid phase.

Nitrocellulose membranes are recommended as the solid phase carrier system.

Here we describe the ELISPOT assay for detection of anti-KLH ISC (*Protocol 3*) and the reverse ELISPOT for enumeration of IgM[a]-secreting cells (*Protocol 4*). Two different substrate systems are used.

Protocol 3. ELISPOT for the detection and enumeration of mouse spleen cells secreting anti-KLH antibodies

Equipment and reagents

- Stereoscopic microscope (Nikon SMZ10)
- 96-well flat-bottom nitrocellulose plates (Millipore, N45 50)
- Tissue culture media (TCM): RPMI 1640 supplemented with 2 mM L-glutamine, 10% (v/v) fetal calf serum, 100 U/ml penicillin, and 100 μg/ml streptomycin
- PBS/Tween: PBS, 0.05% (v/v) Tween 20
- PBS/Tween/BSA: PBS, 0.05% (v/v) Tween 20, 1% (w/v) BSA
- KLH (Calbiochem-Novabiochem 374805)

- Rabbit anti-mouse Ig–HRP conjugate (Dako P0260)
- 3-Amino-9-ethylcarbazole (AEC)—HRP substrate (Sigma A6926)
- Citrate–phosphate buffer: 0.1 M $C_6H_8O_7$, 0.2 M Na_2HPO_4 pH 5
- Substrate solution: dissolve 30 mg of AEC in 500 μl of dimethylformamide (DMF), add 90 ml of citrate–phosphate solution, filter through a 0.45 μm filter, add 15 μl of 30% H_2O_2—use immediately

Method

1. Add 100 μl of KLH[a] (10 μg/ml in PBS) to required wells of a 96 nitro-cellulose-based well plate. Add 100 μl of PBS to control wells.

2. Incubate the plate at 4°C overnight.[b]

3. Add 200 μl of PBS to every well.

4. Aspirate the PBS from the plate.

5. Repeat steps 3–4 twice.

6. Immerse the plate in a container of PBS at room temperature for 5 min.

7. Aspirate the PBS from the plate.

8. Blot dry the nitrocellulose membrane by placing on absorbent paper towel.

9. Add 50 μl of TCM to each well

10. Incubate the plate at 37°C for 1 h.

11. Add 50 μl of mouse spleen cells[c] to wells in at least three different concentrations (2×10^5 cells/well, 5×10^4 cells/well, 1.25×10^4 cells/ml).[d] Test each concentration in duplicate or triplicate.

12. Incubate the plate at 37°C in a 5% CO_2 incubator for 4 h.

13. Repeat steps 3–5.

14. Add 200 μl of PBS/Tween, leave 1 min.

15. Aspirate the PBS/Tween from the plate.

Protocol 3. *Continued*

16. Repeat steps 14–15 three times.

17. Immerse the plate in a container of PBS/Tween at room temperature for 5 min.

18. Aspirate the PBS/Tween from the plate.

19. Blot dry the nitrocellulose membrane by placing on absorbent paper towel.

20. Add 50 μl of HRP-conjugated rabbit anti-mouse Ig (specificity dependent on the isotype being examined) diluted 1/100–300 in PBS/Tween/BSA to each well.

21. Incubate the plate at 4°C overnight.[b]

22. Repeat steps 14–16 and then steps 3–5.

23. Immerse the plate in a container of PBS at room temperature for 5 min.

24. Aspirate the PBS from the plate.

25. Blot dry the nitrocellulose membrane by placing on absorbent paper towel.

26. Add 100 μl of AEC substrate solution.

27. Incubate the plate at room temperature for 10 min.[e]

28. Stop reaction by flooding the plate with tap-water.

29. Aspirate tap-water from the plate.

30. Blot dry the plates.

31. Count the spots using a stereoscopic microscope at low magnification (x 40 to × 200) under direct illumination.

[a] Antigen chosen will depend on cell-secreted product being detected.
[b] Incubation step can be performed at either room temperature (20°C) for 2 h, or at 4°C overnight, whichever is most convenient.
[c] Mouse spleen cells prepared according to *Protocol 1*. Spleen cells can, for example, be derived from mice immunized with two i.p. injections of 100 μg KLH emulsified in 100 μl of Freunds complete adjuvant (Difco), two weeks apart. The mice are sacrificed four days after the second injection.
[d] Cell concentration will vary depending on the cells and the product being assayed. However the concentrations given provide an appropriate starting point.
[e] After 1–3 min reddish spots recognized by the naked eye will appear. The plate is incubated for up to a further 7 min during which time development is monitored carefully. Over-incubation will lead to excessive background staining.

3.3 Protocol for reverse ELISPOT

The most common group of ligands detected in the reverse ELISPOT are cytokines, but the principles for cytokine or total Ig ISC determination are identical. *Protocol 4* describes the use of the reverse ELISPOT to detect transgene encoded IgM[a] splenic ISC in an IgM[b] host mouse.

Protocol 4. ELISPOT for the detection and enumeration of mouse spleen cells secreting IgM[a]

Equipment and reagents

- TCM (see *Protocol 3*)
- PBS/Tween (see *Protocol 3*)
- PBS/Tween/BSA (see *Protocol 3*)
- 0.025 M Tris-buffered PBS pH 8.0: 3.03 g tris(hydroxymethyl)methylamine dissolved in 1 litre of PBS
- Stereoscopic microscope (see *Protocol 3*)
- 96-well flat-bottom nitrocellulose plates (see *Protocol 3*)
- *p*-Nitrophenyl phosphate (NPP) buffer: pH 9.8, 0.03 M NaHCO$_3$, 0.015 M Na$_2$CO$_3$, 0.0001 M MgCl$_2$, 10 mM NaN$_3$

- Bromochloro indolyl phosphate (BCIP) (Bio-Rad, 170–6539)/nitroblue tetrazolium (NBT) (Bio-Rad, 170–6532) substrate solution: 98 ml NPP buffer, 15 mg BCIP dissolved in 1 ml of 100% DMF, 30 mg NBT dissolved in 1 ml of 70% DMF—filter through a 0.45 μm filter
- Affinity purified rabbit anti-mouse Ig (Dako Z0456)
- Biotinylated mouse anti-mouse IgM[a] (Pharmingen 05092D)
- Avidin–AP (Sigma A 7294)

Method

1. Add 100 μl of polyvalent[a] rabbit anti-mouse Ig (5 μg/ml in PBS) to required wells. Uncoated control wells should be set-up (see *Protocol 3*).

2. Incubate the plate at 4°C overnight.[b]

3. Add 200 μl of PBS to every well.

4. Aspirate the PBS from the plate.

5. Repeat steps 3–4 twice.

6. Immerse the plate in a container of PBS at room temperature for 5 min.

7. Aspirate the PBS from the plate.

8. Blot dry the nitrocellulose membrane by placing on absorbent paper towel.

9. Add 50 μl of TCM to each well.

10. Incubate the plate at 37°C for 1 h.

11. Add 50 μl of mouse spleen cells[c] to wells in at least three different concentrations (2 × 10^5 cells/well, 5 × 10^4 cells/well, 1.25 × 10^4 cells/well).[d] Test each concentration in duplicate or triplicate.

12. Incubate the plate at 37°C in a 5% CO$_2$ incubator for 4 h.

13. Repeat steps 3–5.

14. Add 200 μl of PBS/Tween, leave 1 min.

15. Aspirate the PBS/Tween from the plate.

16. Repeat steps 14–15 twice.

17. Immerse the plate in a container of PBS/Tween at room temperature for 5 min.

18. Aspirate the PBS/Tween from the plate.

227

Protocol 4. *Continued*

19. Blot dry the nitrocellulose membrane by placing on absorbent paper towel.

20. Add 100 μl of biotinylated anti-IgM[a] mAb (0.5 μg/ml) diluted in PBS/Tween/BSA to each well.

21. Incubate the plate at 4°C overnight.[b]

22. Repeat steps 3–5 and steps 14–19.

23. Add 100 μl avidin–AP (1 μg/ml in PBS/Tween/BSA).

24. Incubate the plate at room temperature for 1 h.

25. Repeat steps 14–16 and then steps 3–5.

26. Aspirate the PBS from the plate.

27. Immerse the plate in Tris-buffered PBS at room temperature for 2 min.

28. Blot dry the nitrocellulose membrane by placing on absorbent paper towel.

29. Add 100 μl of BCIP/NBT substrate solution.

30. Incubate the plate at room temperature for 30 min–1 h.[e]

31. Stop reaction by flooding the plate with tap-water.

32. Aspirate the tap-water from the plate.

33. Blot dry the plates.

34. Count the spots using a stereoscopic microscope at low magnification (× 40 to × 200) under direct illumination.

---

[a] Antibody chosen will depend on ligand to be detected. In the case of detection of cytokine-secreting cells, mAb are commonly used to coat the solid phase.
[b] See *Protocol 3* footnote *b*.
[c] See *Protocol 1*.
[d] See *Protocol 3* footnote *d*.
[e] Spots appear within 5–10 min but generally 30 min–1 h is required. Monitor to avoid high backgrounds.

The typical appearance of a 96-well nitrocellulose-based plate where the reverse ELISPOT technique has been employed to detect IgM[a] is shown in *Figure 4*.

In *Protocol 3*, HRP is the detection enzyme whereas in *Protocol 4* AP is used. These enzyme systems are the most widely used and there is not a great deal of difference between the two. HRP-based systems may suffer from spontaneous darkening if kept in the light too long, but this is easy to avoid. When using HRP, AEC is more sensitive than *p*-phenylenediamine (20). The incorporation of NBT into the BCIP substrate has been shown to increase the sensitivity of AP-based ELISPOT assays and is recommended (22). We

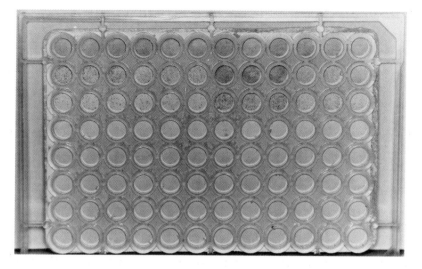

Figure 4. Appearance of spots on a 96-well nitrocellulose-based plate.

favour the use of AP with BCIP/NBT substrate system and there are more reports using the AP system than HRP, however either system is adequate. It is possible to perform the ELISPOT without using an enzyme/substrate system but instead, using immunogold/silver staining techniques (23). This method appears to be effective and sensitive and may be suited to some applications.

The sensitivity of the ELISPOT assay can be increased by a number of techniques which have been reviewed by Czerkinsky and Sedgwick (24) and include peroxidase–anti-peroxidase, avidin–biotin, and the avidin–biotin–anti-peroxidase systems. The use of avidin–biotin techniques are among the most sensitive and sensitivity enhancing procedures that can be incorporated into the ELISPOT assay. Conjugation of antibodies with biotin is simple and yields antibodies that retain virtually all their original activity (25).

3.4 Advantages and disadvantages

The sensitivity of the ELISPOT assay is much greater than the sensitivity of conventional ELISA (26) and this is illustrated clearly in *Table 1*. A further advantage of ELISPOT over the haemolytic PFC technique is its ability to measure low frequency events such as IgE ISC amongst a much greater number of cells secreting other isotypes (17). The technique can also detect ISC specific for antigens that are not readily coupled to erythrocytes, such as particulate antigens like bacteria (22) and other intact cells such as T lymphocytes (27).

In general, antigen concentrations required to achieve appropriate coating

Table 1. Superior sensitivity of the ELISPOT compared to ELISA[a]

Cells/Well	Spots/Well	Absorbance ($\times 10^3$)
0	0 ± 0[b]	62 ± 6[c]
3	2.0 ± 1.2	63 ± 8
6	4.4 ± 1.5	69 ± 8
12	7.6 ± 2.2	70 ± 9
25	15 ± 2	70 ± 7
50	32 ± 6	69 ± 4
100	61 ± 4	68 ± 8
200	119 ± 8	69 ± 10

[a] From ref. 26 with permission..
[b] Interferon-α2a specific mAb-secreting hybridoma cells were added to wells and tested directly by ELISPOT. Mean $\pm$ SD of ten replicate wells.
[c] Supernatants from ten replicate wells were pooled and evaluated for interferon-α2a antibodies by sandwich ELISA. Mean absorbance $\pm$ SD of quadruplicate determinations. Reagent blank was 0.069 ± 0.008.

in the ELISPOT assay are higher than those required for a standard ELISA, at least with plastic solid phase. This problem is particularly acute for low molecular weight proteins such as ovalbumin which require coating concentrations in excess of 100 μg/ml to enable sensitive detection of anti-ovalbumin ISC in plastic wells. Capture antibodies are effective in the reverse ELISPOT at concentrations of around 5–10 μg/ml. Two studies (28,29) employed the avidin–biotin system to circumvent the inadequate antigen binding properties of plastic. In these cases, plastic solid phase was coated with low concentrations of avidin which then captured biotinylated antigen.

A similar method termed cell blotting has been described (30) for enumeration of rat prolactin-secreting cells using computerized image analysis of the cell blots. The technique differs from the ELISPOT in that it relies on passive adsorption of the secreted product to the nitrocellulose or similar membranes rather than active capture.

3.5 Applications of the technique

The species, origin, cell types, and experimental systems used with the ELISPOT are numerous and have been listed elsewhere (3). The major application of the standard ELISPOT is the enumeration of ISC, while for the reverse ELISPOT the major application is the enumeration of cytokine-secreting cells.

4. Cell proliferation

4.1 Background to the technique

The quantitative assessment of cellular proliferation is one of the cornerstones of immunological investigation. *In vitro* cell proliferation in response to antigen is taken to reflect the responsiveness *in vivo* of lymphocytes to antigenic stimuli. The classical way to measure cell-mediated immune responses is by the incorporation of radioactively-labelled [³H]thymidine analogues into DNA during the S phase of the cell cycle. The production of mAbs to the thymidine analogue bromodeoxyuridine (BrdU) (31,32) made it possible to develop rapid quantitative assays of cell proliferation avoiding the use of radioactivity. BrdU, like [³H]thymidine, is incorporated into the DNA when cells enter the S phase of the cell cycle. The measurement of cell proliferation in a cell–ELISA using an antibody against BrdU was first demonstrated by Porstmann (33). They extracted and semi-purified DNA from BrdU pulsed proliferating cells followed by an enzyme immunoassay in a separate vessel. Although the method showed good correlation to the standard ³H assay for cell proliferation, it was laborious. The method was simplified by permeabilizing the cells at the end of the culture period allowing the anti-BrdU mAb to enter the cell and bind BrdU incorporated into the DNA (34). This allows the entire procedure to be performed in a single 96-well microtitre plate.

4.2 Protocol for cell proliferation using BrdU

The main features for determination of cell proliferation using BrdU are provided in *Figure 5*. BrdU is added to wells of microtitre plates during the final 16–24 hours of the culture during which time the BrdU is incorporated into DNA of dividing cells (*Figure 5A*). To allow the incorporated BrdU to be detected by the anti-BrdU mAb the cells must be permeabilized and importantly the DNA denatured by treatment with either formamide or acid to enable mAb binding. mAb is then added to the culture (*Figure 5B*). Bound anti-BrdU mAb is detected by enzyme-linked antibody (*Figure 5C*). The addition of substrate yields a soluble product revealing the presence of incorporated BrdU. Absorbance values indicate the relative proliferation indices (*Figure 5D*).

A method for measuring human peripheral blood mononuclear cells (PBMC) proliferation is described (*Protocol 5*). Note that it is possible to detect BrdU incorporation after pulsing cells for 10 min, however a progressively stronger signal is obtained when the incubation period is increased to 24 h (34). If the denaturation step is omitted it is not possible to detect BrdU uptake. Denaturation with 2 N HCl results in a lower background than when cells are denatured with formamide/heat treatment (35).

4.3 Advantages and disadvantages

The BrdU assay can be performed easily, reasonably rapidly, and most importantly it does not employ radioactive materials or require the sophisticated equipment necessary for radioactive work. The method allows, in addition to cell proliferation, other investigations to be performed on cells at the single

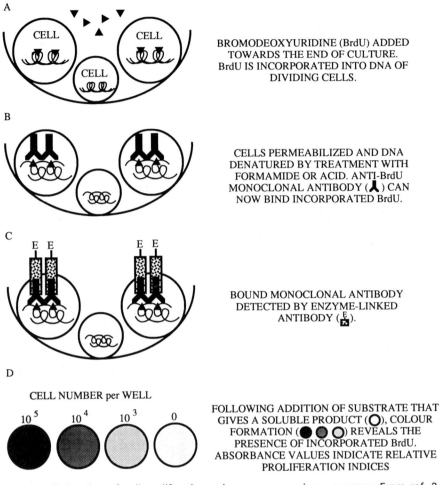

A. BROMODEOXYURIDINE (BrdU) ADDED TOWARDS THE END OF CULTURE. BrdU IS INCORPORATED INTO DNA OF DIVIDING CELLS.

B. CELLS PERMEABILIZED AND DNA DENATURED BY TREATMENT WITH FORMAMIDE OR ACID. ANTI-BrdU MONOCLONAL ANTIBODY (𝗬) CAN NOW BIND INCORPORATED BrdU.

C. BOUND MONOCLONAL ANTIBODY DETECTED BY ENZYME-LINKED ANTIBODY (E).

D. CELL NUMBER per WELL

10^5 10^4 10^3 0

FOLLOWING ADDITION OF SUBSTRATE THAT GIVES A SOLUBLE PRODUCT (○), COLOUR FORMATION (●◉○) REVEALS THE PRESENCE OF INCORPORATED BrdU. ABSORBANCE VALUES INDICATE RELATIVE PROLIFERATION INDICES

Figure 5. Estimation of cell proliferation using an enzyme immunoassay. From ref. 3 with permission

cell level. Possibilities include flow cytometry (36) and histochemical staining (33,37) to directly enumerate positive cells. Obviously in these cases the cells should be extracted from the well before ethanol fixation.

Protocol 5. Detection of human peripheral blood mononuclear cell proliferation by uptake of BrdU

Equipment and reagents

- Heparinized human blood
- 5-Bromo-2′-deoxyuridine (ICN, 100171)
- Ficoll–Paque (Pharmacia, 17–0840–02)

- 2 M HCl
- 0.1 M $Na_2B_4O_7$ (sodium borate)
- Mouse anti-BrdU mAb (Dako, M0744)

- Rabbit anti-mouse IgG (H + L) conjugated to AP (Dako D0314)
- Phytohaemagglutinin (PHA) (ICN 151884)
- PBS/BSA: PBS, 0.2% (w/v) BSA
- PBS/BSA/Az (see *Protocol 1*)
- PBS/Tween (see *Protocol 3*)
- 96-well plate washer (ICN Flow M96V)

- 96-well plate reader (see *Protocol 2*)
- NPP (Sigma 104–0)
- NPP buffer (see *Protocol 4*)
- Microplate carriers (Heraeus 8082) for Heraeus bench-top Megafuge 2.0R centrifuge or microplate carriers (Beckman 362394) for Beckman bench-top GPR centrifuge

Method

1. Add 15 ml of heparinized human blood to a 50 ml centrifuge tube.

2. Add 15 ml of PBS/BSA to the tube.

3. Underlay the blood PBS/BSA mixture with 9 ml of Ficoll–Paque.

4. Spin the tube at 900 g for 30 min with the brake off.

5. Remove the PBMC from the Ficoll–Paque plasma interface and dispense the cells into a clean 50 ml centrifuge tube.

6. Add PBS/BSA to the tube so that the volume in the tube is 50 ml.

7. Centrifuge the PBMC at 400 g for 5 min.

8. Aspirate the PBS/BSA and resuspend the cells in 50 ml of PBS.

9. Centrifuge the PBMC at 400 g for 5 min.

10. Aspirate the PBS/BSA and resuspend the cells in suitable TCM (such as RPMI 1640 supplemented with 2 mM L-glutamine, 100 U/ml penicillin, 100 μg/ml streptomycin, and 5–10% normal human AB serum) at a concentration of 4×10^6 cells/ml. Aliquot 200 μl of cells into triplicate starting wells of a 96-well flat-bottom plate and double dilute 100 μl volumes through five dilution steps to leave 100 μl of cells per well at 4×10^5 cells/well at the highest dilution.[a]

11. Add 2 μl of PHA (at 250 μg/ml) to test wells. PHA is not added to control wells.

12. Incubate the cells in a humidified 5% CO_2 incubator at 37°C for 96 h. 24 h before the end of the culture add 25 μl of BrdU (5×10^{-5} M in PBS) to each well, so that the final concentration in each well is 10^{-5} M.

13. At the end of the 96 h culture, centrifuge the plate at 400 g for 5 min.

14. Aspirate the TCM.

15. Resuspend the cells in 200 μl of PBS.

16. Centrifuge the plate at 400 g for 5 min.

17. Aspirate the PBS.

18. Repeat steps 15–17 twice.

19. Dry the plate under a hair dryer for 2 h.

20. Add 200 μl of 100% ethanol/well to fix and attach the cells to the plate.

21. Leave plate at 20°C for 10 min.

Protocol 5. *Continued*

22. Air dry the plate.

23. Add 200 μl of 2 M HCl to each well.

24. Leave plate at 20°C for 30 min.

25. Add 50 μl of 0.1 M $Na_2B_4O_7$ (sodium borate).

26. Wash the plate once with 200 μl of PBS in an automatic plate washer.[b]

27. Add 200 μl of PBS/BSA/Az to each well.

28. Incubate the plate at 37°C for 1 h.

29. Wash the plate five times with PBS/Tween in an automatic plate washer.

30. Add 50 μl of mouse anti-BrdU mAb diluted 1/100 in PBS/Tween to test wells.

31. Leave plate at 20°C for 30 min.

32. Wash the plate five times with PBS/Tween in an automatic plate washer.

33. Add 50 μl of rabbit anti-mouse IgG (H + L)–AP diluted 1/1000 in PBS/Tween, to each well.

34. Leave plate at 20°C for 30 min.

35. Wash the plate five times with PBS/Tween in an automatic plate washer.

36. Add 100 μl of AP substrate solution (1 mg/ml of NPP in NPP buffer) to each well.

37. Leave plate at 20°C for 30 min.

38. Measure absorbance at 405 nm or if necessary a visual assessment of colour can be made.

[a]There is non-linearity with BrdU methods at high proliferation values. An appropriate level on the linear part of the curve should be found in this way.
[b]Plate can be washed manually if no plate washer is available by adding 200 μl of PBS to each well, then flicking out the PBS.

The method is as sensitive as [³H]thymidine incorporation at low levels of cell proliferation (34,35) (*Figure 6*). However [³H]thymidine incorporation provides a linear measurement of cell proliferation over a broader range of cell concentrations. BrdU incorporation lacks linearity at higher cell concentrations.

The use of anti-Ki-67 mAbs (38) also can be used to measure cell proliferation. The Ki-67 protein is expressed during all active parts of the cell cycle, G_1, S, G_2, and mitosis but is absent in G_0. However the measurement of cell proliferation using these antibodies, like the incorporation of BrdU, lacks

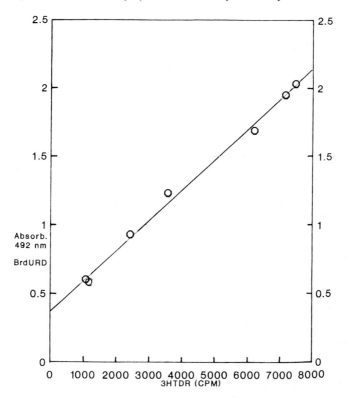

Figure 6. Comparison of cellular proliferation measured by BrdU and [³H]thymidine incorporation demonstrating good correlation at the low end of the cell proliferation curve. From ref. 35 with permission.

linearity over a broad range of cell concentrations. A disadvantage of BrdU is its toxicity mediated by the depletion of cytidine (39). However this can be prevented by the addition of 2-deoxycytidine in equimolar concentrations. Anti-BrdU, anti-Ki-67, and anti-proliferating cell nuclear antigen (PCNA) mAbs are all used extensively to measure cell proliferation in immunohisto-chemistry and also for cytometric analysis. However none of these mAbs have been used extensively to measure *in vitro* cell proliferation.

Other enzymatic assays using 3-(4,5-dimethylthiazol-2-yl)-2,5-diphenylte-trazolium (MTT) to measure dehydrogenase turnover (40,41) and p-nitro-phenol-N-acetyl-β-D-glucosaminide to measure the hexosaminidase reaction (42), detect proliferation by the increase in cell number but lack the sensitivity of either [³H]thymidine or BrdU incorporation (43). The enzyme immuno-assay is probably more convenient than [³H]thymidine incorporation and lends itself to the analysis of large numbers of samples. However, significant im-provements in equipment and procedures leave [³H]thymidine incorporation as the most widely used assay of cell proliferation due to its sensitivity, low

235

background, and most importantly its linear measurement of cell proliferation over a broad range of cell concentrations.

4.4 Applications of the technique

The BrdU incorporation technique has been used to measure proliferation of various types of cells ranging from lymphocytes (34,35), to neural cells (36), and even parasites (44). In general this particular enzyme immunoassay is rarely used, in part due to alternative techniques such as [^{3}H]thymidine incorporation being extremely effective. Probably the most useful application is for studies in the field or in laboratories lacking the specialized equipment necessary for measuring [^{3}H]thymidine incorporation.

References

1. Avrameas, S. and Guilbert, B. (1971). *Eur. J. Immunol.*, **1**, 394.
2. Jerne, N. K., Henry, C., Nordin, A. A., Fuji, H., Koros, A., and Lefkovits, I. (1974). *Transplant. Rev.*, **18**, 130.
3. Sedgwick, J. D. and Czerkinsky, C. (1992). *J. Immunol. Methods*, **150**, 159.
4. Aida, Y., Onuma, M., Kasai, N., and Izawa, H. (1987). *Am. J. Vet. Res.*, **48**, 1319.
5. Schlosser, M., Witt, S., Ziegler, B., and Ziegler, M. (1991). *J. Immunol. Methods*, **140**, 101.
6. Gaffar, S. A., Li, Z., and Epstein, A. L. (1989). *Hybridoma*, **8**, 331.
7. Lansdorp, P. M., Oosterhof, F., Astaldi, G. C., and Zeijlemaker, W. P. (1982). *Tissue Antigens*, **19**, 11.
8. Borzini, P., Tedesco, F., Greppi, N., Rebulla, P., Parravicini, A., and Sirchia, G. (1981). *J. Immunol. Methods*, **44**, 323.
9. Cobbold, S. P. and Waldmann, H. (1981). *J. Immunol. Methods*, **44**, 125.
10. Gay, H., Clark, W. R., and Docherty, J. J. (1984). *J. Histochem. Cytochem.*, **32**, 447.
11. Forster, R., Emrich, T., Voss, C., and Lipp, M. (1993). *Biochem. Biophys. Res. Commun.*, **196**, 1496.
12. Masuda, K., Nagata, S., Harada, S., Hirano, K., and Takagishi, Y. (1992). *Microbiol. Immunol.*, **36**, 873.
13. Jerne, N. K. and Nordin, A. A. (1963). *Science*, **140**, 405.
14. Golub, E. S., Mischell, R. I., Weigle, W. O., and Dutton, R. W. (1968). *J. Immunol.*, **100**, 133.
15. Sedgwick, J. D. and Holt, P. G. (1983). *J. Immunol. Methods*, **57**, 301.
16. Czerkinsky, C. C., Nilsson, L. A., Nygren, H., Ouchterlony, O., and Tarkowski, A. (1983). *J. Immunol. Methods*, **65**, 109.
17. Sedgwick, J. D. and Holt, P. G. (1983). *J. Exp. Med.*, **157**, 2178.
18. Czerkinsky, C. C., Tarkowski, A., Nilsson, L. A., Ouchterlony, O., Nygren, H., and Gretzer, C. (1984). *J. Immunol. Methods*, **72**, 489.
19. Moller, S. A. and Borrebaeck, C. A. (1985). *J Immunol. Methods*, **79**, 195.
20. Czerkinsky, C., Moldoveanu, Z., Mestecky, J., Nilsson, L. A., and Ouchterlony, O. (1988). *J. Immunol. Methods*, **115**, 31.
21. Shirai, A., Sierra, V., Kelly, C. I., and Klinman, D. M. (1994). *Cytokine*, **6**, 329.

22. Franci, C., Ingles, J., Castro, R., and Vidal, J. (1986). *J. Immunol. Methods*, **88**, 225.
23. Walker, A. G. and Dawe, K. I. (1987). *J. Immunol. Methods*, **104**, 281.
24. Czerkinsky, C. and Sedgwick, J. (1993). In *Methods of immunological analysis* (ed. M. R. F. Masseyeff, W. H. Albert, and N. A. Staines), Vol. 3, p. 504. Verlagsgesellschaft, Weinhiem.
25. Hertzman, H. and Richards, R. M. (1974). *Proc. Natl. Acad. Sci. USA*, **71**, 3537.
26. Prummer, O., Cederblad, B., Alm, G., Drees, N., and Porzsolt, F. (1990). *J. Immunol. Methods*, **130**, 187.
27. Klinman, D. M. and Steinberg, A. D. (1987). *J. Immunol. Methods*, **102**, 157.
28. Verheul, A. F., Versteeg, A. A., Westerdaal, N. A., Van, D. G., Jansze, M., and Snippe, H. (1990). *J. Immunol. Methods*, **126**, 79.
29. Schliebs, B., Gattner, H., Naithani, K., Fabry, M., Khalaf, A. N., Kerp, L., *et al.* (1990). *J. Immunol. Methods*, **126**, 169.
30. Kendall, M. E. and Hymer, W. C. (1987). *Endocrinology*, **121**, 2260.
31. Gonchoroff, N. J., Greipp, P. R., Kyle, R. A., and Katzmann, J. A. (1985). *Cytometry*, **6**, 506.
32. Gratzner, H. G. (1982). *Science*, **218**, 474.
33. Porstmann, T., Ternynck, T., and Avrameas, S. (1985). *J. Immunol. Methods*, **82**, 169.
34. Magaud, J. P., Sargent, I., and Mason, D. Y. (1988). *J. Immunol. Methods*, **106**, 95.
35. Huong, P. L., Kolk, A. H., Eggelte, T. A., Verstijnen, C. P., Gilis, H., and Hendriks, J. T. (1991). *J. Immunol. Methods*, **140**, 243.
36. Cheitlin, R. A. and Bodell, W. J. (1988). *Anticancer Res.*, **8**, 471.
37. Muir, D., Varon, S., and Manthorpe, M. (1990). *Anal. Biochem.*, **185**, 377.
38. Key, G., Kubbutat, M. H., and Gerdes, J. (1994). *J. Immunol. Methods*, **177**, 113.
39. Meuth, M. and Green, H. (1974). *Cell*, **2**, 109.
40. Mosmann, T. (1983). *J. Immunol. Methods*, **65**, 55.
41. Mosmann, T. R. and Fong, T. A. (1989). *J. Immunol. Methods*, **116**, 151.
42. Landegren, U. (1984). *J. Immunol. Methods*, **67**, 379.
43. Wemme, H., Pfeifer, S., Heck, R., and Muller, Q. J. (1992). *Immunobiology*, **185**, 78.
44. Doi, H., Ishii, A., and Shimono, K. (1988). *Trans. R. Soc. Trop. Med. Hyg.*, **82**, 190.

12

Immunotoxins

ELAINE J. DERBYSHIRE, CLAUDIA GOTTSTEIN, and
PHILIP E. THORPE

1. Introduction

Immunotoxins (ITs) are cell type-specific cytotoxic agents consisting of an antibody linked to a protein toxin. The antibody component is responsible for directing the toxin selectively to the target cell and the toxin component is responsible for killing the cell (for reviews on ITs see refs 1–5). Because of their potency and selectivity, ITs have several clinical applications. First, ITs have been employed *ex vivo* to kill residual tumour cells in autologous bone marrow grafts, or T cells in allogenic bone marrow grafts for the prevention of graft-versus-host disease (6,7). Secondly, they have been used systematically to treat leukaemias, lymphomas, and solid tumours (8). Thirdly, they have been used to treat solid tumours by intracavity administration (8). They have also been evaluated for the treatment of autoimmune diseases like psoriasis (9).

The most commonly used toxins for forming ITs are the plant toxins ricin and abrin, and the bacterial toxins, diphtheria toxin (DT) and *Pseudomonas* exotoxin (PE). In their active forms, the toxins consist of two polypeptide chains which are joined by a single disulfide bond. Ricin and abrin are synthesized in plants as their active two-chain forms whereas DT and PE are synthesized in bacteria as a single polypeptide chain which is converted by proteolytic cleavage (either post-translationally or inside the target cell) into the active two-chain form. The toxins all act in a similar manner. They bind by means of recognition sites on the B-chain (domain I in PE) to receptor molecules that are on virtually all cell types: ricin and abrin recognize galactose-terminating glycoproteins and glycolipids, PE recognizes an α_2-macroglobulin receptor-like molecule, and DT recognizes an epidermal growth factor receptor-like molecule. The toxins are then endocytosed and are transported to an intracellular compartment where the A-chain (domain III in PE) translocates to the cytosol. Once in the cytosol, the A-chain kills the cell by inhibiting protein synthesis. Ricin and abrin enzymatically inactivate the ribosomal 60S subunits by cleaving off a crucial adenine residue needed for EF2 binding, whereas DT and PE ADP-ribosylate EF2 directly (reviewed in ref. 10).

ITs prepared with intact toxin molecules are invariably highly potent as cytotoxic agents for their designated target cells because, after binding via the antibody component, the toxin can interact with its receptor which directs the IT into the cell via the same route as the native toxin. However, ITs prepared with intact toxins have substantial non-specific toxicity because they bind via the toxin's binding site to non-target cells (11). This 'non-specific' binding can be abolished by removing the toxin's cell binding domain. This is readily accomplished with ricin and abrin by reducing the disulfide bond between the A- and B-chains and attaching the isolated A-chain to the antibody. Similarly, mutated forms of DT and PE, which lack cell binding activity, can be used (12). Immunotoxins lacking the toxin's binding domain have high selectivity for target cells. However, A-chain immunotoxins exhibit unpredictable and variable potency depending on whether the antibody recognizes an antigen that routes the IT to an intracellular compartment favourable for A-chain translocation, a property normally conferred by the B-chain. In an attempt to obtain more consistent potency, 'blocked ricin' ITs have been constructed where the galactose-binding sites of the B-chain are obstructed, but not eliminated (13,14). These ITs have residual galactose recognition that is strong enough to confer correct intracellular routing, but weak enough not to cause significant binding to non-target cells (13).

Plants also produce proteins which resemble the toxin A-chains in size and function. These proteins are termed ribosome-inactivating proteins (RIPs). They naturally lack a binding domain and so are only cytotoxic when conveyed to the cell by an antibody. Gelonin, pokeweed anti-viral protein, and saponin are examples of plant RIPs which have been used in IT construction (15). ITs have also been prepared with fungal RIPs such as α-sarcin (16). Fungal RIPs are smaller in size than plant RIPs, with a molecular weight of approximately 17 kDa, but like plant RIPs they consist of a single polypeptide chain that inactivates ribosomes and lacks a mechanism of binding to the surface of cells.

ITs having good anti-tumour activity have been prepared from monovalent antibody fragments (Fab', Fv), as well as from intact antibodies. There is no clear answer to the question of which is preferable. On the one hand, monovalent fragments of antibody are less immunogenic and, lacking the Fc portion, cannot bind to cells of the reticulo-endothelial system or to other cells expressing Fc receptors. Furthermore, they are smaller than intact antibody and their decreased size should give faster extravasation and penetration of solid tumour masses (17). On the other hand, antibody fragments are more rapidly cleared from the blood and tissues than intact antibodies and so have less time to reach their target cells. Also, Fab' fragments bind to their target cells less avidly than their bivalent counterparts and always form less effective ITs *in vitro* (18).

The traditional way to prepare an IT is to link the antibody chemically to the toxin moiety. This method is still the method of choice for ITs containing plant toxins or RIPs because it is technically difficult to generate a disulfide bond between the antibody and the toxin by recombinant DNA technology (19). In contrast, the bacterial toxins naturally occur as single polypeptide chains with internal sequences which are cleaved intracellularly to allow release of the catalytic domains. Therefore, genetically engineered fusion proteins of antibody fragments and bacterial toxins are also active and hence genetic engineering has emerged as the favoured method for preparing such ITs (20,21).

Recombinant techniques offer several advantages over the older chemical techniques. The molecules which are generated are homogeneous, well-defined, and can be produced in large quantities. Further, they can be engineered to improve their utility. For example, the antibody component can be humanized to reduce its immunogenicity (22) and its affinity can be improved by chain shuffling (23) or by mutagenesis (24). It is also possible to create synthetic antibodies from naive DNA libraries, thereby bypassing the immunization process (25). Modifications to the toxin moiety can also be made to reduce its size (26,27), or to improve potency by adding intracellular signalling sequences, such as KDEL (28), or by generating hybrid toxins with dual enzymatic activity (29,30).

2. Preparation of chemically coupled immunotoxins

2.1 General considerations

The monoclonal antibody and toxin component used in the preparation of the immunotoxin should be as pure as possible. Methods for the purification of monoclonal antibodies, diphtheria toxin, ricin, and abrin, for the separation of the A- and B-chains of these toxins, and for the isolation of RIPs have been described elsewhere (31–34). Extreme care must be taken when handling intact toxins due to their extremely high toxicity. For recommendations on handling procedures see Cumber *et al.* (35).

Heterobifunctional coupling reagents, having two dissimilar reactive groups, are used for coupling the antibody and toxin components (36). The antibody or toxin is reacted with the coupling agent using one of the groups and the second group is activated after addition of the second protein component. The heterobifunctional reagents can also be used to introduce complementary reactive groups into the two protein components which on mixing react to form the conjugate. The use of heterobifunctional reagents favours conjugation between antibody and toxin molecules and prevents the formation of intramolecular cross-links.

The derivatization of antibody or toxin component should not impair the antigen-binding capacity of the antibody or the enzymatic activity of the toxin component. These properties are generally unaffected by the introduction of only one or two coupling groups per molecule and so the

241

procedures outlined in the following section are designed to give such a low level of antibody and toxin modification.

2.2 Conjugation methods

2.2.1 Preparation of ITs by chemical linkage of antibody to isolated toxin A-chains

The A-chains of plant toxins are isolated by reduction of the single disulfide bond which links the A- and B-chain in the native toxin followed by extensive purification. The isolated A-chain contains a single free thiol group which can be used in the attachment to antibody. Since it is possible for the A-chain to form dimers during storage, the A-chain must be freshly reduced immediately prior to conjugation. A-chain ITs are prepared by linking the A-chain to the antibody by means of a reducible disulfide bond. ITs prepared by linking the A-chain to the antibody via irreducible bonds (e.g. thioether bonds) are usually much less active, probably because splitting of the bond between the A-chain and antibody is necessary once inside the cell so that the A-chain can gain access to the cytosol (37).

i. Linkage via SPDP
A generalized scheme for the preparation of A-chain ITs is shown in *Figure 1*.

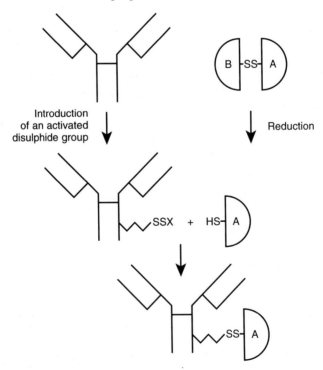

Figure 1. Generalized scheme for preparation of A-chain ITs.

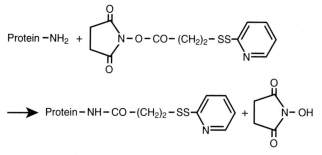

Figure 2. Introduction of 2-pyridyl disulfide groups into protein using SPDP.

a) SPDP: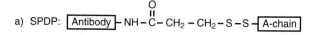

b) SMPT: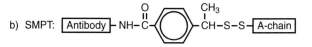

Figure 3. Chemical structures of the linkages in disulfide bonded A-chain ITs.

The procedure can be employed for intact antibodies, F(ab')₂, or Fab fragments since all require the introduction of an activated disulfide bond for conjugation to A-chain. The most frequently used heterobifunctional reagent for IT construction is *N*-succinimidyl-3-(2-pyridyldithio)propionate (SPDP) (*Protocol 1*). This reagent, first described by Carlsson *et al.* (38), reacts with ε-amino groups of lysine residues present on the surface of the antibody molecule to introduce 2-pyridyl disulfide groups (*Figure 2*). The number of 2-pyridyl disulfide groups introduced per antibody can be determined spectrophotometrically (see *Protocol 2*). On reaction with the A-chain, SPDP produces the linkage shown in *Figure 3a*.

Protocol 1.   Attachment of antibody to isolated A-chain

Equipment and reagents

- Low-pressure chromatography equipment consisting of a pump, a UV detector capable of monitoring absorbance at 280 nm, a chart recorder
- Sephadex G-25 (F) column (1.6 cm diameter × 30 cm) (Pharmacia)
- 50 ml ultrafiltration cell equipped with a PM10 membrane (Amicon)
- 0.45 μm low protein binding filters (Whatman)
- Antibody

- SDS–PAGE equipment (e.g. Pharmacia PhastSystem)
- A-chain
- SPDP (Pierce)
- Dry dimethylformamide (DMF) (Sigma)
- Dithiothreitol (DTT) (Sigma)
- Nitrogen
- PBS: 18.8 mM Na₂HPO₄.7H₂O, 1.8 mM KH₂PO₄, 17.1 mM NaCl, 2.7 mM KCl pH 7.4
- PBSe: PBS containing 1 mM EDTA

Protocol 1. *Continued*

Method[a]

1. Prepare 10–20 mg of antibody at a concentration of 10 mg/ml in PBS.

2. Prepare a fresh solution of SPDP[b] at 10 mM in dry DMF.

3. Add a 2.5-fold molar excess of SPDP over antibody to the antibody solution, mix, and incubate the mixture for 30 min at room temperature.

4. Pass the mixture through a 0.45 μm filter to remove aggregates and then apply the mixture to a column (1.6 cm diameter × 30 cm) of Sephadex G-25 (F) equilibrated in PBS.

5. Collect the first peak corresponding to derivatized antibody.

6. Quantify the level of substitution of the antibody with pyridyl disulfide groups (see *Protocol 2*).

7. Concentrate the derivatized antibody to about 3.5 ml by ultrafiltration (Amicon PM10 membrane).

8. Fully reduce the toxin A-chain by incubation in the presence of 50 mM DTT for 1 h at room temperature. A 2.5-fold molar excess of toxin A-chain over antibody is required.

9. Pass the mixture through a 0.45 μm filter and apply the mixture to a column (1.6 cm diameter × 30 cm) of Sephadex G-25 (F) equilibrated with nitrogen-flushed PBSe.

10. Collect the first peak corresponding to reduced toxin A-chain directly into the ultrafiltration cell containing the concentrated antibody.

11. Concentrate the reaction mixture to about 3 ml by ultrafiltration and incubate the mixture at room temperature for 18 h in the ultrafiltration cell in an atmosphere of nitrogen to allow conjugation of the A-chain to the antibody.[c] It is usual for a light protein precipitate to form.

12. Analyse a sample of conjugation mixture by SDS–PAGE to verify that conjugation has occurred. Run the sample under non-reducing conditions along with proteins of known molecular mass on a 4–12% polyacrylamide gradient gel. Detect protein bands using Coomassie blue staining. A prominent band having a molecular mass of 180 kDa should be seen, corresponding to one molecule of IgG and one molecule of A-chain. Minor bands (210 kDa etc.) of ITs having two or more A-chains are usually also visible.

[a] Similar procedures have been used to link antibodies to the A chains of diphtheria toxin and ricin (reviewed in ref. 39) and abrin (40,41).
[b] SMPT is an alternative heterobifunctional reagent that forms conjugates with greater *in vivo* stability.
[c] When using SMPT as the cross-linker, allow conjugation to proceed for 72 h.

Protocol 2. Quantitation of the level of antibody substitution by pyridyl disulfide groups

Equipment and reagents
- Spectrophotometer
- Cuvette (0.6 ml)
- DTT (Sigma)
- PBS (see *Protocol 1*)

Method

1. Add 0.5 ml of the derivatized antibody sample to the cuvette and measure the absorbance of the solution at 280 nm and 343 nm.

2. Prepare a 0.5 M solution of DTT in PBS.

3. Add 5 µl of the DTT solution to the derivatized antibody in the cuvette, mix, and incubate at room temperature.

4. Measure the absorbance of the solution at 280 nm and 343 nm periodically, until the maximal 343 nm absorbance is achieved (about 10 min).

5. Determine the antibody concentration in the cuvette by dividing the absorbance at 280 nm (without DTT) by the molar extinction coefficient of the antibody,[a] i.e.:

$$\text{Concentration of antibody (M)} = \frac{A_{280}^{-DTT}}{2.1 \times 10^5}$$

6. Determine the concentration of released pyridine-2-thione in the cuvette according to the following formula:

$$\text{Concentration of released pyridine-2-thione (M)} = \frac{A_{343}^{+DTT} - A_{343}^{-DTT}}{8.08 \times 10^3}$$

7. Determine the level of antibody substitution by pyridyl disulfide (PDP) groups by dividing the concentration of released pyridine-2-thione by the concentration of antibody.

[a] The molar extinction coefficient for the derivatized antibody can be taken to be the same as for underivatized antibody, provided the SPDP derivatization level is low (≤ two PDP groups per IgG). At higher loadings, correction is needed for the 280 nm absorbance attributable to the PDP groups (38).

---

ii. Linkage via SMPT

Although SPDP produces stably linked conjugates for *in vitro* usage, such conjugates break down *in vivo* diminishing their therapeutic efficacy. To overcome this problem, heterobifunctional reagents have been synthesized which link the A-chain and antibody via a 'hindered' disulfide bond (42,43). An example is 4-succinimidyl-oxycarbonyl-α-methyl-α-(2-pyridyldithio)-toluene (SMPT) which contains a bulky phenyl ring and a methyl group that hinder the reduction of the disulfide bond (44). On reaction with the A-chain,

SMPT produces the linkage shown in *Figure 3b*. SMPT yields more stable ITs with improved anti-tumour activity *in vivo*.

The method of conjugation of A-chain to antibody via SMPT is essentially the same as that for conjugation via SPDP (*Protocol 1*) except that the conjugation reaction is allowed to proceed for 72 hours. This is because the A-chain reacts with the SMPT-derivatized antibody at a slower rate than it does with the SPDP-derivatized antibody.

2.2.2 Preparation of ITs by chemical linkage of antibody to RIP

Unlike isolated A-chains, RIPs do not contain a free thiol group for attachment to antibody. Therefore one must first be introduced using a thiolating agent. SPDP has been used as the thiolating agent for this purpose but thiolation of RIPs with SPDP can decrease their ability to inactivate ribosomes in cell-free systems (45). In contrast, thiolation with 2-iminothiolane hydrochloride (2-IT) can preserve the complete ribosome inactivating activity of the RIP (45), presumably because it reacts with amino groups in proteins to form charged amidine groups, thereby preserving the positive charge of the protein (*Figure 4*).

As with the A-chain ITs, RIP ITs in which the RIP is linked to the antibody via a disulfide bond show a higher cytotoxicity compared with RIP ITs which contain a thioether linkage (45–47). Therefore, the thiolated RIP is usually reacted with antibody which has been derivatized with SPDP to produce a conjugate with the linkage shown in *Figure 5*. The procedure for preparing RIP ITs in this way is outlined in *Protocol 3*. A more stably linked conjugate with greater *in vivo* efficacy can be prepared by using SMPT instead of SPDP.

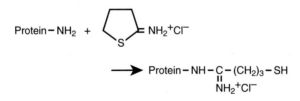

Figure 4. Introduction of thiol groups into protein using 2-IT.

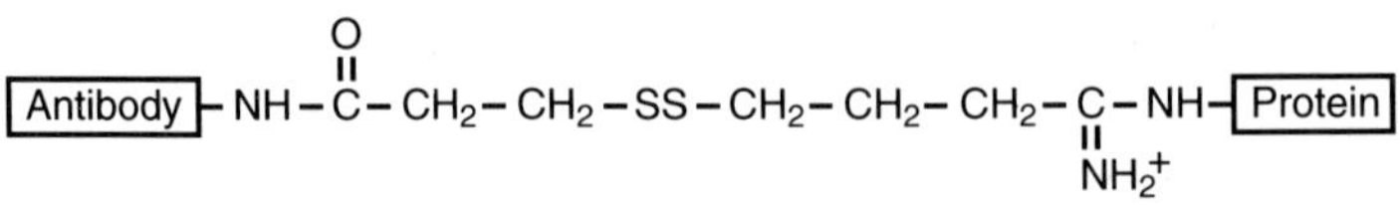

$$\boxed{\text{Antibody}}\!-\!\text{NH}-\overset{\overset{\text{O}}{\|}}{\text{C}}-\text{CH}_2-\text{CH}_2-\text{SS}-\text{CH}_2-\text{CH}_2-\text{CH}_2-\underset{\underset{\text{NH}_2^+}{\|}}{\text{C}}-\text{NH}-\boxed{\text{Protein}}$$

Figure 5. Chemical structure of the linkage in disulfide bonded RIP ITs utilizing SPDP plus 2-IT.

Protocol 3. Attachment of antibody to RIP

Equipment and reagents

- Low-pressure chromatography equipment
- Sephadex G-25 (F) column (1.6 cm diameter × 30 cm) (Pharmacia)
- 50 ml ultrafiltration cell equipped with a PM10 membrane (Amicon)
- 0.45 μm low protein binding filters (Whatman)
- SDS–PAGE equipment (e.g. Pharmacia PhastSystem)
- Antibody

- RIP
- SPDP (Pierce)
- Dry DMF (Sigma)
- 2-IT (Sigma)
- Nitrogen
- PBS (see *Protocol 1*)
- PBSe (see *Protocol 1*)
- Borate buffer: 0.05 M borate, 0.3 M NaCl pH 9.0

Method

1. Derivatize 10–20 mg of antibody with SPDP as described in *Protocol 1*, steps 1–7.

2. Prepare the RIP at a concentration of 10 mg/ml in borate buffer. A 2.5-fold molar excess of RIP over antibody is required.

3. Prepare a fresh solution of 2-IT in borate buffer.

4. Add a sixfold molar excess of 2-IT over RIP to the RIP solution, mix, and incubate the mixture at room temperature for 1 h.

5. Apply the mixture to a column (1.6 cm diameter × 30 cm) of Sephadex G-25 (F) equilibrated with nitrogen-flushed PBSe.

6. Collect the first peak, corresponding to modified RIP, directly into the ultrafiltration cell containing the concentrated derivatized antibody.[a]

7. Concentrate the mixture to about 3 ml by ultrafiltration and incubate the mixture at room temperature in the ultrafiltration cell in an atmosphere of nitrogen for 18 h to allow conjugation of the RIP to the antibody.

8. Subject a sample of the conjugation mixture to SDS–PAGE under non-reducing conditions to verify that conjugation has occurred.

[a] The average number of free sulfydryl groups introduced into the RIP can be quantified spectrophotometrically according to *Protocol 4*.

Protocol 4. Determination of the number of free sulfydryl groups introduced into a protein

Equipment and reagents

- Spectrophotometer
- Cuvette (0.6 ml)
- Tris buffer: 1 M Tris–HCl pH 8

- Ellman's reagent (5,5'-dithio-bis(2-nitrobenzoic acid)) (Sigma)
- PSBe (see *Protocol 1*)

Protocol 4. Continued

Method[a]

1. Freshly prepare a 10 mM solution of Ellman's reagent in Tris buffer (dissolve the Ellman's reagent in a few drops of ethanol before adding the Tris buffer).

2. Add 0.25 ml of Tris buffer, 0.25 ml PBSe, and 10 μl of 10 mM Ellman's reagent to a cuvette, mix, and measure the absorbance of the solution at 412 nm.

3. Add 0.25 ml of Tris buffer and 0.25 ml of derivatized protein in PBSe to a cuvette, mix, and measure the absorbance of the solution at 280 nm.

4. Determine the molar concentration of protein by dividing the absorbance at 280 nm by the molar extinction coefficient of the protein.

5. Add 10 μl of 10 mM Ellman's reagent to the cuvette, mix, and measure the absorbance of the solution at 412 nm when it has reached its maximum value.

6. Calculate the concentration of free sulfydryl groups according to the following formula:

$$\text{Concentration of free sulfydryl groups (M)} = \frac{A_{412}^{\text{Step 5}} - A_{412}^{\text{Step 2}}}{1.36 \times 10^4}$$

7. Calculate the average number of free sulfydryl groups introduced into the protein by dividing the molarity of free sulfydryl groups by the molarity of protein.

[a] The procedure for the quantitation of free sulfydryl groups into proteins was devised by Ellman (48).

2.2.3 Preparation of ITs by chemical linkage of antibody to intact toxins

Immunotoxins containing intact toxins are usually prepared with a thioether linkage which is stable to reduction in order to avoid the release of free toxin molecules into the circulation which could cause non-specific toxicity. The immunotoxin preparation usually consists of multiple molecular species. Those immunotoxin molecules in which the toxin is linked to the antibody via its binding domain are cytotoxic because the catalytic domain can be released by reduction within the cytosol. Those immunotoxin molecules in which the toxin is attached to the antibody via the catalytic domain are probably not biologically active. The linkage of antibody to intact toxin is usually performed by introducing an alkylating group into the toxin and a free thiol group into the antibody and then reacting the two derivatized components

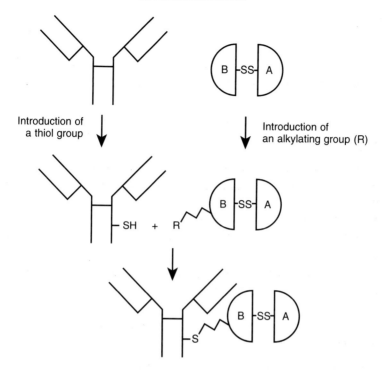

Figure 6. Generalized scheme for preparation of ITs with intact toxins.

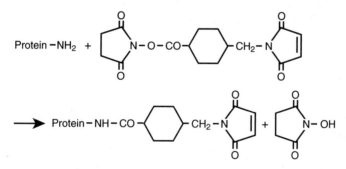

Figure 7. Introduction of alkylating groups into protein using SMCC.

together (*Figure 6*). Several different methods have been employed to pre-pare such conjugates. *Protocol 5* describes the use of *N*-succinimidyl-4-(*N*-maleimidomethylcyclo-hexane)-1-carboxylate (SMCC) to introduce an alkylating group into the toxin (*Figure 7*) and 2-IT to introduce a thiol group into the antibody. The linkage between the antibody and the toxin has the structure shown in *Figure 8*.

Blocked ricin is a derivative of ricin in which the galactose binding sites of

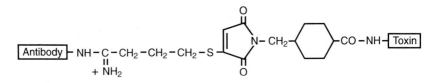

Figure 8. Chemical structures of the linkages in thioether-linked intact toxin ITs formed with 2-IT plus SMCC.

the B-chain are blocked by chemical modification with affinity ligands (14), resulting in a 3500-fold lower binding affinity (49). The molecule can be reacted directly with antibody containing an alkylating group, introduced by SMCC (49) as in *Protocol 5*. An alternative method which does not require access to proprietary affinity ligands is to synthesize an antibody–ricin immunotoxin using a short cross-linker which generates an immunotoxin in which the toxin binding sites are obstructed by the antibody itself (50). These sterically blocked immunotoxins are readily separated by filtration through a galactose-containing absorbent gel, such as Sepharose 4B.

Protocol 5.   Attachment of antibody to intact toxin using 2-IT and SMCC

Equipment and reagents
- Low-pressure chromatography equipment
- Sephadex G-25 (F) column (1.6 cm diameter × 30 cm) (Pharmacia)
- 50 ml ultrafiltration cell equipped with a PM10 membrane (Amicon)
- 0.45 μm low protein binding filters (Whatman)
- SDS–PAGE equipment (e.g. Pharmacia PhastSystem)
- Antibody
- Toxin
- SMCC (Pierce)
- Dry DMF (Sigma)
- 2-IT (Sigma)
- Nitrogen
- PBS (see *Protocol 1*)
- PBSe (see *Protocol 1*)
- Phosphate buffer pH 8: 100 mM sodium phosphate, 2 mM EDTA pH 8.0
- Phosphate buffer pH 7: 100 mM sodium phosphate, 2 mM EDTA, 3 mg/ml NAD, 3 mM EDTA

Method[a]
1. Prepare 10–20 mg solution of toxin at a concentration of 5 mg/ml in phosphate buffer pH 7.
2. Prepare a fresh solution of SMCC at 10 mM in dry DMF.
3. Add a threefold molar excess of SMCC over toxin to the toxin solution, mix, and incubate the mixture at room temperature for 30 min.
4. Filter the mixture through a 0.45 μm filter and apply the mixture to a column of Sephadex G-25 (F) equilibrated in phosphate buffer pH 8.
5. Collect the first peak corresponding to derivatized toxin.

6. Concentrate the toxin to approx. 5 mg/ml in an ultrafiltration cell equipped with a PM10 membrane. Store at 4°C until mixed with the derivatized antibody.

7. Prepare the antibody at a concentration of 10 mg/ml in phosphate buffer pH 8. A fourfold molar excess of toxin over antibody is required.

8. Prepare a fresh solution of 2-IT at 10 mM in borate buffer.

9. Add a threefold molar excess of 2-IT over antibody to the antibody solution, mix, and incubate the mixture for 1 h at room temperature.

10. Filter the mixture through a 0.45 μm filter and apply the mixture to a column of Sephadex G-25 (F) equilibrated in nitrogen-flushed phosphate buffer pH 8.

11. Collect the first peak corresponding to derivatized antibody[b] directly into the ultrafiltration cell containing the concentrated derivatized toxin[c].

12. Concentrate the mixture to about 3 ml by ultrafiltration and incubate at room temperature in the ultrafiltration cell in an atmosphere of nitrogen for 20 h to allow conjugation of the toxin to the antibody. It is usual for a light protein precipitate to form.

13. Analyse a sample of the conjugation mixture by SDS–PAGE to verify that conjugation has occurred.

[a] The method was used by Zangemeister-Wittke *et al.* (51) to link PE to antibody.
[b] The average number of free sulfydryl groups introduced into the antibody can be quantitated spectrophotometrically according to *Protocol 4*.
[c] Yields can be improved by conducting steps 1–6 and 7–10 concurrently using two low-pressure chromatography apparatuses, and mixing the derivatized toxin and the derivatized antibody as soon as they have eluted from the Sephadex G-25 columns (step 11).

2.2.4 Immunotoxins prepared by linking antibody Fab′ fragments to toxins or their catalytic subunits

Antibody Fab′ fragments, prepared by reducing F(ab′)$_2$ fragments, contain one to three thiol groups, depending on the heavy chain subclass. Conjugation to isolated A-chains, thiolated RIPs, or intact toxins containing an alkylating group can be achieved using one of the thiol groups in the Fab′ molecule. Before reaction with A-chain or thiolated RIP, the thiol groups in the Fab′ fragment are converted into activated disulfides by treating the Fab′ fragments with Ellman's reagent (48) (*Figure 9*). The activated disulfide then reacts with the A-chain or thiolated RIP to give an immunotoxin containing a disulfide bond (*Protocol 6*). The bond formed has a stability greater than an SPDP linkage but less stable than an SMPT linkage (8). An alternative approach is to activate the thiol group of the A-chain or RIP and then mix it with the reduced Fab′ fragment to form the IT.

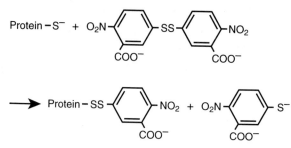

Figure 9. Activation of disulfide groups by reaction with Ellman's reagent (48).

Protocol 6. Attachment of Fab' antibody fragments to isolated A-chain or thiolated RIP

Equipment and reagents

- Low-pressure chromatography equipment
- Sephadex G-25 (F) column (1.6 cm diameter × 30 cm) (Pharmacia)
- 50 ml ultrafiltration cell equipped with a PM10 membrane (Amicon)
- 0.45 μm low protein binding filters (Whatman)

- SDS–PAGE equipment (Pharmacia)
- F(ab')$_2$ fragment
- A-chain
- 2-Mercaptoethanol (Sigma)
- Ellman's reagent (Sigma)
- Sephadex G-25 (F) (Pharmacia)
- Tris buffer: 50 mM Tris–HCl pH 7.4

Method

1. Prepare 20 mg of the F(ab')$_2$ fragment at 10 mg/ml in Tris buffer.

2. Prepare a 20 mM solution of 2-mercaptoethanol in Tris buffer.

3. Add 222 μl of the 2-mercaptoethanol to the F(ab')$_2$ fragment solution and incubate at 37°C for 1 h to reduce the F(ab')$_2$ fragment to the Fab' fragment.

4. Prepare a fresh 50 mM solution of Ellman's reagent in Tris buffer.

5. Add 247 μl of the Ellman's reagent solution to the Fab' fragment and incubate at room temperature for 50 min to convert the free thiol groups of the Fab' fragment to 3-carboxylate-4-nitrophenylthiol groups.

6. Filter the mixture through a 0.45 μm filter to remove aggregates and then apply the mixture to a column of Sephadex G-25 equilibrated in PBS.

7. Collect the first peak corresponding to the Ellman's reagent-modified Fab' fragment.

8. Quantify the average number of 3-carboxy-4-nitrophenylthiol groups introduced per Fab' fragment according to *Protocol 4* (except that the concentration of Fab' is determined instead of toxin).

9. Concentrate the Fab′ fragment to approx. 3 ml in an ultrafiltration cell.

10. Add a 2.5-fold molar excess of freshly reduced toxin A-chain or thiolated RIP to the Fab′ fragment.

11. Concentrate the mixture to approx. 3 ml by ultrafiltration and incubate at room temperature for 72 h.

12. Subject a sample of the conjugation mixture to SDS–PAGE under non-reducing conditions to verify that conjugation has occurred.

3. Purification of chemically coupled immunotoxins

3.1 Gel permeation chromatography

After the coupling of antibody and toxin components as described in Section 2, the conjugate must be purified from other components in the reaction mixture. These consist of unconjugated antibody, unconjugated toxin components, and low molecular weight by-products of the reaction. Gel permeation chromatography on columns of Sephacryl S200 or S300 can resolve conjugate from unconjugate toxin components and low molecular weight by-products (33,46) (see *Protocol 7*). Conjugates of Fab′ fragments and A-chains are best fractionated using Sephadex G-100. Alternatively, purification of conjugates can be performed using TSK gel permeation columns (52,53).

The immunotoxins can consist of one, two, or more toxin components linked to each antibody molecule. This heterogeneity results from the introduction of various numbers of thiol groups per molecule of antibody. Conjugates with the best biological activity contain one or two toxin components per molecule of antibody (46,54,55) and so conjugates lacking high molecular weight species are desirable. Both low- and high-pressure gel filtration columns will separate high molecular weight conjugates containing more than one molecule of antibody or toxin from the (antibody)$_1$–(toxin)$_1$ conjugate. However, the resolution of molecules in the high molecular weight range is generally incomplete and so fractions of the (antibody)$_1$–(toxin)$_1$ conjugates will usually also contain (antibody)$_1$–(toxin)$_2$ conjugate and unconjugated antibody.

Protocol 7. Partial purification of immunotoxins by gel permeation chromatography

Equipment and reagents

- Low-pressure chromatography equipment
- Sephacryl S300 column (1.6 cm diameter × 100 cm) (Pharmacia) or Sephadex G-100 (Pharmacia)
- 5 ml fraction collector tubes (Sarstedt)
- Conjugation mixture

- SDS–PAGE equipment (e.g. Pharmacia PhastSystem)
- 0.45 μm low protein binding filters (Whatman)
- PBS (see *Protocol 1*)

Protocol 7. *Continued*

Method

1. Filter the conjugation mixture through a 0.45 μm filter and load onto a column of Sephacryl S300 equilibrated in PBS (in a volume not exceeding 2% of the column volume).

2. Fractionate the mixture at a flow rate of 12 ml/h and collect the eluate into 2 ml fractions.

3. Subject a sample of each fraction in the main peak (corresponding to conjugate) to SDS–PAGE to determine the molecular composition of the species in each fraction. Run the sample under non-reducing conditions along with proteins of known molecular mass on a 4–12% polyacrylamide gradient gel. Detect protein bands using Coomassie blue staining.

4. Pool fractions containing (antibody)$_1$–(toxin)$_1$ conjugate and (antibody)$_1$–(toxin)$_2$ conjugate but lacking significant amounts of higher molecular weight conjugates.

3.2 Affinity chromatography

Knowles and Thorpe (56) developed a method for separating contaminating antibody from conjugate for ITs containing ricin A-chain or abrin A-chain. The separation is based on the ability of the toxin A-chains to bind to Blue Sepharose CL-6B when applied in low salt buffer and to be released in the presence of 0.5 M salt (see *Protocol 8*).

Protocol 8. Purification of ITs containing ricin A-chain and abrin A-chain using Blue Sepharose CL-6B

Equipment and reagents

- Low-pressure chromatography equipment
- Blue Sepharose CL-6B column (0.5 × 25 cm) (Pharmacia)
- 0.45 μm low protein binding filter (Whatman)
- 5 ml fraction collector tubes (Sarstedt)
- Washing buffer: 0.05 M sodium phosphate pH 7.5
- Elution buffer: 0.05 M sodium phosphate, 0.5 M NaCl pH 7.5

Method

1. Wash the column of Blue Sepharose with 20 column volumes of washing buffer at 0.15 ml/min.

2. Dialyse the partially purified immunotoxin into the washing buffer. Adjust the total protein concentration of the retentate to 1 mg/ml and pass it through a 0.45 μm filter.

3. Apply 1 ml of the solution to the column. Elute the free antibody by washing with the washing buffer at a flow rate of 0.15 ml/min. Collect

1.5 ml fractions. Continue until the absorbance at 280 nm of the eluate returns to zero.

4. Apply the elution buffer[a] to elute the bound conjugate. Collect 1.5 ml fractions. Continue until the absorbance has again returned to zero.

5. Analyse a sample of each fraction having absorbance at 280 nm by SDS–PAGE to verify complete separation of antibody and conjugate. Run the sample under non-reducing conditions on a 4–12% polyacrylamide gradient gel. Detect protein bands using Coomassie blue staining.

6. Pool the fractions containing purified conjugate, and dialyse the conjugate against PBS.

[a] If the antibody also binds to the Blue Sepharose, use a salt gradient from 0–0.5 M NaCl in 0.05 M sodium phosphate to first elute the antibody and then elute the conjugate.

4. Calculation of the composition of chemically prepared ITs

The precise composition of the conjugate preparation can be determined by SDS–PAGE under non-reducing conditions. On 4–15% polyacrylamide gradient gels, $(antibody)_1$–$(toxin)_1$ conjugates are clearly distinguished from $(antibody)_1$–$(toxin)_2$ conjugates and from any contaminating higher molecular weight conjugates or free antibody. Any contaminating free toxin or toxin A-chain are also resolved on 4–15% gradient gels. The proportions of the different molecular species can be quantified by densitometric scanning of Coomassie blue stained gels and dividing the area under each peak by the total area for all peaks.

From knowing the proportions of the different molecular species in the IT preparation, the extinction coefficients and molecular weights of the antibody and toxin components, and the absorbance of the IT preparation at 280 nm, the concentration of antibody and toxin in the IT preparation can be calculated. Alternatively the IT can be analysed by SDS–PAGE under reducing conditions and staining with Coomassie blue. Under these conditions, ITs which are linked via a disulfide bond break down into the toxin component and the antibody heavy and light chains. The reduced sample of the IT is run alongside various concentrations of the unconjugated toxin and antibody components. The concentration of toxin and antibody in the conjugate can be estimated by densitometry relative to standards of known concentration.

5. Preparation of ITs by genetic engineering

It is beyond the scope of this chapter to describe all the different ways of cloning, screening, expression, purification, and modification of immunotoxins.

Here, we will give one example of how a single-chain variable region (scFv)–PE fusion protein can be prepared from a given antibody by fusing the genes of the variable regions of the antibody with that of PE toxin. Alternative methods are reviewed in ref. 57.

The cloning and expression of the scFv-PE fusion protein is performed in several steps:

- extraction of RNA from B cell hybridoma cell line
- preparation of DNA coding for variable regions
- cloning of DNA and assembly of scFv and scFv–immunotoxins
- expression and purification of fusion proteins

5.1 Extraction of RNA from B cell hybridoma cell line

The preparation of and work with RNA always bears the risk of RNA degradation through RNases which are ubiquitously present. Special precautions must be taken to prevent contamination of RNA solutions with RNases. These precautions are outlined in detail in ref. 58. In brief, gloves must be worn at all times. The use of disposable materials is preferred. Non-disposable materials can be decontaminated from RNases either by dry heat sterilization (180°C for 10 h) or by extensive rinsing with DEPC water (*Protocol 9*), followed by extensive rinsing with sterile water, and autoclaving.

It is also important to avoid contamination of the RNA solution with genomic DNA, because the DNA can serve as a false template in the subsequent DNA amplifying reaction. Contaminating DNA can be removed by DNase II treatment or by incorporating an additional purification step with oligo(dT) membranes to extract the mRNA from the total RNA. Alternatively, the acid/phenol RNA extraction procedure (*Protocol 9*) can be used, which exploits the fact that DNA is relatively insoluble at acid pH, leaving the RNA in solution free from contaminating genomic DNA.

The use of aerosol resistant tips (ART-tips), which create a barrier between the pipetter and the solution to be pipetted, is highly recommended to avoid cross-contamination with DNA, which may be present in traces on the pipetter.

Protocol 9. Preparation of total RNA from hybridoma cells[a,b]

Equipment and reagents

- Hybridoma cells secreting the monoclonal antibody, viability ≥ 90%
- Sterile sodium chloride 0.9% (w/v)
- ART-tips (Molecular BioProducts)
- DEPC water:[c] prepare 0.1% (w/v) DEPC (diethyl pyrocarbonate) (Sigma) in distilled water and shake at 37°C for several hours—autoclave

- GTC buffer:[d] 4 M GTC (guanidium thiocyanate) (Sigma), 25 mM sodium citrate pH 7, is prepared as follows. Add to 25 g GTC (in manufacturer's bottle) 31.9 ml DEPC water and 1.76 ml 0.75 M sodium citrate pH 7. The solution can be kept at room temperature for at least three months. Shortly before use, add 7.2 µl 2-mercaptoethanol per ml GTC buffer.

- 2-Mercaptoethanol (Sigma)
- 2 M sodium acetate pH 4, autoclaved
- Acid phenol:[e] crystalline phenol is dissolved in 0.1 M sodium acetate pH 5, and overlayed by distilled water
- CI:[e] chloroform/isoamyl alcohol at a ratio of 49 : 1 (v/v)
- Isopropanol
- 80% ethanol (diluted with DEPC water)

- TE: 10 mM Tris–HCl, 1 mM EDTA pH 8, autoclaved—the Tris and EDTA should be from a newly opened bottle
- 3 M sodium acetate pH 5.3, autoclaved
- Agarose (Gibco)
- Ethidium bromide[f] (Sigma)
- Horizontal gel electrophoresis equipment
- UV transilluminator (or UV lamp)

Method

1. Wash 2×10^7 hybridoma cells once in 0.9% sodium chloride and resuspend in 50 µl 0.9% sodium chloride.

2. Keep the cells on ice. All following steps are performed on ice unless specified otherwise.

3. Add 1 ml GTC buffer to the cells and homogenize by passing through a syringe using first a needle of 20 gauge, then one of 23 gauge, and lastly one of 27 gauge. Ensure that no cell clumps remain in the suspension.

4. Remove 250 µl cell suspension and freeze the remainder at –20 °C for further preparations.

5. Add in this order to the cell suspension:
   - 250 µl GTC buffer
   - 50 µl 2 M Na acetate pH 4
   - 500 µl acid phenol
   - 100 µl CI

   Vortex.

6. Incubate 15 min on ice.

7. Centrifuge at 10 000 g for 20 min at 4 °C.

8. Transfer the aqueous phase (upper phase) into a tube containing 500 µl CI. Vortex.

9. Centrifuge at room temperature at 10 000 g for 5 min.

10. Transfer the aqueous phase (upper phase) into a tube containing 500 µl isopropanol.

11. Allow to precipitate overnight at –20 °C.[g]

12. Centrifuge at 10 000 g for 30 min at 4 °C.

13. Wash precipitate with 80% ethanol, and allow it to dry in the air for 5–10 min.

14. Resuspend in 50 µl TE.

Protocol 9. *Continued*

15. Run 3 μl of the solution on a 1.2% agarose gel containing 0.15 μl/ml ethidium bromide to ensure that the RNA is intact. Two distinct bands at 28S and 18S (ribosomal RNA) and an additional band at 5S (tRNA) should be clearly visible on a UV transilluminator.[h]

[a] This method is a modification from Chomczynski *et al.* (59).
[b] Alternatively, various RNA extraction kits are available from different vendors. In our laboratory, the Trizol reagent kit from Gibco worked well, but kits from other vendors may be of equal value.
[c] Do not use DEPC water together with a Tris-based buffer, since DEPC carboxymethylates amino groups.
[d] Care should be taken when handling GTC since it is a hazardous material.
[e] Phenol and chloroform containing solutions must be dispensed under a chemical hood.
[f] Ethidium bromide is a carcinogenic reagent and must be handled with care.
[g] Precipitation can also be performed at –80°C for 4 h.
[h] Working with UV light requires appropriate protection of eyes, e.g. by glasses or glass/plastic shields.

5.2 Preparation of DNA coding for variable regions

The preparation of double-stranded DNA coding for the variable regions of the heavy (V_H) and the light chain (V_L) is performed in two steps: In the first step, RNA is transcribed with reverse transcriptase to a single-stranded cDNA (*Protocol 10*). This cDNA is then amplified in a polymerase chain reaction (PCR, *Protocol 11*) to yield sufficient DNA for cloning into a vector. To avoid contamination of DNA, which will later be used for cloning, sterile solutions are required and gloves should be worn at all times.

Protocol 10. cDNA synthesis

Equipment and reagents

- Template RNA (*Protocol 9*)
- Thermocycler and PCR tubes (Perkin Elmer)[a]
- 1 M Tris–HCl pH 8.3, autoclaved
- 1 M KCl, autoclaved
- Oligo(dT) primer at 100 μM (Gibco)
- 50 mM MgCl₂, autoclaved

- 0.1 M DTT (Sigma)
- Deoxynucleotide mix (dNTP): 2 mM dATP, 2 mM dGTP, 2 mM dCTP, and 2 mM dTTP (Gibco)
- Autoclaved distilled H₂O (dH₂O)
- RNase inhibitor (Gibco)
- Reverse transcriptase (Gibco)

Method

1. To denature RNA in the presence of primers mix the following:
   - 1.6 μl 1 M Tris–HCl pH 8.3
   - 1.6 μl 1 M KCl
   - 1 μl oligo(dT) primer
   - 15.8 μl RNA

   and heat for 2 min at 96°C and for 8 min at 70°C.

2. Put immediately on ice.

3. Add the following:

  - 1 μl 1 M Tris–HCl pH 8.3
  - 1 μl 1 M KCl
  - 6 μl 50 mM MgCl$_2$
  - 5 μl 0.1 M DTT
  - 12.5 μl dNTP
  - 3 μl dH$_2$O
  - 0.5 μl RNase inhibitor
  - 1 μl reverse transcriptase

4. Spin down in a microcentrifuge at 10 000 *g* for 5 sec.

5. Transcribe at 42°C for 60 min, then heat to 70°C for 5 min.

6. Store at –20°C.

[a] If a thermocycler is used which heats the tubes from the bottom instead of from the top, the solution in the PCR tube must be overlayed with mineral oil to avoid evaporation of the fluid.

Protocol 11. PCR reaction with hot start

Equipment and reagents

- Template cDNA (*Protocol 10*)
- Thermocycler and PCR tubes (Perkin Elmer)[a]
- Primers for amplification of V$_H$ and V$_L$ (see *Figure 10*) at 20 μM
- Autoclaved distilled H$_2$O (dH$_2$O)
- dNTP (see *Protocol 10*)
- *Taq* polymerase (Gibco)

- 10 × PCR buffer:[b] 200 mM Tris–HCl, 500 mM KCl, 15 mM MgCl$_2$ pH 8.3
- Agarose (Gibco)
- Ethidium bromide[c] (Sigma)
- Horizontal gel electrophoresis equipment
- DNA marker, e.g. hundred base pair ladder (Gibco)

Method

1. Mix the following:

  - 2.5 μl cDNA
  - 1 μl forward primer
  - 1 μl back primer
  - 15.5 μl H$_2$O

2. Denature at 96°C for 5 min, cool down to primer annealing temperature,[b] e.g. 55°C, and hold the temperature.

3. Add the following: 2.5 μl dNTP, 2.5 μl 10 × PCR buffer, 0.2 μl *Taq* polymerase.

4. Continue PCR program with:

  - 72°C, 1 min

259

Protocol 11. *Continued*

then five cycles of:
- 94 °C, 30 sec
- 55 °C, 30 sec (primer annealing temperature)[b]
- 72 °C, 30 sec

then 25–30 cycles of:
- 94 °C, 30 sec
- 57 °C, 30 sec (polymerase annealing temperature)
- 72 °C, 30 sec

then 72 °C, 5 min.

5. Store the DNA at –20 °C.

6. Run a sample (2 μl) of the PCR reaction alongside a DNA marker on a 2% agarose gel containing 0.15 μg/ml ethidium bromide to confirm the amplification of the DNA. The V_H and V_L chains have a size of 320–350 bp.

[a] See *Protocol 10*, footnote *a*.
[b] It might be necessary to vary and optimize the concentrations for Tris–HCl and MgCl$_2$ as well as the primer annealing temperature.
[c] See *Protocol 9*, footnote *f*.

5.3 Cloning of DNA and assembly of DNA encoding scFv and scFv–immunotoxin

In this protocol a vector system is used which was developed by Wels *et al.* (64); it allows direct cloning of V_H and V_L fragments into a cloning vector which contains a synthetic (Gly$_4$Ser)$_3$ linker (*Protocol 12*). Alternatively, V_H and V_L chains can also be linked together by an additional PCR step; in this step, called overlap extension, primers overlapping for the 3′ end of the V_H chain and for the 5′ end of the linker and, secondly, for the 5′ end of the linker and the 3′ end of the V_L chain are used in a PCR reaction to amplify the 750 bp scFv fragment.

Once the scFv fragment is assembled, it can be (sub)cloned into an expression vector (*Protocol 14*). In the system described here, subcloning into expression vectors allows expression of a scFv (psw50) or a scFv–PE–immunotoxin (psw202), since the expression vector psw202 contains the gene for the truncated toxin. Cloning of the toxins PE and DT has been described elsewhere (65–67). The vector maps are given in *Figure 11*.

The cloning of the V_H and V_L fragments is performed in two steps. First the V_H chain is cloned into the vector pwwl52 or pwwc1; then the V_L chain is cloned into the plasmid containing the heavy chain (pwwl52–V_H or pwwc1–V_H) (see *Protocol 12*, footnote *a*). The procedures are the same for both steps, except that different restriction enzymes are used. A *Hind*III/

a) Sites of annealing of the primers

Back-primer   MVH-Back
                  MVL-Back

VH/VL: | FR1 | CDR1 | FR2 | CDR2 | FR3 | CDR3 | FR4 |

Forward-primer   MVH-For
                                                  MVL-For

b) Sequences of the primers[a] containing the restriction sites needed for cloning in the vector system described in this chapter

	HindIII
MVH Back	ATC GAT AAG CTT CAG GTS MAA CTG CAG SAG TCW GG
	Bst EII
MVH For	TGA GGA GAC GGT GAC CGT GGT CCC TTG GCC CC

MVL Back (κ)[b,c]	1	GAY ATT GTG MTG ACM CAG WCT CCA
	2	GAT GTT TTG ATG ACC CAA ACT CCA
	3	GAC ATT GTG CTR ACC CAG TCT CCA
	4	GAC ATC CAG ATG ACN CAG TCT CCA
	5	CAA ATT GTT CTC ACC CAG TCT CCA
	6	GAA AAT GTG CTC ACC CAG TCT CCA

	XbaI
MVL For(κ)[c]	ACT AGT CTA GAC GTT TSA KYT CCA GCT T
	BglII
MVL Back (λ)	ACT CAG AGA TCT GCA CTC ACC ACA TCA
	XbaI
MVL For(λ)	TAC CTC TAG ACC TAG GAC AGT CAG YTT GGT TCC

Figure 10. Primers used for amplification of heavy and light chain variable regions. [a]: The primer sequences have been determined by comparison with known V gene sequences from ref. 60 or modified from published primer sequences (61,62). [b]: Note that the MVL back (κ) primers do not contain a restriction site for cloning. Because FR1 in κ-chains has a high variability, it is necessary to find the best combination out of a set of primers first. Once the best primer for this specific κ-chain is determined, a PvuII site (CAG CTG) can be introduced in codons 3 and 4. [c]: Amplification of light chain variable regions from a hybridoma mRNA source bears the risk of accidently amplifying the light chain of the myeloma fusion partner. Therefore it is advisable to do a restriction analysis with an enzyme cutting frequently in V_L chains, e.g. BstNI, and compare the pattern to the expected pattern of the myeloma light chain BstNI digest (63).

BstEII digestion is performed to clone the V_H chains and a PvuII/XbaI (Vκ) or BglII/XbaI (Vλ) digestion is performed to clone the V_L chains. The enzymes for the second step, i.e. cloning of V_L are given in parentheses.

The transformation protocol describes a heat shock transformation of DNA into the E. coli host. Alternatively, a transformation by electroporation can be performed which generally yields higher transformation efficiencies. Other E. coli strains than XL1 blue and CC118 can also be used. In that case IPTG concentrations and incubation times for the growth of the bacteria have to be adapted, when the protein is expressed.

Several screening methods can be applied to verify the successful cloning into the cloning vector, e.g. PCR screening using the same primers as in the

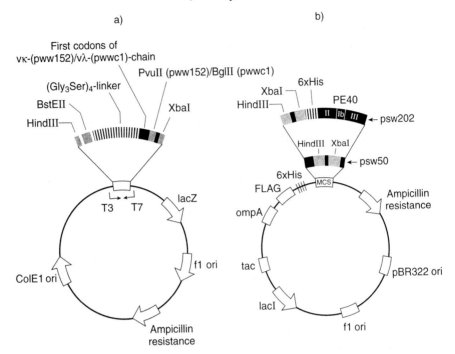

Figure 11. Cloning and expression vectors for direct cloning of V_H and V_L PCR fragments. (a) Heavy chains can be cloned into pww152/pwwc1 after *Hind*III/*Bst*EII digestion of plasmid and PCR fragment. κ-light chains are cloned into pww152–V_H by *Pvu*II/*Xba*I digestion, λ-light chains into pwwc1–V_H by *Bgl*II/*Xba*I digestion. Pww152 is a Blueskript II SK+ (Stratagene) derived cloning vector which was constructed by Wels *et al.* (64). Pwwc1 is a modification of pww152 for the cloning of λ-chains and was constructed by Gottstein (UTSW Medical Center, Dallas, unpublished). (b) The scFv is subcloned as a *Hind*III/*Xba*I fragment into psw50 for expression of a scFv or into psw202 for expression of a scFv–PE40–immunotoxin. Psw50 and psw202 are based on the FLAG expression vector (IBI) and were constructed by Wels *et al.* (64).

PCR amplification step or preparation of the DNA of bacterial colonies in combination with a restriction analysis. The latter will be described in *Protocol 13*.

Protocol 12. Cloning of PCR fragments encoding V_H and V_L regions into a cloning vector

Equipment and reagents

- DNA encoding for V_H and V_L containing the appropriate restriction sites (*Protocol 11*)
- Gel running buffer: 0.5 × TBE, prepared from a 5 × TBE stock (0.45 M Tris, 0.45 M borate, 10 mM EDTA pH 8)
- Ethidium bromide[b] (Sigma)

- Agarose (Gibco)
- Horizontal gel electrophoresis equipment
- Gel extraction kit (based on solubilization of agarose with sodium perchlorate and selective adsorption of DNA onto beads; Qiagen)

- Distilled H$_2$O (dH$_2$O), autoclaved
- Plasmid pWW152 or pwwc1 (see *Figure 11*)
- Restriction enzymes: *Hind*III, *Bst*EII, *Pvu*II, *Bgl*II, *Xba*I at 10 U/μl (New England Biolabs)
- 10 × restriction buffers 2 and 3 (New England Biolabs)
- 10 × BSA solution (1 mg/ml), sterile
- Alkaline phosphatase at 20 U/μl (Boehringer Mannheim)
- T4 ligase supplied with ligation buffer 10 × (Boehringer Mannheim)

- 1 M NaCl, autoclaved
- *E. coli* strain XL1 blue supercompetent (Stratagene)
- Petri dishes (Baxter) 100 × 15 mm for selection plates
- LB medium: 10 g bactotryptone (Gibco), 5 g bactoyeast (Gibco), 10 g NaCl, distilled H$_2$O to a volume of 1 litre pH 7, autoclaved
- LB–amp agar solution: add 15 g agar noble (Gibco) per litre LB medium, autoclave, cool to 50°C, add ampicillin to a final concentration of 20 μg/ml

A. *Digestion of PCR fragments and cloning vector*

1. Gel purify the PCR product for V$_H$ (V$_L$) on a 2% agarose gel using a gel extraction kit according to the manufacturer's recommendations. Resuspend the purified DNA in 20 μl dH$_2$O.

2. Run the purified PCR product and the cloning vector on a 1.5% agarose gel alongside a known standard to estimate the concentrations.

3. Add to each 1 μg DNA of the PCR product V$_H$ (V$_L$) and of the plasmid pww152/pwwc1 (pww152–V$_H$/pwwc1–V$_H$):

   - 1 μl *Hind*III (1 μl *Pvu*II plus 1 μl *Xba*I for Vκ, or 1 μl *Xba*I only for Vλ)
   - 2 μl 10 × BSA solution
   - 2 μl restriction buffer 2
   - dH$_2$O to a total volume of 20 μl^c

   Incubate for 1 h at 37°C.

4. Only for *Hind*III/*Bst*EII (*Bgl*II/*Xba*I) digestion, add to the DNA solution:

   - 1.5 μl *Bst*EII (1 μl *Bgl*II)
   - 0.5 μl restriction buffer 3
   - 1 μl 1 M NaCl solution
   - 2.5 μl dH$_2$O (3 μl dH$_2$O)

   Incubate for 3 h at 60°C (2 h at 37°C for *Bgl*II).

5. Heat inactivate at 65°C for 10 min (not necessary for *Hind*III/*Bst*EII digestion).

6. Only for the plasmids add 0.5 μl alkaline phosphatase and incubate for 30 min at 37°C. (The digested PCR fragment can be kept on ice during this time.)

7. Gel purify the digested PCR fragment on a 2% agarose gel and the digested plasmid on a 1% agarose gel using a gel extraction kit. Resuspend in 20 μl dH$_2$O.

B. *Ligation of PCR fragments to linearized cloning vector*

1. Run the digested and purified fragments and plasmids on a 1.5% agarose gel alongside a known standard to estimate the concentrations.

Protocol 12. *Continued*

2. Mix together:
   - 100 ng plasmid
   - 30 ng fragment
   - 1 μl 10 × ligation buffer
   - 1 μl ligase
   - dH$_2$O to a total volume of 10 μl[d]

   As a control, set-up a second mixture, omitting the fragment.

3. Incubate at 16°C for 5 h.

C. *Heat shock transformation of ligation into E. coli hosts*

1. Thaw *E. coli* bacteria on ice.

2. Transfer 100 μl *E. coli* suspension into a 1.5 ml reaction tube and add 5 μl ligation mixture. Mix well. Incubate for 30 min on ice. Meanwhile pour selection plates with LB–amp agar solution (twice as many as the number of the ligation samples).

3. Transfer the reaction tubes gently into a 42°C water-bath and incubate for 45 sec. Place immediately back on ice and incubate for another 2 min on ice.

4. Add 800 μl LB medium to each tube. Shake at 37°C for 1 h.

5. Spin down the tubes at 250 *g* in a microcentrifuge for 10 min.

6. Discard the supernatant and resuspend the bacteria in 200 μl fresh LB medium.

7. Plate the bacterial suspensions on LB–amp selection plates in two different volumes (50 μl and 150 μl).

8. Incubate at 37°C for 20 h.[e]

---

[a] If the V_H and V_L chains contain restriction sites which are used for cloning, it may be necessary to perform the steps in a different order, i.e. first clone the V_L chain into pww152, and then clone the V_H chain into pww152–V_L. It is also possible to clone the V_H and V_L chain separately into pww152 and then subclone, e.g. the V_L into pww152–V_H.

[b] See *Protocol 9*, footnote f.

[c] If the total volume exceeds 20 μl, the amounts of BSA solution (10 × stock) and restriction buffer (10 × stock) have to be adjusted to the higher volume.

[d] If the total volume exceeds 10 μl, the amount of ligation buffer (10 × stock) has to be adjusted to the higher volume. If the total volume exceeds 15 μl, it is advisable to first precipitate and resuspend the DNA to bring it to a higher concentration and then carry out the ligation in a 10 μl volume.

[e] If other *E. coli* strains than XL1 blue are used the temperature and time may have to be varied.

Protocol 13. Screening of bacterial colonies by restriction analysis

For this screening method, a small amount of DNA is prepared from the transformed bacteria by alkaline lysis and subsequent purification and then a restriction analysis is performed. The preparation of the DNA can be done according to ref. 58 or with a DNA miniprep kit, as described here.

Equipment and reagents

- *E. coli* colonies transformed with V_H, V_L, or scFv (from *Protocol 12* or *14*)
- LB–amp medium: LB medium (see *Protocol 12*) containing ampicillin at 20 μg/ml
- DNA minipreparation kit (Qiagen: Qiaprep Spin plasmid kit)[a]
- Distilled H_2O (dH_2O), autoclaved

- Restriction enzymes: *Hind*III, *Xba*I at 10 U/μl (New England Biolabs)
- 10 × restriction buffer 2 (New England Biolabs)
- Horizontal gel electrophoresis equipment
- Agarose (Gibco)
- Ethidium bromide[b] (Sigma)

Method

1. Inoculate 2 ml LB–amp medium with one single colony of transformed *E. coli* and shake at 37°C overnight.

2. Prepare DNA from the bacterial suspension with the DNA preparation kit according to the manufacturer's recommendations. Resuspend in 100 μl TE (typical yield: 3–5 μg).

3. Mix 2 μl DNA with 0.1 μl *Hind*III, 0.1 μl *Xba*I, 1.5 μl restriction buffer 2, and 11.3 μl dH_2O. Incubate at 37°C for 3 h.

4. Run samples (15 μl each) on a 2% agarose gel containing 0.15 μg/ml ethidium bromide. The excised inserts have a size of approximately 450 bp (V_H, V_L) or 750 bp (scFv).

[a] Other vendors sell similar DNA miniprep kits which may be equally suitable.
[b] See *Protocol 9*, footnote *f*.

Protocol 14. Subcloning of scFv into expression vectors

Equipment and reagents

- *E. coli* transformed with scFv in pww152 or pwwc1 (from *Protocol 12*)
- Expression plasmids psw50 and psw202 (see *Figure 11*)
- *E. coli* strain CC118[a]
- Agarose (Gibco)
- Ethidium bromide[b] (Sigma)
- Gel running buffer: 0.5 × TBE (see *Protocol 12*)

- Horizontal gel electrophoresis equipment
- Restriction enzymes: *Hind*III, *Xba*I at 10 U/μl (New England Biolabs)
- 10 × restriction buffer 2 (New England Biolabs)
- 10 × BSA solution (1 mg/ml), sterile
- Alkaline phosphatase at 20 U/μl (Boehringer Mannheim)
- Gel extraction kit (Qiagen)

Protocol 14. *Continued*

- T4 ligase (Boehringer Mannheim) supplied with 10 × ligation buffer
- Petri dishes (Baxter) 100 × 15 mm for selection plates
- LB–amp agar solution and LB medium (see *Protocol 12*)
- DNA minipreparation kit (see *Protocol 13*)

Method

1. Prepare DNA from an overnight culture of the transformed DNA as described in *Protocol 13* (until end of step 2).

2. Mix 1 µg of expression plasmids (psw50, psw202) with the following:
   - 0.25 µl *Hind*III
   - 0.25 µl *Xba*I
   - 1 µl BSA solution
   - 1 µl restriction buffer 2
   - dH$_2$O, to a total volume of 10 µl

   Mix 35 µl (≈ 1–3 µg DNA) of pww152–scFv DNA solution with the following:
   - 0.25 µl *Hind*III
   - 0.25 µl *Xba*I
   - 5 µl BSA solution
   - 5 µl restriction buffer 2
   - dH$_2$O, to a total volume of 50 µl

   Incubate the mixtures at 37°C for 1 h.

3. Only for expression plasmids. Add 0.5 µl alkaline phosphatase and incubate for 30 min at 37°C, while keeping the digested scFv plasmids on ice.

4. Gel purify the digested expression plasmids on a 1% agarose gel and the scFv (750 bp band) on a 2% agarose gel using a gel extraction kit.

5. Run a 1.5% agarose gel for estimation of concentrations of purified DNA.

6. Follow *Protocol 12*, part B and C for ligation and transformation, and *Protocol 13* for screening of the colonies.

7. Confirm by sequencing that the plasmid contains the desired DNA.[c]

[a] This strain has been used by Manoil *et al.* (68). Other *E. coli* strains may also be suitable.
[b] See *Protocol 9*, footnote f.
[c] Protocols for double-stranded and single-stranded sequencing are given in refs 69 and 70, respectively.

5.4 Expression and purification of fusion proteins

In contrast to the cloning of DNA, where general protocols yield the desired product in most cases, the expression and purification protocols for proteins have to be tailored to the protein being purified. Once it has been confirmed by sequencing that the DNA to be expressed does not contain any mutations or deletions which will affect the translation, there is still the problem of misfolding, denaturing, and formation of aggregates during the expression and purification procedures. Changing a single amino acid can necessitate changes in purification conditions to attain maximal yield and quality.

ScFv and their fusion proteins can be extracted from bacteria and purified in several ways. If the expression system contains a leader sequence, the proteins are secreted into the periplasm of the bacteria from which they can be released by osmotic shock and subsequently purified (71). The advantage of this method is that the proteins are present in their native form, and problems of misfolding are much reduced. The disadvantage is that the yield of the proteins is typically low (up to 2 mg/litre of bacterial culture). Proteins which lack a leader sequence can be purified from inclusion bodies where they are present in a denatured form. This requires that a denaturing–renaturing procedure be developed to refold the protein, but yields are typically high (72). Here we describe a method for extraction and purification of proteins from the cytoplasm under denaturing conditions.

After extracting the scFv or fusion protein by freeze-fracturing transformed bacteria, the protein is usually purified by a combination of affinity chromatography and gel filtration. The affinity chromatography can be based on binding of an antibody or chelating agent on the column to a specific tag encoded in the expression plasmid. Such tags can be the FLAG epitope on FLAG-derived plasmids (see *Figure 11*) or a histidine cluster. Histidine clusters in the protein and Ni^{2+} in the chromatography gel will form chelating complexes; the protein can then be eluted by high concentrations of imidazole. In addition to this purification step a gel chromatography can be performed, an example of which is described in *Protocol 9*. For purification of scFv (27 kDa) use a Sephadex G-75 column (Pharmacia) and for purification of scFv–PE (67 kDa) use a Sephadex S200 column (Pharmacia).

Protocol 15.   Expression of fusion proteins and extraction from the cytoplasm[a]

Equipment and reagents

- Transformed CC118 *E. coli* colonies (*Protocol 14*)
- LB medium (see *Protocol 12*)
- LB–amp: LB medium containing 100 μg/ml ampicillin
- LB–amp–glc: LB–amp containing 0.6% glucose
- IPTG (isopropyl-β-D-thiogalactopyranoside) (Boehringer Mannheim)
- Lysis buffer: 100 mM Tris–HCl pH 8.0, 1 mg/ml lysozyme, 0.3 mM PMSF[b] (phenylmethylsulfonyl fluoride) (Sigma), 1 μg/ml aprotinin (Boehringer Mannheim), 0.5 μg/ml leupeptin (Boehringer Mannheim), 1 μg/ml DNase I (Gibco)
- PBS (see *Protocol 1*)
- Guanidine hydrochloride, 6 M in PBS
- Ultracentrifuge

267

Protocol 15. *Continued*

Method

1. Inoculate 30 ml LB–amp–glc with a single colony of transformed *E. coli* and shake at 37 °C overnight.

2. Add 1 litre LB–amp and grow at 37 °C with shaking until the A_{550} has reached 0.5.

3. Add IPTG to a final concentration of 0.5 mM and grow at room temperature with shaking for 45 min.[c]

4. Spin down at 10 000 *g* for 10 min at 4 °C and resuspend the pellet in 0.5 ml lysis buffer. Lyse the cells by several cycles of freezing in liquid nitrogen and subsequent thawing.

5. Add 5 ml 6 M guanidine hydrochloride per gram of pellet and incubate with stirring for 30 min at room temperature.

6. Centrifuge at 100 000 *g* for 25 min at room temperature.

7. Transfer the supernatant into a fresh tube and add 1 vol. PBS. The protein solution can now be purified by chromatography (see *Protocol 16*).

[a] This method was used by Schmidt *et al.* (73).
[b] PMSF is highly toxic and must be handled with care.
[c] The concentration of IPTG as well as the temperature and time of growing the bacteria may have to be optimized, especially if strains other than CC118 are used.

Protocol 16. Purification of fusion proteins using a nickel column[a]

Equipment and reagents

- Protein solution (*Protocol 15*)
- Low-pressure chromatography equipment
- Column loaded with 5 ml Ni–NTA–Sepharose (Qiagen)
- Elution buffer: 250 mM imidazole in PBS
- Running buffer: 3 M guanidine hydrochloride, 20 mM imidazole in PBS (PBS, see *Protocol 1*)
- 5 ml fraction collector tubes (Sarstedt)
- SDS–PAGE equipment

Method

1. Load the protein solution on a Ni–NTA column equilibrated in the running buffer. Run at 1 ml/min.

2. Wash the column with running buffer until the absorbance at 280 nm returns to zero.

3. Elute with 250 mM imidazole in PBS. Collect 1 ml fractions.

4. Analyse a sample of each fraction in the main peak (corresponding to conjugate) by SDS–PAGE. Run the sample under non-reducing conditions on a 4–12% polyacrylamide gradient gel (IT) or a 8–25% gel

(scFv). Detect protein bands using Coomassie blue staining. The scFv has a molecular size of 27 kDa, the scFv–immunotoxin of 67 kDa.

5. Pool fractions containing the scFv or the scFv–immunotoxin and dialyse against PBS in three steps: two times for 3 h and once overnight.[b]

[a] This method was used by Schmidt *et al.* (73).
[b] To avoid precipitation of the proteins by too rapid renaturation it may be necessary to reduce the guanidine hydrochloride concentration gradually; e.g. by dialysing against PBS containing decreasing concentrations of guanidine hydrochloride. Alternatively, the refolding can be performed on the Ni–NTA column by running a 6 M–1 M urea gradient, before eluting with imidazole.

6. Biochemical characterization of ITs

Before ITs are analysed for their ability to kill cells *in vitro* or *in vivo* it should be ensured that both the antibody component and the toxin component are functional. The ability of the antibody component to retain antigen binding activity can be assessed by RIA, ELISA, or flow cytometry (74). There is sometimes a small (about twofold) increase in the concentration of IT that is required to give the same binding signal as the unmodified antibody. Competition assays can also be used, where IT and antibody are compared in competition with labelled antibody for binding sites to antigens on cell surfaces.

The enzymatic activity of the toxin component of the conjugate is measured in a cell-free system in comparison to the unmodified toxin component. Toxins or toxin components which inactivate ribosomes can be tested for enzymatic activity in a rabbit reticulocyte lysate system (see *Protocol 17*).

Protocol 17.   Cell-free assay of protein synthesis

Equipment and reagents

- Liquid scintillation counter, scintillation fluid, and vials (ICN)
- 2.5 cm filter paper discs (Whatman)
- 1 ml glass tubes (Sarstedt)
- Test IT
- Corresponding unmodified toxin, A-chain, or RIP
- Rabbit reticulocyte lysate (Sigma)
- Trichloroacetic acid (TCA) (Sigma)
- Methanol (Aldrich)
- Acetone (Aldrich)
- PBS–BSA: PBS (*Protocol 1*) containing 0.1 mg/ml bovine serum albumin (Sigma)
- 0.4 M 2-Me buffer: 0.4 M 2-mercaptoethanol (Sigma) in PBS–BSA

- Hydrogen peroxide (Sigma)
- 0.2 M 2-Me buffer: 0.2 M 2-mercaptoethanol (Sigma) in PBS–BSA
- Enzyme mixture:[a] 330 mg ammonium acetate, 250 mg creatine phosphate (Sigma), 31 mg adenosine 5′-triphosphate (Sigma), 21 mg magnesium acetate (Sigma), 5.5 mg guanosine 5′-triphosphate (Sigma), 3 mg creatine phosphokinase (Sigma), 0.8 mg L-glutamine (Sigma), 0.5 mg alanine (Sigma), 0.5 mg cysteine (Sigma), dissolved in 5 ml of PBS–BSA
- [³H]leucine, 65 Ci/mmol (Amersham)
- Incubation mixture: 51 vol. PBS–BSA, 12 vol. enzyme mixture, 4 vol. [³H]leucine

Protocol 17. *Continued*

Method[b]

1. Make up the test IT and the corresponding toxin at 2×10^{-9} M with respect to toxin in PBS–BSA.

2. Incubate 100 µl of each sample with 100 µl of 0.4 M 2-Me buffer for 1 h at 37°C to reduce the disulfide bond between the A-chain and B-chain or antibody.

3. Dilute both samples six times by threefold serial dilutions in 0.2 M 2-Me buffer.

4. Add 13 µl aliquots from each dilution to 67 µl of incubation mixture dispensed into 1 ml glass tubes, in triplicate.

5. Freshly thaw the rabbit reticulocyte lysate, dilute 1/2.5 with PBS–BSA, and keep on ice.

6. Start the reaction by mixing 50 µl of the lysate with the contents of each tube at 15 sec intervals.

7. After incubation for 25 min at room temperature, remove 30 µl of the reaction mixture from each tube onto a 2.5 cm diameter filter paper disc and immediately immerse the disc into the 5% TCA solution to stop the reaction.

8. Gently stir the discs in the TCA solution for 1 h and then wash the discs in methanol for 1 min, methanol/acetone (1:1) for 1 min, and acetone for 1 min.

9. Bleach the discs in hydrogen peroxide/water (1:3) for 10 sec, wash the discs with copious quantities of water, and then allow the discs to dry.

10. Measure [³H]leucine incorporation in a liquid scintillation counter.

11. Determine the mean incorporation of [³H]leucine of samples as a percentage of controls without toxin. Plot this against the toxin concentration of the IT and unmodified toxin. Compare the two lines to determine the ribosome inactivating activity of the IT.

[a] The enzyme mixture can be frozen in small volumes for future use.
[b] The method was used by Wawrzynczak *et al.*(16) and was based on that used by Forrester *et al.* (40).

Acknowledgements

We thank Karen Schiller for excellent secretarial assistance.

References

1. Blakey, D. C., Wawrzynczak, E. J., Wallace, P. M., and Thorpe, P. E. (1988). In *Progress in allergy (monoclonal antibody therapy)* (ed. H. Waldmann), pp. 50–90. Karger, Basel, Switzerland.
2. Wawrzynczak, E. J. (1992). *Anti-Cancer Drug Des.*, **7**, 427.
3. Vitetta, E. S., Thorpe, P. E., and Uhr, J. W. (1993). *Immunol. Today*, **14**, 148.
4. Brinkmann, U. and Pastan, I. (1994). *Biochim. Biophys. Acta*, **1198**, 27.
5. Gottstein, C., Winkler, U., Bohlen, H., Diehl, V., and Engert, A. (1994). *Ann. Oncol.*, **5**, s97.
6. Vallera, D. A. (1988). *Cancer Treat. Res.*, **37**, 515.
7. Salzman, D. and LeMaistre, F. (1993). *Cancer Treat. Res.*, **68**, 133.
8. Ghetie, V. and Vitetta, E. (1994). *Pharmacol. Ther.*, **63**, 209.
9. Estis, L., Krueger, J., Nichols, J., and Parker, K. (1995). *Fourth International Symposium on Immunotoxins*, June 8–10, 167. Myrtle Beach, SC.
10. Wawrzynczak, E. J. (1991). *Br. J. Cancer*, **64**, 679.
11. Thorpe, P. E., Cumber, A. J., Williams, N., Edwards, D. C., Ross, W. C., and Davies, A. J. (1981). *Clin. Exp. Immunol.*, **43**, 195.
12. Pastan, I. and FitzGerald, D. (1991). *Science*, **254**, 1173.
13. Wawrzynczak, E. J., Watson, G. J., Cumber, A. J., Henry, R. V., Parnell, G. D., Reiber, E. P., *et al.* (1991). *Cancer Immunol. Immunother.*, **32**, 289.
14. Lambert, J. M., McIntyre, G., Gauthier, M. N., Zullo, D., Rao, V., Steeves, R. M., *et al.* (1991). *Biochemistry*, **30**, 3234.
15. Lambert, J. M., Blattler, W. A., McIntyre, G. D., Goldmacher, V. S., and Scott, C. F. Jr. (1988). In *Immunotoxins* (ed. A. E. Frankel), pp. 175–213. Kluwer Academic Publishers, Norwell, MA.
16. Wawrzynczak, E. J., Henry, R. V., Cumber, A. J., Parnell, G. D., Derbyshire, E. J., and Ulbrich, N. (1991). *Eur. J. Biochem.*, **196**, 203.
17. Nakamura, R. M., Spiegelberg, H. L., Lee, S., and Weigle, W. O. (1968). *J. Immunol.*, **100**, 376.
18. Ghetie, M. A., May, R. D., Till, M., Uhr, J. W., Ghetie, V., Vitetta, E. S., *et al.* (1993). *Cancer Res.*, **48**, 2610.
19. O'Hare, M., Brown, A. N., Hussain, K., Gebhardt, A., Watson, G., Roberts, L. M., *et al.* (1990). *FEBS Lett.*, **273**, 200.
20. Woodworth, T. G. and Nichols, J. C. (1993). *Cancer Treat. Res.*, **68**, 145.
21. FitzGerald, D. and Pastan, I. (1993). *Ann. N Y Acad. Sci.*, **685**, 740.
22. Winter, G. and Harris, W. J. (1993). *Immunol. Today*, **14**, 243.
23. Marks, J. D., Griffiths, A. D., Malmqvist, M., Clackson, T. P., Bye, J. M., and Winter, G. (1992). *Bio/Technology*, **10**, 779.
24. Hawkins, R. E., Russell, S. J., and Winter, G. (1992). *J. Mol. Biol.*, **226**, 889.
25. Hoogenboom, H. R. and Winter, G. (1992). *J. Mol. Biol.*, **227**, 381.
26. Kuan, C.-T. and Pastan, I. (1995). *Fourth International Symposium on Immunotoxins*, June 8–10, 54. Myrtle Beach, SC.
27. Sheldon, K. and Raso, V. (1995). *Fourth International Symposium on Immunotoxins*, June 8–10, 50. Myrtle Beach, SC.
28. Zhan, J. B., Stayton, P., and Press, O. W. (1995). *Fourth International Symposium on Immunotoxins*, June 8–10, 51. Myrtle Beach, SC.

29. Ramakrishnan, S., Mohanraj, D., and Lu, B.-Y. (1995). *Fourth International Symposium on Immunotoxins*, June 8–10, 41. Myrtle Beach, SC.

30. Lord, J. M., Pitcher, L. M., Roberts, S., Fawell, S., Zdanofsky, A. G., and FitzGerald, D. J. (1995). *Fourth International Symposium on Immunotoxins*, June 8–10, 57. Myrtle Beach, SC.

31. Lambert, J. M. and Blattler, W. A. (1988). *Cancer Treat. Res.*, **37**, 323.

32. Barbieri, L. and Stirpe, F. (1982). *Cancer Surveys*, **1**, 489.

33. Collier, R. J., and Kandel, J. (1971). *J. Biol. Chemistry*, **246**, 1496.

34. Fulton, R. J., Blakey, D. C., Knowles, P. P., Uhr, J. W., Thorpe, P. E., and Vitetta, E. S. (1986). *J. Biol. Chem.*, **261**, 5314.

35. Cumber, A. J., Forrester, J. A., Foxwell, B. M., Ross, W. C., and Thorpe, P. E. (1985). In *Drug and Enzyme Targeting. Part A* (ed. K. J. Widder), p. 207. Academic Press, Orlando.

36. Wawrzynczak, E. J. and Thorpe, P. E. (1987). In *Immunoconjugates: antibody conjugates in radioimaging and therapy of cancer* (ed. C. W. Vogel), pp. 28–55. Oxford University Press, New York.

37. Masuho, Y., Kishida, K., Saito, M., Umeto, N., and Hara, T. (1982). *J. Biochem.*, **91**, 1583.

38. Carlsson, J., Drevin, H., and Axen, R. (1978). *Biochem. J.*, **173**, 723.

39. Thorpe, P. E. (1985). In *Monoclonal antibodies, 1984: biological and clinical applications* (ed. A. Pinchera, G. Doria, F. Dammacco, and A. Bargellesi), pp. 475–512. Editrice Kurtis, Milano, Italy.

40. Forrester, J. A., McIntosh, D. P., Cumber, A. J., Parnell, G. D., and Ross, W. C. (1984). *Cancer Drug Delivery*, **1**, 283.

41. Hwang, K. M., Foon, K. A., Cheung, P. H., Pearson, J. W., and Oldham, R. K. (1984). *Cancer Res.*, **44**, 4578.

42. Worrell, N. R., Cumber, A. J., Parnell, G. D., Mirza, A., Forrester, J. A., and Ross, W. C. (1986). *Anti-Cancer Drug Design*, **1**, 179.

43. Thorpe, P. E., Wallace, P. M., Knowles, P. P., Relf, M. G., Brown, A. N. F., Watson, G. J., *et al.* (1987). *Cancer Res.*, **47**, 5924.

44. Thorpe, P. E., Wallace, P. M., Knowles, P. P., Relf, M. G., Brown, A. N. F., Watson, G. J., *et al.* (1988). *Cancer Res.*, **48**, 6396.

45. Lambert, J. M., Senter, P. D., Yau-Young, A., Blattler, W. A., and Goldmacher, V. S. (1985). *J. Biol. Chem.*, **260**, 12035.

46. Thorpe, P. E. and Ross, W. C. (1982). *Immunol. Rev.*, **62**, 119.

47. Ramakrishnan, S. and Houston, L. L. (1984). *Cancer Res.*, **44**, 201.

48. Ellman, G. L. (1959). *Arch. Biochem. Biophys.*, **82**, 70.

49. Lambert, J. M., Goldmacher, V. S., Collinson, A. R., Nadler, L. M., and Blattler, W. A. (1991). *Cancer Res.*, **51**, 6236.

50. Thorpe, P. E., Ross, W. C., Brown, A. N., Myers, C. D., Cumber, A. J., Foxwell, B. M., *et al.* (1984). *Eur. J. Biochem.*, **140**, 63.

51. Zangemeister-Wittke, U., Collinson, A. R., Frosch, B., Waibel, R., Schenker, T., and Stahel, R. A. (1994). *Br. J. Cancer*, **69**, 32.

52. Youle, R. J. and Neville, D. M. Jr. (1980). *Proc. Natl. Acad. Sci. USA*, **77**, 5483.

53. Bjorn, M. J., Ring, D., and Frankel, A. (1985). *Cancer Res.*, **45**, 1214.

54. Ross, W. C., Thorpe, P. E., Cumber, A. J., Edwards, D. C., Hinson, C. A., and Davies, A. J. (1980). *Eur. J. Biochem.*, **104**, 381.

55. Ghetie, V., Swindell, E., Uhr, J. W., and Vitetta, E. S. (1993). *J. Immunol. Methods*, **166**, 117.
56. Knowles, P. P. and Thorpe, P. E. (1987). *Anal. Biochem.*, **160**, 440.
57. Kreitman, R. J. and Pastan, I. (1994). In *Advances in pharmacology* (ed. J. T. August, M. W. Anders, F. Murad, and J. T. Coyle), pp. 203–10. Academic Press, San Diego.
58. Sambrook, J., Fritsch, E. F., and Maniatis, T. (ed.) (1989). *Molecular cloning, a laboratory manual*. Cold Spring Harbor Laboratory Press.
59. Chomczynski, P. and Sacchi, N. (1987). *Anal. Chem.*, **162**, 156.
60. Kabat, E. A., Wu, T. T., Perry, H. M., Gottesman, K. S., and Foeller, C. (ed.) (1991). *Sequences of proteins of immunological interest*. US Department of Health and Human Services, National Institutes of Health, Bethesda.
61. Orlandi, R., Gussow, D. H., Jones, P. T., and Winter, G. (1989). *Proc. Natl. Acad. Sci. USA*, **86**, 3833.
62. Ward, E. S., Gussow, D., Griffiths, A. D., Jones, P. T., and Winter, G. (1989). *Nature*, **341**, 544.
63. Clackson, T., Hoogenboom, H. R., Griffiths, A. D., and Winter, G. (1991). *Nature*, **352**, 624.
64. Wels, W., Beerli, R., Hellmann, P., Schmidt, M., Marte, B. M., and Kornilonn, E. S. (1995). *Int. J. Cancer*, **60**, 137.
65. Hwang, J., FitzGerald, D. J., Adhya, S., and Pastan, I. (1987). *Cell*, **48**, 129.
66. Greenfield, L., Bjorn, M. J., Horn, G., Fong, D., Buck, G. A., Collier, R. J., *et al.* (1983). *Proc. Natl. Acad. Sci. USA*, **80**, 6853.
67. Bishai, W. R., Rappuoli, R., and Murphy, J. R. (1987). *J. Bacteriol.*, **169**, 5140.
68. Manoil, C. and Beckwith, J. (1985). *Proc. Natl. Acad. Sci. USA*, **82**, 8129.
69. Murphy, G. and Ward, E. S. (1989). In *Nucleic acids sequencing: a practical approach* (ed. C. J. Howe and E. S. Ward), pp. 99–115. IRL Press, Oxford.
70. Howe, C. J. and Ward, E. S. (1994). In *Essential molecular biology: a practical approach* (ed. T. A. Brown), pp. 157–82. Oxford University Press, Oxford.
71. Koshland, D. and Bottstein, D. (1980). *Cell*, **20**, 749.
72. Buchner, J., Pastan, I., and Brinkmann, U. (1992). *Anal. Biochem.*, **205**, 263.
73. Schmidt, M., Hynes, N. E., Groner, B., and Wels, W. (1995). *Cancer Res.*,
74. Klaus, G. G. B. (ed.) (1987). *Lymphocytes: a practical approach*. IRL Press pp. 35–42 and 169.

A1

List of suppliers

Agar Scientific Ltd., 66a Cambridge Road, Stansted, Essex CM24 8DA, UK.

Aldrich Chemical Co. Inc., 1001 W. St. Paul Avenue, Milwaukee, Wisconsin 53233, USA.

Aldrich Chemical Co., The Old Brickyard, Gillingham, Dorset SP8 4JL, UK.

America A/S, Winthersmollevej 1, DK 7700 Thisted, Denmark.

American National Can, Greenwich, CT 06836, USA.

American Type Culture Collection, 12301 Parklawn Drive, Rockville, Maryland 20852-1776, USA.

Amersham

Amersham International plc., Lincoln Place, Green End, Aylesbury, Buckinghamshire HP20 2TP, UK.

Amersham Corporation, 2636 South Clearbrook Drive, Arlington Heights, IL 60005, USA.

Amicon Inc., Cherry Hill Drive, Beverly, MA 01915, USA.

Amicon Ltd., Upper Mill, Stonehouse, Gloucestershire GL10 2BJ, UK.

Amplify, Amersham, Little Chalfont, Buckinghamshire, UK.

Anachem, Charles Street, Luton, Bedfordshire LU2 0EB, UK. (Also suppliers of Gilson pipettes)

Anderman

Anderman and Co. Ltd., 145 London Road, Kingston-Upon-Thames, Surrey KT17 7NH, UK.

Avanti Polar Lipids, 700 Industrial Park Drive, Alabaster, Alabama 35007, USA.

Bachem (Switzerland), Hauptstrasse 144, CH-4416 Bubendorf, Switzerland.

Baxter Diagnostics Inc., 1430 Waukegan Road, McGaw Park, IL 60085-6787, USA.

Baxter Diagnostics Inc., Wallingford Road, Compton, Newbury, Berkshire, UK.

Beckman Instruments

Beckman Instruments UK Ltd., Progress Road, Sands Industrial Estate, High Wycombe, Buckinghamshire HP12 4JL, UK.

Beckman Instruments Inc., PO Box 3100, 2500 Harbor Boulevard, Fullerton, CA 92634, USA.

Becton Dickinson, European HQ, Denderstraat 24, B-9320 Erenbodegem-Aalst, Belgium.

Becton Dickinson and Co., Between Towns Road, Cowley, Oxford OX4 3LY, UK.

Becton Dickinson and Co., 2 Bridgewater Lane, Lincoln Park, NJ 07035, USA.

Behringwerke, Marburg, Germany.

BIAcore Ltd., Davy Avenue, Knowlhill, Milton Keynes MK5 8PH, UK.

Bio

Bio 101 Inc., c/o Statech Scientific Ltd., 6163 Dudley Street, Luton, Bedfordshire LU2 0HP, UK.

Bio 101 Inc., PO Box 2284, La Jolla, CA 92038/2284, USA.

Biogenesis, 7 New Fields, Stinsford Road, Poole, Dorset BH17 0NF, UK.

Biological Detection Systems, Pittsburgh, PA, USA.

Biomedicals, Rheinstrasse 28-32, CH-4302 Augst, Switzerland.

bio-Merieux, 69280 Marcy-l'Etoile, France.

Bio-Rad Laboratories

Bio-Rad Laboratories Ltd., Bio-Rad House, Maylands Avenue, Hemel Hempstead HP2 7TD, UK.

Bio-Rad Laboratories, Division Headquarters, 3300 Regatta Boulevard, Richmond, CA 94804, USA.

Biotecx Laboratories Inc., 6023 South Loop East, Houston, TX 77033-9980, USA.

Boehringer Mannheim

Boehringer Mannheim UK (Diagnostics and Biochemicals) Ltd., Bell Lane, Lewes, East Sussex BN17 1LG, UK.

Boehringer Mannheim Corporation, Biochemical Products, 9115 Hague Road, PO Box 504, Indianopolis, IN 46250-0414, USA.

Boehringer Mannheim Biochemica, GmbH, Sandhofer Str. 116, Postfach 310120 D-6800 Ma 31, Germany.

British Drug Houses (BDH) Ltd., Poole, Dorset, UK.

Calbiochem–Novabiochem (UK) Ltd., 3 Heathcoat Building, Highfields Science Park, University Boulevard, Nottingham NG7 2QJ, UK.

Calbiochem–Novabiochem Corporation, 10394 Pacific Court Centre, San Diego, CA 92121, USA.

Celltech Therapeutics Ltd., 216 Bath Road, Slough, Berkshire SL1 4EN, UK.

Central Labs Netherlands Red Cross (CNB), Plesmanlaan 125, 1066 CX Amsterdam, The Netherlands.

Chiron Mimotopes, 11055 Roselle St, San Diego, CA 92121, USA.

Chiron Mimotopes, (now Chiron Technologies) PO Box 1415, Clayton South, Victoria 3169, Australia.

Chiron Mimotopes, 10 rue Chevreul, 92150 Suresnes, France.

Cinnabiotecx—OWENS

Conair Churchill, Uxbridge, Middlesex, UK.

Coulter Electronics Ltd., Northwell Drive, Luton, Bedfordshire LU3 3RH, UK.

CP Pharmaceuticals Ltd., Ash Road North, Wrexham, Clwyd LL13 9UF, UK.

DAKO A/S, Produktionsvej 42, DK-2000 Glostrup, Denmark.

DAKO (UK) Ltd., 16 Manor Courtyard, Hughden Avenue, High Wycombe, Buckinghamshire H13 5RE, UK.

DAKO Corporation, 6392 Via Real, Carpinteria, CA 93013, USA.

Dakopatts A/S, Produktionsvej 42, Postbox 1359, DK 2600 Glostrup, Denmark.

Difco Laboratories

Difco Laboratories Ltd., PO Box 14B, Central Avenue, West Molesey, Surrey KT8 2SE, UK.

Difco Laboratories, PO Box 331058, Detroit, MI 48232-7058, USA.

Du Pont

Dupont (UK) Ltd., Industrial Products Division, Wedgwood Way, Stevenage, Hertfordshire SG1 4QU, UK.

Du Pont Co. (Biotechnology Systems Division), PO Box 80024, Wilmington, DE 19880-002, USA.

Dynatech Laboratories Ltd., Daux Road, Billingshurst, West Sussex RH14 9SJ, UK.

European Collection of Animal Cell Culture, Division of Biologics, PHLS Centre for Applied Microbiology and Research, Porton Down, Salisbury, Wiltshire SP4 0JG, UK.

Falcon (Falcon is a registered trademark of Becton Dickinson and Co.)

Fisher Scientific Co., 711 Forbest Avenue, Pittsburgh, PA 15219-4785, USA.

Flow Cytometry Standards, PO Box 1336, 2302 Blt Leiden, The Netherlands.

Flow Laboratories, Woodcock Hill, Harefield Road, Rickmansworth, Hertfordshire WD3 1PQ, UK.

Fluka

Fluka–Chemie AG, CH-9470, Buchs, Switzerland.

Fluka Chemicals Ltd., The Old Brickyard, New Road, Gillingham, Dorset SP8 4JL, UK.

Gibco BRL

Gibco BRL (Life Technologies Ltd., Trident House, Renfrew Road, Paisley PA3 4EF, UK.

Gibco BRL (Life Technologies Inc.), 3175 Staler Road, Grand Island, NY 14072-0068, USA.

Halocarbon, PO Box 661, River Edge, NH 07661, USA.

C.A. Hendley (Essex) Ltd., Oakwood Hill Industrial Estate, Loughton, Essex, UK.

Arnold R. Horwell, 73 Maygrove Road, West Hampstead, London NW6 2BP, UK.

Hybaid

Hybaid Ltd., 111–113 Waldegrave Road, Teddington, Middlesex TW11 8LL, UK.

Hybaid, National Labnet Corporation, PO Box 841, Woodbridge, NJ 07095, USA.

HyClone Laboratories, 1725 South HyClone Road, Logan, UT 84321, USA.

ICN Biomedicals Inc., Costa Mesa, CA, USA.

ICN Pharmaceuticals Inc., 3300 Hyland Avenue, Costa Mesa, CA 92626, USA.

ICN Pharmaceuticals Ltd., Thame Park Business Centre, Wenman Road, Thame, Oxfordshire OX9 3XA, UK.

IGS, Nikon, Nippon Kogaku ICK, Tokyo, Japan.

International Biotechnologies Inc., 25 Science Park, New Haven, Connecticut 06535, USA.

International Blood Group Reference Laboratory (IBGRL), Dagger Lane, Elstree, Hertfordshire WD6 3BX, UK.

Invitrogen Corporation

Invitrogen Corporation, 3985 B Sorrenton Valley Building, San Diego, CA 92121, USA.

Invitrogen Corporation, c/o British Biotechnology Products Ltd., 410 The Quadrant, Barton Lane, Abingdon, Oxfordshire OX14 3YS, UK.

Jackson Immuno Research Laboratories Inc., 61–63 Dudley Street, Luton, Bedfordshire LU2 0NP, UK.

Jackson Immuno Research Laboratories, 827 W. Baltimore Pike, West Grove, PA 19390, USA.

Kabi Pharmacia Diagnostics, Gydevangen 21, 3450 Allerod, Denmark.

Kirkegaard and Perry Labs (KPL), 2 Cessna Court, Gaithersburg, MD 20879, USA.

Kodak: Eastman Fine Chemicals, 343 State Street, Rochester, NY, USA.

Labsystems Oy, PO Box 8, FIN-0-881 Helsinki, Finland.

Leo Pharmaceutical Products BV, Pampusalaan 186, 1382 JS Weesp, The Netherlands.

Life Sciences International (UK) Ltd., Unit 5, The Ringway Centre, Edison Road, Basingstoke, Hampshire RG21 6ZZ, UK. (Suppliers of Finnpipettes and Labsystems microtitre plates).

Life Technologies Inc., 8451 Helgerman Court, Gaithersburg, MN 20877, USA.

Litex A/S, Copenhagen, Denmark.

Merck Ltd. (BDH), Hunter Boulevard, Magna Park, Lutterworth, Leicestershire LE17 4XN, UK.

Merck Industries Inc., 5 Skyline Drive, Nawthorne, NY 10532, USA.

Merck, Frankfurter Strasse, 250, Postfach 4119, D-64293, Germany.

Millipore

Millipore (UK) Ltd., The Boulevard, Blackmoor Lane, Watford, Hertfordshire WD1 8YW, UK.

Millipore Corp./Biosearch, PO Box 255, 80 Ashby Road, Bedford, MA 01730, USA.

Molecular Bio-Products Inc., 9888 Waples Street, San Diego, CA 92121, USA.

Molecular Bio-Products Inc., distributor in the UK: Merck Ltd., BDH Laboratory Supplies, Poole BH15 1TDF, UK.

Molecular Devices Corporation, 3180 Porter Drive, Palo Alto, CA 94304, USA.

Molecular Probes, Eugene, OR, USA.

Molecular Probes Europe BV, Poort Bebrouw, Rijnsburgerweg 10, 2333 AA Leiden, The Netherlands.

NE Technology Ltd., Sighthill, Edinburgh, UK.

New England Biolabs (NBL)

New England Biolabs (NBL), 32 Tozer Road, Beverley, MA 01915-5510, USA.

New England Biolabs (NBL), c/o CP Labs Ltd., PO Box 22, Bishops Stortford, Hertfordshire CM23 3DH, UK.

Nikon Corporation, Fuji Building, 23 Marunouchi 3-chome, Chiyoda-ku, Tokyo, Japan.

Nordic Immunological Laboratories BV, Langestraat 55-61, PO Box 22, 5000 AA Tilburg, The Netherlands.

Novabiochem, Calbiochem-Novabiochem AG, Weidenmattweg 4, Postfach, CH-4448 Laufelfingen, Switzerland.

Nucleopore Inc., 7035 Commerce Circle, Pleasanton, California 94566, USA.

Nunc A/S, Postbox 280, Kamstrup, DK-4000 Roskilde, Denmark.

Nycomed UK Ltd., Nycomed House, 2111 Coventry Road, Sheldon, UK.

Omega Optical, Brattleboro, VT, USA.

Oxoid, Basingstoke, Hampshire, UK.

Perkin-Elmer

Perkin-Elmer Ltd., Maxwell Road, Beaconsfield, Buckinghamshire HP9 1QA, UK.

Perkin-Elmer Ltd., Post Office Lane, Beaconsfield, Buckinghamshire HP9 1QA, UK.

Perkin Elmer-Cetus (The Perkin-Elmer Corporation), 761 Main Avenue, Norwalk, CT 0689, USA.

Pharmacia Biotech Europe Procordia EuroCentre, Rue de la Fuse-e 62, B-1130 Brussels, Belgium.

Pharmacia LKB Biotechnology AB, S-75182 Uppsala, Sweden.

Pharmingen, 10975 Torreyana Road, San Diego, CA 92121, USA.]

Phase Sep, Deeside, Clwyd, UK.

Pierce, 3747 N Meridian Road, PO Box 117, Rockford, IL 61105, USA.

Pierce Europe BV, PO Box 1512, 3260 BA and Beijerland, Holland.

Polysciences Inc., Warrington, PA, USA.

Promega

Promega Ltd., Delta House, Enterprise Road, Chilworth Research Centre, Southampton, UK.

Promega Corporation, 2800 Woods Hollow Road, Madison, WI 53711-5399, USA.

Qiagen

Qiagen Inc., c/o Hybaid, 111–13 Waldegrave Road, Teddington, Middlesex TW11 8LL, UK.

Qiagen Inc., 9259 Eton Avenue, Chatsworth, CA 91311, USA.

Quidel, 10165 McKellar Court, San Diego, CA 92121, USA.

Rathburn Chemicals Ltd., Caberston Road, Walkerburn, Peebleshire, Scotland EH13 6AU, UK.

Rudolf Brand GmbH, PO Box 310, D-6980 Wertheim/Main, Germany.

Sarstedt Inc., PO Box 468, Newton, NC 28658, USA.

Sarstedt Inc., 68 Boston Road, Beaumont Leys, Leicester LE1 AQ, UK.

Schleicher and Schuell

Schleicher and Schuell Inc., Keene, NH 03431A, USA.

Schleicher and Schuell Inc., D-3354 Dassel, Germany. Schleicher and Schuell Inc., c/o Andermann and Company Ltd.

Scott Smith Electronics, Wimborne, Dorset, UK.

Semat Technical Ltd., St Albans, UK.

Seralab, Crawley Down, Sussex RH10 4FF, UK.

Serotec, 22 Bankside, Station Approach, Kidlington, Oxford OX5 1JE, UK.

Serotec distributed by Harlan Bioproducts for Science USA, PO Box 29171, Indianapolis, IN 46163, USA.

Shandon Scientific Ltd., Chadwick Road, Astmoor, Runcorn, Cheshire WA7 1PR, UK.

Sherwood Medical, Crawley, Sussex RH11 7YQ, UK.

Sherwood Medical, St Louis, MO 63103, USA.

Sigma Chemical Company

Sigma Chemical Company (UK), Fancy Road, Poole, Dorset BH17 7NH, UK.

Sigma Chemical Company, 3050 Spruce Street, PO Box 14508, St Louis, MO 63178-9916, USA.

SLT Labinstruments GmbH, Unterbergstrasse 1 A, 5082 Grodig, Austria.

Sorvall DuPont Company, Biotechnology Division, PO Box 80022, Wilmington, DE 19880-0022, USA.

Southern Biotechnology, Birmingham, AL, USA.

Statens Seruminstitut, Artillerivej 5, DK-2300 Copenhagen, Denmark.

Stratagene

Stratagene Ltd., Unit 140, Cambridge Innovation Centre, Cambridge Science Park, Milton Road, Cambridge CB4 4GF, UK.

TCS Biologics Ltd., Botolph Claydon, Buckingham, UK.

Vector Laboratories, Burlingame, CA, USA.

Ventana Medical Systems, Tucson, Arizona, USA.

Wallac Co. (UK), EG&G Instruments Ltd., 20 Vincent Avenue, Crownhill Business Centre, Crownhill, Milton Keyne MK8 0AB, UK.

Wallac Inc. (USA), 9238 Gaither Road, Baithesburg, Maryland 20877, USA.

Wallac Oy, PO Box 10, FIN-20101 Turku, Finland.

Whatman International, Whatman House, St Leonard's Road, Maidstone, Kent ME16 0LS, UK.

Whatman LabSates, PO Box 1359, Hillsboro, OR 97123-9981, USA.

Index

abrin 239, 241
affinity chromatography 109, 254
alcohol dehydrogenase, yeast 187
alkaline phosphatase-based systems 105, 152, 159, 178
amperometric systems 179
antibody(ies)
  anti-melanoma 83–8
  biotin conjugated 104, 165, 229
  CDR regions 35
  chelate-labelled 211
  chemically engineered 1–24
    buffers 8–10
    equipment 10–12
  chicken 89–90
  chimeric 1
  enzyme-labelled 105, 152–5, 177
  Eu^{3+}-labelled 199–202, 205
  fragments 27, 49
  genetically engineered 27–52
    human variable domains, see engineered human variable domains construction
  human monoclonal 55–88
  recombinant, see recombinant antibodies expression
antibody (B cell) epitope mapping with synthetic peptides 127–34
  analoguing with detailed epitope analysis 134
  confirmation of relevance of binding 132–5
    choice of controls 132
    competition test 132–3
    elution test 133
    homologous competition test with peptide 133–4
  direct binding of antibodies on pins 128–30
  direct binding of biotinylated peptides 130–1
  ELISA on pin-bound (non-cleavable) peptides 128–30
  ELISA with biotinylated peptides 130–1
antibody-mediated cytotoxic cells 209
antibody-ricin immunotoxin 250
antigen-presenting cells 134
anti-melanoma antibodies specificity test 87–8
autoimmune diseases 239
2,2-azino-bis(3-ethylbenzthiazoline)-6-sulfonic acid (ABTS) 154

bacterial β-galactose 218
B cells (lymphocytes) 56, 64–9
  antigen-specific 56

Epstein–Barr virus infection 56, 64–5
  isolation 64–9
  proliferation 64–9
bifunctional chelating agent, protein labelling using 200–2
biotinylated oligonucleotide probe 179
biotinylated peptides 127
bismaleimide linkers 7
bovine haemoglobin 110
5-bromo-4-chloro-3-indoyl phosphate 153
bromodeoxyuridine (BrdU) 230–6
  advantages 231–6
  applications of technique 236
  detection of peripheral mononuclear cell proliferation 232–4
  disadvantages 231–6
  protocol for cell proliferation 231

casein 156
cell–ELISA (CELISA) 215
  advantages 221
  background to technique 216
  disadvantages 221
  production of single cell suspension from rat spleen 218–19
  protocol 216–21
  technique applications 221–2
  titrating mouse mAb against rat lymphocyte cell surface antigens 219–21
cell fusion 69–73
cell proliferation 230–6
  estimation using enzyme immunoassay 232 (fig.)
  see also bromodeoxyuridine
chickens 90
  antibodies 89–90
  use of 101–4
  husbandry 91, 92 (fig.)
  immunization 91–3
    antibody response 93–5
  immunoglobulin 90 1
  immunoglobulin G, see immunoglobulin G, chickens
chimeric FabFc$_2$ 21
  preparation 21–2
chlamydia 179
cleavable pin kit 123 (table)
Clostridium kluyveri 179, 185
cytokine standards 113 (table)
cytomegalovirus 63
cytoxic T cell epitopes 122

dehydrogenase turnover measurement 235
DELFIA 194, 195–211
  cytotoxic release assays 209–11
  FIA assay optimization 205–9
  IFMA assay optimization 205–9
  lathanide ions measurement 203
  microtitration plate based 198
  multilabel measurement 204–5
  receptor binding assay 209
  two-site sandwich assay 206–9
dextran sulfate 97–8, 211
diamino benzidine 153
o-diansidine (3,3 -diemethoxybenzidine) 154
diaphorase 179, 185, 187
dicarboxylic acid derivative, esterified 211
3,3-diemethoxybenzidine (o-diansidine) 154
dihydrofolate reductase 46–7
β-diketone 202
β-diketone-based co-fluorescence
    enhancement 204
diketopiperazine 138
3-(4,5-dimethylthiazol-2-yl)2,5
    diphenyltetrazolium (MTT) 235
diphtheria toxin 239, 240, 241
4,4-dipyridyl disulfide 4, 5 (fig.)
discontinuous epitopes 127
DKP-ended peptides 122

eggs 89–105
  antibodies in 90–1
electrofusion 71–3
  lymphoblastoid cell lines 72–3
ELISA 215; *see also* enzyme-linked
    immunoassays
  amplification systems 177–90
  antibody-coated plate (two-site/sandwich)
    149
  antigen-coated plate 147–9
    for detection of rat IgG antibody to
      ovalbumin 166–7
  application of fluorescence to enzyme
    amplification 84–9
    alkaline phosphatase substrate 187
    buffers 186–7
    cycling enzyme concentrations balance
      187–9
    enzymes preparation 187
    proinsulin assay 189
    water sources 186–7
  cell-based system 86 (fig.)
  class capture 149
  competitive 149–52
    for antibody 151
    for antigen 52, 169–70
  enzyme-amplified assay 179–84
    TSH microtitre plates (comparison with
      unamplified detection) 183–4

use of enzyme amplification reagents
    182–3
  non-competitive 147–9
  quantification 170–2
  two-site, for detection of IFN-γ 168–9
  two-site, for detection of IgE 167–8
ELISPOT, *see* enzyme-linked immunospot
engineered human variable domains
    construction 35–41
  CDR replacement 35–7
  construction of engineered variable region
    genes 40–1
  sequence assembly 38–40
  surface residue replacement 37–8
  worked example 38, 39 (fig.)
enzyme amplification 177–9
enzyme labels 152–4, 159, 161–5, 177
enzyme-linked chemiluminescence
    immunoassay 193
enzyme-linked immunoassays 147–73,
    215–36
  buffers 160–1
  coating of solid phase 155–6
  coefficient of variation 173
  enzyme labelling of proteins 162–5
    biotinylation of antibodies 165
    one-step glutaraldehyde method 163
    periodate method 162–5
    two-step glutaraldehyde method 163–4
  enzymes 152–4
  equipment 159–60
  format 147–52
  optimization 155
  purification of conjugate 154–5
  quality controls 172–3
  substrates 152–4, 161–2
    alkaline phosphatase 162
    peroxidase 161–2
enzyme-linked immunospot (ELISPOT)
    215–16, 222–30
  advantages 229–30
  applications 230
  avidin–biotin–antiperoxidase system 229
  avidin–biotin system 229
  disadvantages 229–30
  peroxidase–antiperoxidase system 229
  protocol 223–4
  reverse 226–9
    application 230
Epstein–Barr virus 56, 69, 73
  B cell infection 56, 64–5
  handling in laboratory 65
  infection with 74
  laboratory growth 66
Escherichia coli 27–8, 48–52, 153
  antibody fragments production 51
  heterologous proteins production 49
  outer membrane protein A 49

recombinant antibody fragments 49
secreting cloning vectors 48–9
vector design 48–50
Eu–cryptate 211
Eu–DTPA 211
cell labelling 209–10
Eu-labelling reagent 200
Eu release test 210

Fab$_2$Fc$_4$ 8 (fig.)
Fab′ fragments 240, 251–3
F(ab′)$_2$ thioether-linked bispecific chimeric
derivatives 23–4
F(ab′)$_2$ thioether-linked potential hinge SH
groups 22–3
Fab′γ module 1
derivatives 21–4
murine 18–20
Fab′γ (-maleimide) 7
F(ab′γ)$_2$ preparation 18–19
Fab′ (–SH) 23
Fab′ (SH)$_1$ preparation 20
Fab′γ-SS-Py 4
Fab-SS-pyridine preparation 20
Fc-SS-pyridine preparation 16–18
Fcγ module, human 1, 12–19
derivatives 21–4
Fcγ1 preparation 14–16
Fc–maleimide preparation 16–18
FLAG-derived plasmids 267
fluorescence 193–212
fluorescent chelates 211–12
fluorescent end-product of enzymatic reaction
193
fluoroimmunoassays 193, 194
fluorometry 194
formazan 178 (fig.), 178
fusion proteins 266–7
Fv fragments 240
Fvs construction 41–4
disufide-linked Fvs 42–4
single-chain Fvs 41–2

gadolinium 195
β-D-galactosidase 154
gelonin 240
genes
engineered variable region construction
40–1
guanosine phosphoribosyl transferase 45
V(D)J 89
German measles (rubella) 149
glutamine synthetase 47
(Gly$_4$Ser) synthetic oligomers 41
guanosine phosphoribosyl transferase gene 45

hapten 209
antibodies to 156
hepatitis 179
hepatitis B vaccination 63
hepatitis C 13
heterohybridomas 70
histidine clusters 267
HIV 179
horse-radish peroxidase 153–4, 218, 229
fluorogenic substrates 154
human monoclonal antibodies 55–88
hybridization assays 213
hybridoma 57, 58, 73, 77
BALB/c origin 2
production from lymph node lymphocytes
83–4

2-iminothiolane hydrochloride 246
immunoassays 213, *see also* ELISA
immunochemical procedures standardization
110–19
antibodies epitope specificity 119
availability of standard preparations 111–12
cytokine standards 113 (table)
immunological standards 114 (table)
International Collaborative Study 112
matrix effects 111, 115
multiple standards 115–19
nature of standard preparations 11–12
parallel line analysis 110, 111 (fig.)
parallel line approach 112
unitage of standard preparations 112–15
use of standards 110–11
WHO International Standards 112
immunochemical reagents 109–10
standardization 109–10
immunoglobulin(s) 27–8
chicken 90–1
intrinsic properties 27
modules 1
reduced, SS-interchange on 6–7
immunoglobulin G 12–13
antibodies from ascites/culture fluid 18
Fraction II precipitate 12, 13
preparation 13–14
immunoglobulin G, chicken 90
cleavage by pepsin 3, 5 (fig.)
immunoelectrophoresis 101–3
labelling with alkaline phosphatase 105
labelling with biotin 104, 165, 229
labelling with lanthanide ions 199–202
purification by ammonium sulfate
precipitation 100
purification from yolk 97–101
by PEG precipitation 101
by precipitation with sodium sulfate
98–9

immunoglobulin G, chicken (*cont.*)
  removal of lipids by dextran sulfate
    precipitation 97–8
  removal of yolk lipids by euglobulin
    precipitation 97
  SDS–PAGE 99 (fig.)
immunoglobulin-secreting cells 222
immunoglobulin Y 89
immunoradiometric assays 118 (fig.)
immunotoxins 239–70
  A-chain 240
  affinity chromatography 254–5
  antibody ricin 250
  anti-tumour activity 240
  B-chain 240
  biochemical characterization 269
  cell-free asays of protein synthesis 269–70
  chemically coupled 241–55
    heterobifunctional coupling reagents
      241
    purification 253–5
  chemically prepared, composition
    calculation 255
  fusion proteins expression/purification
    266–9
  non-specific toxicity 240
  preparation 240–69
  preparation by chemical linkage of antibody
    to intact toxins 248–51
  preparation of chemical linkage of antibody
    to isolated toxin A-chains 242–6
  preparation of chemical linkage of antibody
    to RIP 246–8
  preparation by genetic engineering 255–69
  preparation by linking antibody Fab
    fragments to toxins/catalytic subunits
    251–3
  attachment of Fab antibody fragments to
    isolated A-chain/thiolated RIP 252–3
  sterically blocked 250
interchain disulfide (–SS) bonds 1, 2–3
interleukin-1α, IRMA specific for 116 (fig.)
interleukin-4 117 (fig.)

lanthanides 193, 195
  chelates 194
    dissociative fluorescence enhancement
      202
    measurement 203–4
  DELFIA measuremnt 205
  dissociative fluorescence enhancement
    202
  labelling of immunoglobulins with
    lanthanide ions 199–202
  protein labelling 200–2
leucyl–leucine–methyl ester 68
  cytotoxic cells depletion 68–9

lymphoblastoid cell lines 73–5
  bulk, fusion of 74–5
  electrofusion 72–3
  fusion 70
  selected, fusion of 73

macroimaging 194
matrix effects 115
melanoma 83
  native cells, antibodies reacting with 86–7
melanoma-associated antigens, antibodies
    selection against 85
L-methionine sulfoximine 47
methotrexate 46
4–methylumbelliferyl-β-D-galactoside 154
microimaging 194
Milstein bottle 80
miniaturized multiparametric assays 194
monoclonal antibodies (Mabs) 27, 55–81
  cloning 75–9
    limiting dilution 76–8
    soft agar 78–9
  culture 56–8
    basic requirements 56–7
    media 57
    supplements 57–8
  detection assays 56–63
  donor selection 63–4
  ELISA 59–61
    direct for specific antibody 60–1
    indirect for human Ig 59–60
    indirect for specific antibody 61
  haemagglutination assays 61–3
  human 55–88
  Milstein–Kohler production 69
  production strategies 73–5
    anti-CD40 75
    bulk of LCLs fusion 74–5
    Epstein–Barr virus only 73
    selected LCLs fusion 73–4
  roller bottle culture 79–80
  scale up of production 79–81
  therapeutic agents 27
  tumour-specific, identification of 84–8
multiple standards 115–16
myelomas 44, 47
  cell fusion 56
  fusion partners 33, 70

p-nitro-phenol-N -acetyl-β-D-glucosaminide
    235
p-nitrophenyl-β-D-galactoside 154
p-nitrophenyl phosphate 153
NS0 cell lines, amplified antibody-producing
    construction 47–8
nystatin 58

ouabain 70

papain 3, 4 (fig.), 101
    activation 14–15
pectate lyase B 49
Pepscan 127
pepsin 101
peptides 127–34
    assay in CTL killing assay 140–1
    biotinylated 127
    discontinuous 127
    helper T cell clone response 139
    homologous competition test 133–4
    synthetic, *see* antibody (B cell) epitope
        mapping with synthetic peptides;
        synthetic peptides
peripheral blood mononuclear cells 145
    isolation from human blood 65
    mapping with T helper epitopes 141–4
    proliferation detection by uptake of BrdU
        232–4
o-phenylene diamine 154
o-phenylene-disuccinimidyl 7
phosphofructokinase 177
phospho-immunoassay 193
photoluminescence immunoassays 193–213
pokeweed anti-viral protein 240
pokeweed mitogen 56
polyacrylamide gradient gels 255
polycarboxylate-based complexes 195
polyethylene glycol fusion 70–1
polymerase chain reaction (PCR) cloning of
        immunoglobulin variable domains
        28–35
Protein-A–Sepharose 11
Pseudomonas exotoxin 239, 240
psoriasis 239
pyridyl disulfides 3–6
4-pyridylthiol 5
pyruvate kinase, *E.coli* type 1 isoenzyme 177

rabbit reticulocyte lysate system 269
radioimmunoassays 193
radioluminescence 193
receptor–ligand interactions 213
recombinant antibodies expression 44–52
    E.coli 48–52
        factors affecting level of expression 50–2
        vector design 48–50
    mammalian cells 44–8
        low copy vectors 44–6
        vector amplification 46–8
    transient expression in COS cells 44–5
resazurin 179, 185–6, 189, 190 (fig.)
resorufin 185
Rh(D) blood group determinant 55, 63

ribosomal 60S subunits 239
ribosome-inactivating proteins 240
    fungal 240
ricin 239, 241
    blocked 249–50

saponin 240
α-sarcin 240
sensitive fibre sensors 194
solid tumours 239
SPDP 246
SS-interchange on reduced Ig modules 6–7
streptavidin, chelate-labelled 211–12
streptavidin–alkaline phosphatase conjugate
        179
streptomycin 58
N-succinidimidyl-4–(*N*-maleimidomethycyclo-
        hexane)-1-carboxylate 249
4–succinimidyl-oxycarbonyl-α-methyl-α-
        (2–pyridyldithio)-toluene 245–6
N-succinimidyl-3-(2-pyridyldithio) propionate
        243
sulfydryl group 1, 3–7
    alkylation for blocking/cross linking 7
synthetic peptides 121–45
    cleavage 124–5
    handling 126–7
    linear 121
    multiple peptide synthesis on peptides
        121–2
        choice of ending 122
        cleaved/non-cleavable 122
    peptide pooling, decoding, narrowing to
        minimal epitope 144–5
    redissolving 126–7
    side chain deprotection 124–5
    storage 126–7
    synthesis 122–5
        on pins 123–4

TAG-72 38
T cell (lymphocyte) 229, 239
    cytotoxic 64
        killing assay, peptide assay 140–1
    cytotoxic epitopes 122, 138
    depletion 66–8
T cell clones
    cytotoxic 140
    epitope mapping with 138
T cell epitope mapping with synthetic peptides
        134–45
    design of peptide sets 135
        to scan protein sequence 135–6
    design of truncations/analogues of epitope
        137
tetanus vaccination 63

3,3′,5,5′-tetramethyl benzidine 54
T helper cell clones 138
    response to peptide 139
T helper epitopes 122, 138
    mapping with peripheral blood
        mononuclear cells 141–4
thiolating agent 246
4–thiopyridone 5
time-resolved fluroimmunoasay 193
time-resolved flurometric assays with stable
    fluorescent chelates 211–12

time-resolved flurometry 194
tissue autofluorescence 211
trinitrophenyl phosphate 156
tri-*n*-octylphosphineoxide

V(D)J genes 89
vitamin B12 209

Wallac customer labelling service 200
World Health Organization 112